METHODS IN MOLECULAR BIOLOGY

For further volumes:
http://www.springer.com/series/7651

Bone Morphogenetic Proteins

Methods and Protocols

Edited by

Melissa B. Rogers

New Jersey Medical School, Rutgers Biomedical and Health Sciences, Newark, NJ, USA

Humana Press

Editor
Melissa B. Rogers
New Jersey Medical School
Rutgers Biomedical and Health Sciences
Newark, NJ, USA

ISSN 1064-3745 ISSN 1940-6029 (electronic)
Methods in Molecular Biology
ISBN 978-1-4939-8903-4 ISBN 978-1-4939-8904-1 (eBook)
https://doi.org/10.1007/978-1-4939-8904-1

Library of Congress Control Number: 2018959761

This Humana Press imprint is published by the registered company Springer Science+Business Media, LLC part of Springer Nature.
The registered company address is: 233 Spring Street, New York, NY 10013, U.S.A.

Preface

The 31 members of the transforming growth factor β superfamily profoundly influence animal biology from the earliest stages of germ cell development through life. Those classified within the bone morphogenetic protein (BMP) branch are the most evolutionarily ancient members. Although first identified by their osteogenic properties, these proteins are critical signals in all developing and adult invertebrate and vertebrate tissues. BMP orthologs from distant species such as flies and humans are functionally interchangeable and play key roles in fundamental metazoan processes such as dorso-ventral axis determination, epithelial-mesenchymal interactions, and programmed cell death to name just a few. Disruptions of BMP regulation and signaling cause or influence the entire panoply of human clinical conditions—cancers, cardiovascular and lung diseases, osteogenic diseases involving thin or overly dense bones and pathological soft tissue calcification, etc. Consequently, the hope of discovering new clinical strategies for both common diseases such as atherosclerosis and rare diseases such as fibrodysplasia ossificans progressiva (FOP) drives vigorous research into the biology of BMPs.

This volume reviews methods to study BMPs in cell culture and in both invertebrate and vertebrate animals, and therapeutic applications of BMPs and BMP modulators. The topics have been chosen to extend and complement the recent *Methods in Molecular Biology* volume entitled *TGF-β Signaling* [1]. Specifically, a greater focus has been placed on genetic approaches and model organisms including both invertebrates and nonmammalian vertebrates.

The book begins with a cell and molecular focus before moving into invertebrate models and then ending with vertebrate models. The first three chapters discuss primary and immortalized cell lines used for in vitro differentiation studies as well as the delivery of BMPs for cell-based gene therapy. Cutting-edge methodologies addressing transcription and receptor mechanisms using genome editing and high-throughput biosensors are then presented. The powerful genetic approaches to understanding BMP function in *Caenorhabditis elegans* and *Drosophila melanogaster* are presented along with imaging techniques. A chapter on the newly developed methods to culture the chordate Amphioxus provides the transition between invertebrate and vertebrate systems. Amphibians and fishes are represented by three chapters discussing Xenopus and zebrafish embryology techniques and a new FOP model in zebrafish. Several chapters discuss approaches using genetically engineered mouse models and drugs to study BMP receptor function and activities at the molecular and microscopic levels. The three terminal chapters discuss preclinical approaches to understanding BMP function in bone regeneration and cancers and using a mouse model of FOP.

Finally, I wish to express my gratitude to the authors who contributed great effort and valuable time to this volume. Their cooperation, hard work, and the multitude of useful experimental hints should make this book a helpful BMP resource for many years. I am truly indebted to each contributor. I also must thank Dr. John M. Walker, first for the invitation to edit this volume and then for his patience and advice. I am particularly grateful for his willingness to wait for "just one more exciting protocol." May this collection promote greater understanding of these essential proteins.

Newark, NJ, USA *Melissa B. Rogers, Ph.D.*

Reference

1. Feng, X.-H., Xu, P., Lin, X. (Eds.) (2016) TGF-β signaling: methods in molecular biology. Springer, New York

Contents

Contributors

MAAIKE ALKEMA • *Department of Cell and Chemical Biology, Leiden University Medical Center, Leiden, The Netherlands*

SENEM AYKUL • *Department of Biochemistry and Molecular Biology, Michigan State University, East Lansing, MI, USA*

JEAN-DENIS BÉNAZET • *Department of Orofacial Sciences and Program in Craniofacial Biology, University of California, San Francisco, San Francisco, CA, USA*

GEOFFREY A. BOCOBO • *Division of Cardiovascular Medicine, Department of Medicine, Brigham and Women's Hospital, Harvard Medical School, Boston, MA, USA*

POLYXENI BOZATZI • *MRC Protein Phosphorylation and Ubiquitylation Unit, Sir James Black Centre, University of Dundee, Dundee, Scotland, UK*

ELIZABETH A. BRETT • *Department of Plastic and Hand Surgery, Technical University Munich, Ismaninger Strasse, Munich, Germany*

DONYA BURGESS • *Biological Sciences, Seton Hall University, South Orange, NJ, USA*

SALIN A. CHAKKALAKAL • *Department of Orthopaedic Surgery, Perelman School of Medicine at the University of Pennsylvania, Philadelphia, PA, USA; Center for Research in FOP and Related Disorders, Perelman School of Medicine at the University of Pennsylvania, Philadelphia, PA, USA*

JAN L. CHRISTIAN • *Division of Hematology and Hematologic Malignancies, Department of Neurobiology, Anatomy and Internal Medicine, University of Utah School of Medicine, Salt Lake City, UT, USA*

JESSICA ANN COTTRELL • *Biological Sciences, Seton Hall University, South Orange, NJ, USA*

ALAN R. DAVIS • *Center for Cell and Gene Therapy, Baylor College of Medicine, Texas Children's Hospital and Houston Methodist Hospital, Houston, TX, USA; Section of Hematology/Oncology, Department of Pediatrics, Baylor College of Medicine, Houston, TX, USA; Department of Orthopedic Surgery, Baylor College of Medicine, Houston, TX, USA*

ELEANOR L. DAVIS • *Center for Cell and Gene Therapy, Baylor College of Medicine, Texas Children's Hospital and Houston Methodist Hospital, Houston, TX, USA*

AUSTIN DICKERSON • *Center for Cell and Gene Therapy, Baylor College of Medicine, Texas Children's Hospital and Houston Methodist Hospital, Houston, TX, USA*

TERESA DINTER • *Division of Cardiovascular Medicine, Department of Medicine, Brigham and Women's Hospital, Harvard Medical School, Boston, MA, USA*

RYAN J. GLEASON • *Department of Biology, Johns Hopkins University, Baltimore, MD, USA*

MARIE-JOSÉ GOUMANS • *Department of Cell and Chemical Biology, Leiden University Medical Center, Leiden, The Netherlands*

YAN HUANG • *Department of Agricultural and Biological Engineering, Purdue University, West Lafayette, IN, USA*

DEBORAH A. HURSH • *Division of Cell and Gene Therapy, Center for Biologics Evaluation and Research, Food and Drug Administration, Silver Spring, MD, USA*

LUKE D. HUTCHINSON • *MRC Protein Phosphorylation and Ubiquitylation Unit, Sir James Black Centre, University of Dundee, Dundee, Scotland, UK*

DEBOLEENA KANJILAL • *Rutgers University, Newark, NJ, USA*

HYUNG-SEOK KIM • *Division of Hematology and Hematologic Malignancies, Department of Neurobiology, Anatomy and Internal Medicine, University of Utah School of Medicine, Salt Lake City, UT, USA*

IRYNA KOZMIKOVA • *Institute of Molecular Genetics of the ASCR, Prague, Czech Republic*

BOUDEWIJN P. T. KRUITHOF • *Department of Cell and Chemical Biology, Leiden University Medical Center, Leiden, The Netherlands; Department of Cardiology, Leiden University Medical Center, Leiden, The Netherlands; Netherlands Heart Institute, Utrecht, The Netherlands*

MELISSA LABONTY • *Program in Cell, Molecular, and Developmental Biology, Sackler School of Graduate Biomedical Sciences, Tufts University School of Medicine, Boston, MA, USA; Department of Orthodontics, Division of Craniofacial and Molecular Genetics, Tufts University School of Dental Medicine, Boston, MA, USA*

JOHN LANGENFELD • *Division of Cardiothoracic Surgery, Rutgers Cancer Institute of New Jersey, Rutgers Robert Wood Johnson Medical School, Rutgers, The State University of New Jersey, New Brunswick, NJ, USA*

JUN LIU • *Department of Molecular Biology and Genetics, Cornell University, Ithaca, NY, USA*

MICHAEL T. LONGAKER • *Division of Plastic and Reconstructive Surgery, Department of Surgery, Stanford University School of Medicine, Stanford, CA, USA; Hagey Laboratory for Pediatric Regenerative Medicine, Department of Surgery, Stanford University School of Medicine, Stanford, CA, USA*

THOMAS MACARTNEY • *MRC Protein Phosphorylation and Ubiquitylation Unit, Sir James Black Centre, University of Dundee, Dundee, Scotland, UK*

MARCELO ROCHA MARQUES • *Department of Morphology, Area of Histology and Embryology, Piracicaba Dental School, University of Campinas, Campinas, SP, Brazil; Department of Orofacial Sciences and Program in Craniofacial Biology, University of California, San Francisco, San Francisco, CA, USA*

CLEMENT D. MARSHALL • *Division of Plastic and Reconstructive Surgery, Department of Surgery, Stanford University School of Medicine, Stanford, CA, USA*

ERIK MARTINEZ-HACKERT • *Department of Biochemistry and Molecular Biology, Michigan State University, East Lansing, MI, USA*

AUTUMN MCKNITE • *Division of Hematology and Hematologic Malignancies, Department of Neurobiology, Anatomy and Internal Medicine, University of Utah School of Medicine, Salt Lake City, UT, USA*

SARAH MICHAELS • *Biological Sciences, Seton Hall University, South Orange, NJ, USA*

YUJI MISHINA • *Department of Biologic and Materials Sciences, School of Dentistry, University of Michigan, Ann Arbor, MI, USA*

ALESSANDRA L. MOORE • *Division of Plastic and Reconstructive Surgery, Department of Surgery, Stanford University School of Medicine, Stanford, CA, USA; Department of Surgery, Brigham and Women's Hospital, Boston, MA, USA*

MARY MULLINS • *Department of Cell and Developmental Biology, University of Pennsylvania Perelman School of Medicine, Philadelphia, PA, USA*

RACHEL NEMOYER • *Department of Surgery, Rutgers Robert Wood Johnson Medical School, Rutgers, The State University of New Jersey, New Brunswick, NJ, USA*

JENNA NEWMAN • *Rutgers Cancer Institute of New Jersey, New Brunswick, NJ, USA; Department of Surgery, Section of Surgical Oncology Research, Rutgers Robert Wood Johnson Medical School, Rutgers, The State University of New Jersey, New Brunswick, NJ, USA*

ELIZABETH A. OLMSTED-DAVIS • *Center for Cell and Gene Therapy, Baylor College of Medicine, Texas Children's Hospital and Houston Methodist Hospital, Houston, TX, USA; Section of Hematology/Oncology, Department of Pediatrics, Baylor College of Medicine, Houston, TX, USA; Department of Orthopedic Surgery, Baylor College of Medicine, Houston, TX, USA*

RICHARD W. PADGETT • *Department of Molecular Biology and Biochemistry, Waksman Institute, Rutgers University, Piscataway, NJ, USA; Cancer Institute of New Jersey, Rutgers University, New Brunswick, NJ, USA*

MELISSA B. ROGERS • *Microbiology, Biochemistry, and Molecular Genetics, Rutgers New Jersey Medical School, Newark, NJ, USA*

GOPAL P. SAPKOTA • *MRC Protein Phosphorylation and Ubiquitylation Unit, Sir James Black Centre, University of Dundee, Dundee, Scotland, UK*

CATHY SAVAGE-DUNN • *Department of Biology, Queens College, CUNY, Flushing, NY, USA; PhD Programs in Biology and Biochemistry, The Graduate Center, CUNY, New York, NY, USA*

EILEEN M. SHORE • *Department of Orthopaedic Surgery, Perelman School of Medicine at the University of Pennsylvania, Philadelphia, PA, USA; Center for Research in FOP and Related Disorders, Perelman School of Medicine at the University of Pennsylvania, Philadelphia, PA, USA; Department of Genetics, Perelman School of Medicine at the University of Pennsylvania, Philadelphia, PA, USA*

CORINNE SONNET • *Center for Cell and Gene Therapy, Baylor College of Medicine, Texas Children's Hospital and Houston Methodist Hospital, Houston, TX, USA*

BRIAN G. STULTZ • *Division of Cell and Gene Therapy, Center for Biologics Evaluation and Research, Food and Drug Administration, Silver Spring, MD, USA*

FRANCESCA TUAZON • *Department of Cell and Developmental Biology, University of Pennsylvania Perelman School of Medicine, Philadelphia, PA, USA*

DAVID UMULIS • *Department of Agricultural and Biological Engineering, Purdue University, West Lafayette, IN, USA; Weldon School of Biomedical Engineering, Purdue University, West Lafayette, IN, USA*

DERRICK C. WAN • *Division of Plastic and Reconstructive Surgery, Department of Surgery, Stanford University School of Medicine, Stanford, CA, USA*

JINGWEN YANG • *Department of Biologic and Materials Sciences, School of Dentistry, University of Michigan, Ann Arbor, MI, USA; State Key Laboratory Breeding Base of Basic Science of Stomatology (Hubei-MOST) and Key Laboratory for Oral Biomedicine of Ministry of Education (KLOBM), School and Hospital of Stomatology, Wuhan University, Wuhan, Hubei, China*

PAMELA C. YELICK • *Program in Cell, Molecular, and Developmental Biology, Sackler School of Graduate Biomedical Sciences, Tufts University School of Medicine, Boston, MA, USA; Division of Craniofacial and Molecular Genetics, Department of Orthodontics, Tufts University School of Dental Medicine, Boston, MA, USA*

LUOK WEN YONG • *Institute of Cellular and Organismic Biology, Academia Sinica, Taipei, Taiwan*

JR-KAI YU • *Institute of Cellular and Organismic Biology, Academia Sinica, Taipei, Taiwan*
PAUL B. YU • *Division of Cardiovascular Medicine, Department of Medicine, Brigham and Women's Hospital, Harvard Medical School, Boston, MA, USA*
JOSEPH ZINSKI • *Department of Cell and Developmental Biology, University of Pennsylvania Perelman School of Medicine, Philadelphia, PA, USA*
ANDREW ZLOZA • *Rutgers Cancer Institute of New Jersey, New Brunswick, NJ, USA; Department of Surgery, Section of Surgical Oncology Research, Rutgers Robert Wood Johnson Medical School, Rutgers, The State University of New Jersey, New Brunswick, NJ, USA*

Chapter 1

Common Cell Lines Used to Study Bone Morphogenetic Proteins (BMPs)

Jessica Ann Cottrell, Donya Burgess, Sarah Michaels, and Melissa B. Rogers

Abstract

Many research methods exist to elucidate the functions of BMPs during osteogenesis. This chapter briefly reviews common immortalized mesenchymal cell types used to measure the efficacy of osteogenic factors like BMP-2. Detailed information regarding media and culture conditions are provided. Parameters relevant to experimental reproducibility and cell line authentication are discussed.

Key words BMP-2, Mycoplasma

1 Introduction

BMPs function by stimulating mesenchymal cells to proliferate and differentiate into multiple cell types [1–4]. For example, exposure to BMP-2 prevents myogenesis and promotes osteogenesis, chondrogenesis, or adipogenesis in a concentration-dependent manner [4–9]. Elucidation of the molecular mechanisms by which BMPs exert a cellular response and preliminary dose estimates can be assessed in cell culture prior to initiating time-consuming and expensive animal studies. Primary cell lines used for osteogenesis studies include embryonic stem cells (ESCs), mouse embryo fibroblasts (MEFs), pluripotent mesenchymal stem cells (pMSCs), adipose-derived stem cells (ADSCs), bone-derived mesenchymal stem cells (BMSCs), and knee synovium-derived mesenchymal stem cells (SMSCs). While primary cell types potentially mimic in vivo processes more precisely, their limited lifespan makes establishing multiple batches of cells necessary, thus increasing experimental variability. For historical reasons and for experimental consistency, several immortalized cell lines are commonly used to

Donya Burgess and Sarah Michaels are contributed equally to this work.

Melissa B. Rogers (ed.), *Bone Morphogenetic Proteins: Methods and Protocols*, Methods in Molecular Biology, vol. 1891, https://doi.org/10.1007/978-1-4939-8904-1_1,

assay of the osteogenic activity of BMP preparations and for tests of molecular mechanisms. This chapter discusses five cell types often used to investigate BMP phenomena in vitro: C3H10T$_{1/2}$ (mouse fibroblasts [3]), C_2C_{12} (mouse myoblast [4]), MC3T3-E1 (mouse pre-osteoblasts [10–12]), NIH3T3 (mouse fibroblasts [13]), and ATDC5 (mouse chondrocytes [14]).

"Lack of reproducibility" in preclinical studies is now widely recognized. The unintentional, but common, use of misidentified and contaminated cell lines contributes to this problem [15]. Consequently, the National Institutes of Health now requires formal "Authentication of Key Biological and/or Chemical Resources" (https://grants.nih.gov/grants/guide/notice-files/NOT-OD-17-068.html) in grant applications. Attention to culture conditions and good laboratory practices is a critical component of work with BMP-responsive pluripotent cell types. For example, BMP-2 is induced by no less than nine growth factors and hormones, by physical conditions such as stretch, by many drugs and botanical compounds, and by bacterial contamination [16]. Indeed, mycoplasma contamination, which frequently goes undetected, is a potent inducer of BMP-2 synthesis [17]. Consequently, we include suggestions that address the reproducibility and rigor of these cell culture studies.

2 Materials

1. Cell line sources (*see* **Note 1**) and commonly used media formulations (*see* **Note 2** and Table 1).
2. Sera: fetal bovine serum (FBS) also known as fetal calf serum (FCS) (*see* **Note 3**).
3. Commonly used antibiotics: 100 units/mL penicillin and/or 100 μg/mL streptomycin (*see* **Note 4**).
4. Trypsin: 0.05% (*see* **Note 5**).
5. Phosphate-buffered saline (PBS) or Hanks' Balanced Salt solution (HBSS).
6. Mycoplasma testing kit.
7. Infuse® rhBMP-2 (Medtronic Inc., Minneapolis, MN) (*see* **Note 6**).
8. BMP-2 buffer: 5 mM glutamate, 5 nM sodium chloride, 2.5% glycine, and 0.5% sucrose.

3 Cell Passage Procedures on a 10 cm Cell Culture Dish

1. Obtain cells at 75% confluency (*see* **Note** 7).
2. Aspirate the media from the plate.

Table 1
Cell lines commonly used for cell culture studies with media requirements

Cell line	Source	Base media	Supplements
C3H10T1/2	ATTC: CCL-226	α-MEM with 2 mM L-glutamine, 1.5 g/L sodium bicarbonate, and Earle's BSS	10% FBS
MC3T3-E1	ATTC: CRL-2593	α-MEM with ribonucleosides, deoxyribonucleosides, 2 mM L-glutamine, and 1 mM sodium pyruvate, without ascorbic acid	10% FBS
C2C12	ATTC: CRL-1772	DMEM with 4 mM L-glutamine, 4.5 g/L glucose, 1 mM sodium pyruvate, and 1.5 g/L sodium bicarbonate	10% FBS
NIH3T3	ATTC: CRL-1658	DMEM with 4 mM L-glutamine, 4.5 g/L glucose, 1 mM sodium pyruvate, and 1.5 g/L sodium bicarbonate	10% FBS
ATDC5	Sigma	DMEM: Ham's F12 (1:1) and 2 mM L-glutamine	5% FBS

3. Rinse the plate with 3–4 mL of HBSS or PBS (*see* **Note 8**).
4. Add a small amount of trypsin (2–3 mL) to the plate so that it covers the surface of the dish.
5. Incubate at 37 °C for 2–3 min (*see* **Note 9**).
6. Add 7–8 mL of media to the plate.
7. Pipette solution up and down to disrupt any cells that are clustered together.
8. Count the cells (*see* **Note 10**).
9. Calculate how many cells to plate according to your needs (*see* Table 2) (*see* **Note 11**). Differentiation experiments normally require confluent cells switched to the appropriate media (*see* Table 3). The process of differentiation is dependent on cell type and culture conditions but commonly takes place between days 3 and 14.

4 Notes

1. Cell lines should be acquired from reliable sources such as the American Type Culture Collection (ATCC), Manassas, VA, USA. Other sources of authenticated cell lines are listed by the International Cell Line Authentication Committee (http://iclac.org/). The ICLAC also curates a register of lines known to be misidentified. At the time of this writing, these five lines are not known to be frequently misidentified. However, research groups should periodically authenticate all cell lines by chromosomal analysis or short tandem repeat (STR)

Table 2
Useful information for various sizes of cell culture dishes and plates

Plate	Diameter (cm)	Area (cm^2)	Working volumes (mL)	Approximate number of cells at confluence[a]
15 cm dish	13.7	148	20	269.4×10^5
10 cm dish	8.6	55	10	100.1×10^5
6 cm dish	5.2	21	5	38.2×10^5
3.5 cm dish	3.3	8	3	14.6×10^5
6-well plate	3.5	9.6	3	17.5×10^5
12-well plate	2.2	3.8	2	6.9×10^5
24-well plate	1.6	2.0	1	3.6×10^5
48-well plate	1.0	0.75	0.5	1.4×10^5
96-well plate	0.68	0.36	0.2	0.65×10^5

[a]Approximately 1.82×10^5 cells per cm^2

Table 3
Media commonly used for osteoblast differentiation studies with sources

Cell type	Osteoblast differentiation media components	References
C3H10T1/2	α-MEM with 2 mM L-glutamine, 15 g/L sodium bicarbonate, and Earle's BSS, 10% FBS, and BMP-2-conditioned media (0.1–100 ng/mL)	Osteogenesis, chondrogenesis, adipogenesis (Katagiri et al. [3]; Wang et al. [12]; Rawadi et al. [24]; Date et al. [25]; Tang et al. [26]
MC3T3-E1	α-MEM with ribonucleosides, deoxyribonucleosides, 2 mM L-glutamine, and 1 mM sodium pyruvate, 10% FBS, and 50 μg/mL of ascorbic acid	Osteogenesis, chondrogenesis (Sudo et al. [10]; Hiraki et al. [27]; Quarles et al. [11]; Wang et al. [12]; Luppen et al. [28]; Date et al. [25])
C2C12	DMEM with 4 mM L-glutamine, 4.5 g/L glucose, 1 mM sodium pyruvate, and 1.5 g/L sodium bicarbonate, 2% equine sera, and 1 mg/mL of insulin *or* DMEM with 4 mM L-glutamine, 4.5 g/L glucose, 1 mM sodium pyruvate, and 1.5 g/L sodium bicarbonate, 15% FBDm, and 300 ng/mL BMP-2	Osteogenesis, myogenesis (Katagiri et al. [4]; Lawson and Purslow [29]; Rawadi et al. [24]; Almodovar et al. [30])
NIH3T3	DMEM/F12 media, 10% FBS, 39.3 ng/mL dexamethasone, 50 μg/μL ascorbic acid-2-phosphate, and 756 μg/mL β-glycerol	Osteogenesis, chondrogenesis (Li et al. [13]; Rawadi et al. [24])
ATDC5	DMEM: Ham's F12 (1:1) and 2 mM L-glutamine and 1 μg/mL insulin *This media is used for chondrocyte differentiation	Chondrogenesis (Atsumi et al. [14]; Shukunami et al. [31]; Yao et al. [32])

profiling. Commercial services and kits for authentication are widely available.

2. Many of the classical media formulations were developed decades ago. Over time, labs have developed dozens of variations many of which are readily available commercially. Glucose concentration and the presence or absence of glutamine, pyruvate, ascorbic acid, and other components differ. Care should be taken to use the recommended formulation, because changes in components can alter cell growth, gene expression, and behavior significantly.
3. Serum, usually bovine fetal, is the most common supplement to media. However, as a natural product, serum is highly variable [18]. Both source and batches within the same company often differ greatly. To ensure consistent results, multiple batches should be tested on the desired cells in the desired assay. Reputable companies will provide samples of different batches for testing while reserving a large quantity of each lot for later purchase. Sufficient serum to complete each

comparable study should be purchased. Although many labs traditionally raised the temperature of serum to 56 °C for 30 min to destroy complement, such heat inactivation is unnecessary for most cells, except some immune cells, and may damage serum components such as growth factors. An investigator contemplating eliminating or adding heat inactivation of serum should test its impact on the desired cells and assay.

4. Antibiotics are required for preparing primary cell lines. However, the routine use of antibiotics for immortalized cell lines may mask poor sterile technique and is not recommended. Furthermore, although rampant mold or bacterial contamination may be controlled, common antibiotics are ineffective against viruses and mycoplasma that may induce BMPs or alter their actions. Contamination of cultured cells with the small parasitic mycoplasma bacteria is a long-recognized problem. Mycoplasma contamination broadly affects the expression of many genes, including essential cytokines, growth factors, and oncogenes [19, 20]. Indeed, mycoplasma strongly induces BMP-2 [17]. Cultures should be routinely tested for mycoplasma contamination by any standard method [21]. The MycoAlert™ Mycoplasma Detection Kit is a reliable and rapid biochemical for at least 44 strains of mycoplasma and other mollicutes. Contaminated cell cultures should be discarded and fresh stocks obtained. If replacement is not possible, then companies sell reagents that may eliminate the contamination. However, because the elimination process is likely to change the properties of the cells, avoiding contamination by practicing good technique is strongly preferable.

5. The percentage of trypsin used is cell type dependent. C3H10T1/2, C2C12, ATDC5, and NIH3T3 detach readily with 0.05% trypsin. We use 0.25% trypsin to detach MC3T3-E1 cells.

6. BMP-2 and other BMPs are commercially available from different companies. The suggested Medtronic product is FDA approved for clinical use. The effective dose of BMPs depends on additional variables including the length of exposure and media components. Indeed, the short BMP-2 half-life has prompted much research into strategies to extend in vitro and in vivo activity [22, 23]. Furthermore, BMP concentration profoundly affects the type of cells resulting from treatment [4–9]. Dosages of BMP-2 used in vitro commonly range from 2 to 500 ng/mL. The duration of treatment can range from hours to 21 days and may include a single or reoccurring dose. However, cells are generally committed to a particular path within the first few days of exposure.

7. Cells should be viewed under a microscope daily to assess morphology, a lack of contamination, and to estimate

confluency (the density of adherent cells). Cells should be passaged before they become fully confluent (~75–80% coverage). Passaging cells that are too sparse will cause a growth lag. Highly confluent cultures may spontaneously differentiate.

8. FBS inhibits trypsin and must be removed.
9. Observe the cells during the dissociation process to avoid loss of cell viability due to over-trypsinization. Detached cells appear rounded.
10. Automated cell counters that save time and are accurate are available commercially. Alternatively, manual counting using a hemocytometer is inexpensive and straightforward. Clean a glass hemocytometer and the heavy coverslip provided with the hemocytometer with water and then 70% alcohol before use. Moisten the coverslip with water and affix to the hemocytometer. Swirl the container to suspend the trypsinized cells, and sterilely remove a small aliquot. Fill each counting chamber with detached cells. Count cells in the four outside square sections on both sides (eight squares total). Each square is 1×1 mm and the depth is 0.1 mm. To calculate the number of trypsinized cells per mL of media, multiply the average number of cells in each counted square by 10,000. To increase accuracy, count the number of boxes required to total at least 100 cells. Dilute the trypsinized cells with media if they are too concentrated to count.
11. In mesenchymal cells, the concentration of BMP-2 or other BMPs influences myogenesis, adipogenesis, chondrogenesis, and osteogenesis. Differentiation is also influenced by cell density and the presence of soluble and matrix factors. Typically, cells are grown to confluence and then switched to specialized differentiation media with BMPs and other factors. Table 3 provides typical differentiation media compositions and references of a sampling of studies describing differentiation.

References

1. Urist M (1965) Bone: formation by autoinduction. Science 150:893–899
2. Wozney JM et al (1988) Novel regulators of bone formation: molecular clones and activities. Science 242:1528–1534
3. Katagiri T et al (1990) The non-osteogenic mouse pluripotent cell line, C3H10T1/2, is induced to differentiate into osteoblastic cells by recombinant human bone morphogenetic protein-2. Biochem Biophys Res Commun 172(1):295–299
4. Katagiri T et al (1994) Bone morphogenetic protein-2 converts the differentiation pathway of C2C12 myoblasts into the osteoblast lineage. J Cell Biol 127(6 Pt 1):1755–1766
5. Wang EA et al (1993) Bone morphogenetic protein-2 causes commitment and differentiation in C3H10T1/2 and 3T3 cells. Growth Factors 9(1):57–71
6. Denker AE et al (1999) Chondrogenic differentiation of murine C3H10T1/2 multipotential mesenchymal cells: I. Stimulation by bone morphogenetic protein-2 in high-density micromass cultures. Differentiation 64 (2):67–76
7. Sottile V, Seuwen K (2000) Bone morphogenetic protein-2 stimulates adipogenic differentiation of mesenchymal precursor cells in synergy with BRL 49653 (rosiglitazone). FEBS Lett 475(3):201–204

8. Cheng SL et al (2003) MSX2 promotes osteogenesis and suppresses adipogenic differentiation of multipotent mesenchymal progenitors. J Biol Chem 278(46):45969–45977
9. Rosen ED, MacDougald OA (2006) Adipocyte differentiation from the inside out. Nat Rev Mol Cell Biol 7(12):885–896
10. Sudo H et al (1983) In vitro differentiation and calcification in a new clonal osteogenic cell line derived from newborn mouse calvaria. J Cell Biol 96(1):191–198
11. Quarles LD et al (1992) Distinct proliferative and differentiated stages of murine MC3T3-E1 cells in culture: an in vitro model of osteoblast development. J Bone Miner Res 7(6):683–692
12. Wang D et al (1999) Isolation and characterization of MC3T3-E1 preosteoblast subclones with distinct in vitro and in vivo differentiation/mineralization potential. J Bone Miner Res 14(6):893–903
13. Li G et al (2005) Differential effect of BMP4 on NIH/3T3 and C2C12 cells: implications for endochondral bone formation. J Bone Miner Res 20(9):1611–1623
14. Atsumi T et al (1990) A chondrogenic cell line derived from a differentiating culture of AT805 teratocarcinoma cells. Cell Differ Dev 30 (2):109–116
15. Lorsch JR, Collins FS, Lippincott-Schwartz J (2014) Cell biology. Fixing problems with cell lines. Science 346(6216):1452–1453
16. Rogers MB, Shah TA, Shaikh NN (2015) Turning bone morphogenetic protein 2 (BMP2) on and off in mesenchymal cells. J Cell Biochem 116(10):2127–2138
17. Jiang S et al (2007) Mycoplasma infection transforms normal lung cells and induces bone morphogenetic protein 2 expression by post-transcriptional mechanisms. J Cell Biochem 104(2):580–594
18. Baker M (2016) Reproducibility: respect your cells! Nature 537(7620):433–435
19. Miller CJ et al (2003) Mycoplasma infection significantly alters microarray gene expression profiles. BioTechniques 35(4):812–814
20. Olarerin-George AO, Hogenesch JB (2015) Assessing the prevalence of mycoplasma contamination in cell culture via a survey of NCBI's RNA-seq archive. Nucleic Acids Res 43(5):2535–2542
21. Uphoff CC, Drexler HG (2014) Detection of mycoplasma contamination in cell cultures. Curr Protoc Mol Biol 106:28.4.1–28.4.14
22. Zhao B et al (2006) Heparin potentiates the in vivo ectopic bone formation induced by bone morphogenetic protein-2. J Biol Chem 281(32):23246–23253
23. Kim HY et al (2016) Development of porous beads to provide regulated BMP-2 stimulation for varying durations: in vitro and in vivo studies for bone regeneration. Biomacromolecules 17(5):1633–1642
24. Rawadi G, Vayssiere B, Dunn F, Baron R, Roman-Roman S (2003) BMP-2 controls alkaline phosphatase expression and osteoblast mineralization by a Wnt autocrine loop. J Bone Miner Res 18:1842–1853
25. Date T, Doiguchi Y, Nobuta M, Shindo H (2004) Bone morphogenetic protein-2 induces differentiation of multipotent C3H10T1/2 cells into osteoblasts, chondrocytes, and adipocytes in vivo and in vitro. J Orthop Sci 9:503–508
26. Tang QQ, Otto TC, Lane MD (2004) Commitment of C3H10T1/2 pluripotent stem cells to the adipocyte lineage. Proc Natl Acad Sci U S A 101:9607–9611
27. Hiraki Y, Inoue H, Shigeno C, Sanma Y, Bentz H, Rosen DM et al (1991) Bone morphogenetic proteins (BMP-2 and BMP-3) promote growth and expression of the differentiated phenotype of rabbit chondrocytes and osteoblastic MC3T3-E1 cells in vitro. J Bone Miner Res 6:1373–1385
28. Luppen CA, Smith E, Spevak L, Boskey AL, Frenkel B (2003) Bone morphogenetic protein-2 restores mineralization in glucocorticoid-inhibited MC3T3-E1 osteoblast cultures. J Bone Miner Res 18:1186–1197
29. Lawson MA, Purslow PP (2000) Differentiation of myoblasts in serum-free media: effects of modified media are cell linespecific. Cells Tissues Organs 167:130–137
30. Almodovar J, Guillot R, Monge C, Vollaire J, Selimovic S, Coll JL et al (2014) Spatial patterning of BMP-2 and BMP-7 on biopolymeric films and the guidance of muscle cell fate. Biomaterials 35:3975–3985
31. Shukunami C, Shigeno C, Atsumi T, Ishizeki K, Suzuki F, Hiraki Y (1996) Chondrogenic differentiation of clonal mouse embryonic cell line ATDC5 in vitro: differentiation-dependent gene expression of parathyroid hormone (PTH)/PTH-related peptide receptor. J Cell Biol 133:457–468
32. Yao Y, Zhai Z, Wang Y (2014) Evaluation of insulin medium or chondrogenic medium on proliferation and chondrogenesis of ATDC5 cells. Biomed Res Int 2014:569241

Chapter 2

In Vitro and In Vivo Osteogenic Differentiation of Human Adipose-Derived Stromal Cells

Clement D. Marshall, Elizabeth A. Brett, Alessandra L. Moore, Derrick C. Wan, and Michael T. Longaker

Abstract

Adipose-derived stromal cells (ASCs) are a promising population of cells that may be useful for the regeneration of human tissue defects. ASCs are capable of forming bone tissue in vitro and in vivo. Further work is required to determine the optimal conditions that will allow human ASCs to regenerate tissue in clinically significant tissue defects. Here we present three experimental protocols that are indispensable for the study of ASC osteogenic activity.

Key words Adipose-derived stromal cell, Osteogenesis, Osteogenic differentiation, Bone defect, Calvarial defect

1 Introduction

For decades, bone grafting has been used successfully to treat bone defects caused by infection, congenital conditions, traumatic injuries, and tumors. The advantage of this technique is the use of autologous bone tissue, which eliminates the risk of immune rejection or disease transmission [1]. However, the utility of bone grafting may be limited in patients in whom bone defects are so large or numerous that sufficient autologous bone is not available to fill the defects. For cases such as these, there is interest in the use of synthetic bone substitutes. The advantages of these substitutes are that they are available in unlimited quantities and there is no need for a donor site harvest, which carries its own risks and morbidities [2]. Despite their obvious benefits, synthetic bone substitutes carry a risk of foreign body reaction and infection. Also, their suitability is limited in children, who will rapidly outgrow the implant.

To address the shortcomings of autologous bone grafts and synthetic bone substitutes, there is great interest in the

Melissa B. Rogers (ed.), *Bone Morphogenetic Proteins: Methods and Protocols*, Methods in Molecular Biology, vol. 1891, https://doi.org/10.1007/978-1-4939-8904-1_2,

development of regenerative medicine techniques that utilize a patient's own cells or tissue to form new functional tissue. One of the most promising areas of regenerative research involves the use of adipose-derived stromal cells (ASCs). These are a population of multipotent cells that normally reside in the stromal vascular fraction (SVF) of adipose tissue [3]. They have the capacity to differentiate along adipocyte, chondrocyte, and osteoblast lineages by displaying markers associated with these cell types and by exhibiting phenotypes similar to them [4]. However, it remains controversial whether ASCs ever truly transform into adipocytes, chondrocytes, and osteoblasts or simply mimic certain aspects of these cells.

The term ASC is sometimes inaccurately used to refer to all cells of the SVF [4]. In addition to ASCs, the SVF contains many other cell types, including endothelial cells, erythrocytes, lymphocytes, macrophages, and smooth muscle cells [5]. To be precise, ASCs refer specifically to those cells of the SVF that possess multipotent differentiation capacity and that adhere to the plastic culture dish surface [6]. Additionally, ASCs should display the cell surface markers CD34, CD90, CD73, CD105, and CD44 and should be negative for CD45 and CD31 [6].

ASCs hold a number of advantages compared with other methods for filling bone defects. The first is that using a patient's own cells obviates any concern for immune rejection. Also, the risk of a reaction to foreign material is minimized because the ASCs may be delivered using scaffolds that rapidly degrade. ASCs can be easily isolated from fat tissue that is removed from a human patient during a liposuction or abdominoplasty procedure [7]. It is possible to purify ASCs in several hours without exposing the cells to cell culture, which avoids the risk of contaminating the cells with viruses, bacteria, or other human cells [8]. Harvesting of ASCs is invasive but is minimally morbid and could theoretically be performed in multiple sessions over time, if more cell applications were required.

Bone morphogenetic protein (BMP) is a class of proteins that are critical agents of bone formation and maintenance [9]. Not surprisingly, BMPs and BMP analogs as well as other agents that modulate the BMP pathway can increase the osteogenic activity of ASCs [10]. In the future, it may be useful to deliver drugs that affect BMP pathways along with ASCs in order to maximize new bone growth. A typical experiment exploring the effect of BMP-related pathways on ASC function would involve the isolation of ASCs followed by an assay of their osteogenic activity in the presence or absence of a BMP-modulating drug.

Here we present a protocol for ASC isolation and both in vivo and in vitro testing for osteogenic activity in three parts: (1) isolation of ASCs from human tissue and selection for ASCs bearing specific cell surface markers, (2) use of an in vitro assay to measure ASC osteogenic activity, and (3) use of an in vivo assay to measure new bone formation by ASCs. These techniques can be used together or in combination with other methods to understand the role that BMP-related drugs may have in altering ASC osteogenic activity and bone growth.

2 Materials

1. Dissection and cell culture.
2. 100 μm cell strainers.
3. 15-blade scalpel.
4. 4 mm diamond-coated trephine drill bit.
5. 40 μm cell strainer.
6. Alizarin powder.
7. Antibodies against CD34, CD31, and CD34 that are optimized for flow cytometry.
8. Ascorbic acid.
9. *C. hemolyticum* collagenase type II powder.
10. DAPI (4′,6-diamidino-2-phenylindole).
11. Dulbecco's Modified Eagle Medium.
12. Fetal bovine serum (FBS).
13. Fine-toothed forceps.
14. Glass FACS tubes.
15. β-Glycerophosphate.
16. 10,000 U/mL penicillin-streptomycin.
17. Phosphate-buffered saline.
18. PLGA (poly[lactic-co-glycolic acid]) scaffolds coated with hydroxyapatite. Alternately, collagen-based scaffolds that support cell survival may be used.
19. Poloxamer 188.
20. Povidone-iodine solution.
21. Red blood cell lysis buffer.
22. Sodium hydroxide.
23. Sterile surgical drapes.
24. Small sharp scissors.
25. Surgical drill.
26. Veterinary eye ointment.
27. *FACS buffer*: Add 10 mL fetal bovine serum (FBS), 5 mL poloxamer 188, and 5 mL of 10,000 U/mL penicillin-

streptomycin solution to 500 mL sterile phosphate-buffered saline (PBS).

28. *Digest mixture*: Add 0.375 g *C. hemolyticum* collagenase type II powder and 5 mL poloxamer 188 to 500 mL sterile Dulbecco's Modified Eagle Medium (DMEM).
29. *Standard medium*: Add 50 mL fetal bovine serum (FBS) and 5 mL of 10,000 U/mL penicillin-streptomycin solution to 500 mL Dulbecco's Modified Eagle Medium.
30. *Alizarin red stain*: Add 1 g alizarin powder to 500 mL distilled water. Use sodium hydroxide to bring the pH to 6.36–6.4. This solution may be stored in the dark for 1 month at 4 °C.
31. *Osteogenic differentiation medium (ODM)*: Prepare standard medium as described above, and add to the mixture 100 mg/mL ascorbic acid and 10 mM β-glycerophosphate.

3 Methods

3.1 Isolation of ASCs from Human Tissue and Selection for Specific Cell Surface Markers

A typical ASC experiment begins with the isolation of ASCs from human tissue. Traditional protocols for isolating ASCs from fresh fat tissue involve culturing the cells for several days in order to keep only those cells that adhere to plastic, which by definition should be ASCs [6]. However, it is clear that the time spent on cell culture plates drastically alters gene expression and cell surface marker profiles in ASCs and other adherent cell types [11]. For this reason, we employ an ASC isolation strategy that takes advantage of fluorescence-activated cell sorting (FACS) for markers known to identify ASCs within the SVF [12]. Specifically, this protocol gates for cells that are positive for CD34 and negative for both CD31 and CD45, which mark hematopoietic and endothelial cells. Erythrocytes are eliminated prior to FACS sorting using erythrocyte-specific lysis buffer. This protocol yields ASCs within several hours of obtaining fat tissue and avoids cell culture. However, cell phenotypes and cell surface marker expression profiles may be altered by the process of sorting them. Indeed, a consensus definition of ASCs based on cell surface marker profile has not been established, and different techniques for isolating ASCs may yield slightly different populations of cells [13].

1. Obtain human abdominal, flank, or thigh subcutaneous fat from a healthy donor undergoing elective liposuction. Fat from other anatomic sources has not been used routinely by our group. Keep the fat in a closed container, such as a plastic suction canister.
2. Add an equal volume of PBS (1:1 ratio) to the fat and mix by gentle agitation for 30 s. After the aqueous and fat layers have

separated (aqueous layer to the bottom, fat layer to the top), use a pipette to aspirate and discard the aqueous layer.

3. Decant the fat into a 1 L plastic container, and add an equal volume of the digest mixture. Close the container and clean the outside with 70% ethanol to avoid contaminating laboratory surfaces with human cells. Seal the cap with paraffin film to prevent spillage.
4. Agitate this mixture using an orbital shaker set to 180 revolutions per minute (rpm) at 37 °C for 30 minutes (min).
5. Neutralize the digestion by adding an equal volume of standard FBS-containing medium.
6. Distribute the mixture equally into plastic conical centrifuge tubes, and centrifuge the mixture at 300 × *g* for 20 min at 4 °C.
7. Aspirate and discard the supernatant, taking care to leave the pellet intact. Resuspend each pellet in 5 mL standard medium.
8. Combine the cell suspension samples by filtering them through a 100 μm cell strainer into one 50 mL conical centrifuge tube.
9. Centrifuge at 300 × *g* for 15 min at 4 °C.
10. Aspirate and discard the supernatant and resuspend the pellet in 5 mL room temperature red blood cell lysis buffer. Allow the mixture to sit for 5 min at room temperature, and then bring the volume to 15 mL with PBS. Centrifuge at 300 × *g* for 15 min at room temperature.
11. Aspirate the supernatant and resuspend the pellet in 15 mL standard medium. Again, filter through a 100 μm cell strainer into a single 50 mL conical centrifuge tube.
12. Count the cells using a hemocytometer and determine the total cell number.
13. Centrifuge the cell suspension at 300 × *g* for 5 min at 4 °C. Resuspend the cells in FACS buffer at a concentration optimized for your FACS antibodies. A typical concentration is 1 million cells per 100 μL buffer.
14. Transfer 20 μL of the suspension to a glass FACS tube labeled "Unstained" and bring the volume up to 200 μL with FACS buffer. Set aside on ice. Optionally, set aside a certain volume of cell suspension into a tube labeled "Unsorted." These cells will not undergo staining or FACS sorting and will serve as a control in subsequent experiments (*see* **Note 1**).
15. Label the tube containing the remaining cells as "Stained," and add FACS antibodies against CD34, CD31, and CD45. The CD31 and CD45 antibodies may be conjugated to the same fluorophore, but the CD34 antibody must be conjugated to a distinct fluorophore. If the CD31 and CD45 antibodies are conjugated to Pacific Blue, eFluor 450, or a similar color, this

allows for a "dump channel" when used with DAPI as a viability marker (*see* **Note 2**).

16. Once the antibodies are added to the "Stained" cells, pipette the suspension up and down several times to mix thoroughly. Place the cells on ice and cover the ice bucket with aluminum foil. Allow the cell/antibody mixture to incubate for 20 min with gentle agitation (or according to the instructions of the antibody manufacturer).
17. When the incubation period is complete, bring the total volume in the "Stained" tube to 2 mL with FACS buffer and centrifuge at 300 × *g* for 5 min at 4 °C. Aspirate and discard the supernatant. Again, bring the total volume to 2 mL with FACS buffer, and centrifuge at 300 × *g* for 5 min at 4 °C. Aspirate and discard the supernatant. Resuspend the cells in FACS buffer to a final concentration that is optimized for your FACS machine. We typically use 2 million cells per mL. Filter the cell suspension through a 40 μm cell strainer into a glass FACS tube labeled "Stained." The use of a FACS tube with a cell strainer built into the cap is helpful.
18. On the FACS machine, use the unstained cells to define a negative gate for the fluorescent markers. Use a viability stain to exclude nonviable cells (*see* **Note 2**). Then, using a 100 μm nozzle, sort viable $CD34^{+}CD31^{-}CD45^{-}$ cells into a tube containing standard medium. Ideally the collection tube should be chilled to 4 °C during the sort.

3.2 Alizarin Red Assay for In Vitro ASC Osteogenic Activity

Alizarin red is an in vitro assay performed on cultured cells that quantifies calcification on the surface of culture plates. Cells such as osteoblasts and ASCs with inherent osteogenic behavior will typically produce calcific nodules that adhere to the culture plate. The alizarin red assay is therefore a measurement of the osteogenic activity of cells in culture. Alizarin red staining was first described in the early twentieth century for the histological staining of skeletal specimens [14]. It is currently considered a critical part of characterizing ASCs and other osteogenic cell types. In this protocol, the cultured cells are stained with alizarin red, images are obtained, and the extent of staining is quantified.

1. Following isolation of ASCs, plate the cells in standard medium at a concentration of 100,000 cells per plate onto standard 10 cm polystyrene cell culture plates.
2. When the cells have reached 80% confluency, which typically takes several days, change to osteogenic differentiation medium (ODM). Change the medium three times per week for 14 days. The cells will begin to form bone nodules on the culture plate surface (*see* **Note 3**).

3. After 14 days in ODM, wash the cell layer three times with PBS.
4. Fix the cells in 100% ethanol for 15 min at room temperature. Then wash three times with distilled water.
5. Stain the cell layer with alizarin red solution for 60 min at room temperature. Use 10 mL for a 10 cm plate.
6. Wash with PBS five times. Make sure to wash away any stain that is adherent to the plastic but not marking bone nodule tissue. Be careful not to dislodge bone nodules during staining or washing. Remove the PBS by aspirating with a pipette. Place the plates upside down over a paper towel for 2 min to remove the remaining PBS (*see* **Note 4**).
7. Obtain images of the stained plates using a high-quality computer scanner. Load the images into ImageJ (NIH) or a similar image manipulation program. Convert the images to black and white. Define a threshold that includes all stained bone nodules but excludes background staining. Quantify the surface area of bone nodule in each plate.

3.3 Calvarial Defect Model for Assessing In Vivo Bone Regeneration

While the alizarin red assay measures the osteogenic activity of cells in culture, the calvarial defect model quantifies the ability of cells to assist in closing a bone defect, either by creating new bone tissue themselves or by enhancing endogenous bone growth through paracrine signaling. In vitro assays such as alizarin red can suggest that a cell population is osteogenic, but the use of an in vivo assay such as the calvarial defect model is critical in order to prove that the cells are capable of assisting with bone formation in the complex and hostile environment of a living organism. An added advantage of this model is that human or mouse cells can be used, as long as an appropriately immunosuppressed host mouse is selected. For human ASCs, the athymic CD-1 nude mouse is ideal. If the ASCs are of mouse origin, a syngeneic mouse should be used as the host. We typically use female mice to prevent the risk of fighting between animals.

1. Centrifuge the sorted cells at 300 × *g* for 5 min at 4 °C. Resuspend the cells in standard medium to a concentration of 1 million cells per 250 μL.
2. Obtain 4 mm circular PLGA (poly[lactic-co-glycolic acid]) scaffolds coated with hydroxyapatite [15]. Place the scaffolds individually into the wells of a standard 6-well culture plate. Cover each scaffold with 50 μL of cell suspension containing 200,000 cells. (A different cell number may be chosen based on the needs of your experiment.) Cover the plate with its lid, and incubate in a cell culture incubator at 37 °C for 30 min.

3. Anesthetize an 8–12-week-old female CD-1 nude mouse using inhaled isoflurane gas at 2% in 100% oxygen. Apply eye ointment to prevent desiccation of the cornea. Administer 1 mg/kg buprenorphine SR subcutaneously. Providing analgesia prior to the surgery results in superior pain relief (*see* **Note 5**).
4. Sterilize the skin of the dorsal aspect of the skull with povidone-iodine solution followed by 70% ethanol three times. Drape with sterile drapes, leaving the surgical site exposed.
5. Using a 15-blade scalpel, make a sagittal midline incision that extends over the majority of the dorsal skull. Using fine-toothed forceps, retract the skin on the right side of the incision to expose the right parietal bone. It may be necessary to use forceps and a spreading motion with small sharp scissors to clear away loose connective tissue and develop a plane between the periosteum of the skull and the skin (*see* **Note 6**).
6. Using a drill with an autoclaved 4 mm diamond-coated trephine drill bit, gently drill a circular defect through the right-sided parietal bone. It is important to stop drilling once the drill has traversed the bone. Do not extend the defect past bone into the dura mater layer, as this will result in irreversible brain damage. With fine forceps, carefully remove the circle of bone (*see* **Note 7**).
7. Place a scaffold containing cells into the defect, and then close the skin incision with running nylon suture.
8. Monitor the mouse and provide standard postoperative care according to institutional guidelines (*see* **Note 8**).
9. Shortly after the procedure, and then at 2, 4, 6, and 8 weeks, use micro CT scanning to measure the rate of calvarial defect closure. At 8 weeks, bone tissue may be excised and formalin fixed, decalcified, and then embedded in paraffin. Sections may be stained with Movat's pentachrome stain and histologically examined to assess the microscopic quality of the new bone tissue.

4 Notes

1. It is easy to forget to set aside an unstained tube prior to initiating antibody staining of the cells. However, not having the unstained cells will make accurate sorting impossible.
2. It is important to use a viability stain during FACS sorting in order to exclude dead or dying cells from the sort. 4′,6-diamidino-2-phenylindole (DAPI) and propidium iodide (PI) are two common, easy-to-use viability stains. We typically use DAPI, which is added to the cell suspension to a final concentration of 3 μM just before sorting the cells. Cells that are

stained by DAPI are considered to be nonviable and fluoresce strongly in the Pacific Blue channel. If the CD45 and CD31 antibodies are conjugated to Pacific Blue, a "dump channel" can be used to exclude $C45^+$, $CD31^+$, and nonviable cells simultaneously.

3. Be careful not to allow the cells to become more confluent than 80% before starting osteogenic differentiation. Cells that approach 100% confluency are prone to lifting off of the cell culture plate. When this happens, you will typically find a ball of cell material floating in the medium with no cells adhering to the surface of the culture plate.
4. It is easy to wash too vigorously, which will cause the stained cell layer to lift from the culture plate surface. This makes accurate quantification impossible. The best way to wash away excess alizarin red staining is with multiple gentle washes of PBS.
5. Carefully monitor the respiratory rate of the animal during surgery. Since most of the animal is covered with a drape it can be easy to forget to monitor this parameter. Failure to monitor the respiratory rate can result in the animal expiring during the surgery due to hypoventilation. If the respiratory rate slows, decrease the fraction of inspired isoflurane. Do not reduce the oxygen flow rate.
6. It may be necessary to spend 1–2 min developing the plane between the skin and periosteum by clearing away the loose connective tissue that normally connects these layers. Placement of the scaffold and cells into the defect will be much easier if the periosteal surface is free of connective tissue and the skin lifts freely away from it.
7. Creation of the calvarial defect is perhaps the most critical and challenging step in this protocol. Application of too much force will allow the drill but to move past the bone and into the dura mater, likely resulting in brain damage. One way to avoid this is to repeatedly apply light pressure to the skull with the drill bit in the same spot in a gentle tapping motion. If you are not sure whether the drill is all the way through the bone, try lifting the bone circle with the forceps. If it does not lift, then more drilling needs to be done.
8. While the mouse is recovering from anesthesia, actively warm it either under a warming lamp or on a warming pad. Mice under anesthesia are very prone to hypothermia. Monitor the mouse until it is able to ambulate independently.

References

1. de Boer HH (1988) The history of bone grafts. Clin Orthop Relat Res 226:292–298
2. Kolk A, Handschel J, Drescher W, Rothamel D, Kloss F, Blessmann M, Heiland M, Wolff KD, Smeets R (2012) Current trends and future perspectives of bone substitute materials—from space holders to innovative biomaterials. J Craniomaxillofac Surg 40(8):706–718. https://doi.org/10.1016/j.jcms.2012.01.002
3. Zuk PA, Zhu M, Ashjian P, De Ugarte DA, Huang JI, Mizuno H, Alfonso ZC, Fraser JK, Benhaim P, Hedrick MH (2002) Human adipose tissue is a source of multipotent stem cells. Mol Biol Cell 13(12):4279–4295. https://doi.org/10.1091/mbc.E02-02-0105
4. Casteilla L, Planat-Benard V, Laharrague P, Cousin B (2011) Adipose-derived stromal cells: their identity and uses in clinical trials, an update. World J Stem Cells 3(4):25–33. https://doi.org/10.4252/wjsc.v3.i4.25
5. Gimble JM, Bunnell BA, Frazier T, Rowan B, Shah F, Thomas-Porch C, Wu X (2013) Adipose-derived stromal/stem cells: a primer. Organogenesis 9(1):3–10. https://doi.org/10.4161/org.24279
6. Bourin P, Bunnell BA, Casteilla L, Dominici M, Katz AJ, March KL, Redl H, Rubin JP, Yoshimura K, Gimble JM (2013) Stromal cells from the adipose tissue-derived stromal vascular fraction and culture expanded adipose tissue-derived stromal/stem cells: a joint statement of the International Federation for Adipose Therapeutics and Science (IFATS) and the International Society for Cellular Therapy (ISCT). Cytotherapy 15(6):641–648. https://doi.org/10.1016/j.jcyt.2013.02.006
7. Zielins ER, Luan A, Brett EA, Longaker MT, Wan DC (2015) Therapeutic applications of human adipose-derived stromal cells for soft tissue reconstruction. Discov Med 19 (105):245–253
8. Lo DD, Hyun JS, Chung MT, Montoro DT, Zimmermann A, Grova MM, Lee M, Wan DC, Longaker MT (2012) Repair of a critical-sized calvarial defect model using adipose-derived stromal cells harvested from lipoaspirate. J Vis Exp (68). https://doi.org/10.3791/4221
9. Sanchez-Duffhues G, Hiepen C, Knaus P, Ten Dijke P (2015) Bone morphogenetic protein signaling in bone homeostasis. Bone 80:43–59. https://doi.org/10.1016/j.bone.2015.05.025
10. Fan J, Im CS, Guo M, Cui ZK, Fartash A, Kim S, Patel N, Bezouglaia O, Wu BM, Wang CY, Aghaloo TL, Lee M (2016) Enhanced osteogenesis of adipose-derived stem cells by regulating bone morphogenetic protein signaling antagonists and agonists. Stem Cells Transl Med 5(4):539–551. https://doi.org/10.5966/sctm.2015-0249
11. Walmsley GG, Rinkevich Y, Hu MS, Montoro DT, Lo DD, McArdle A, Maan ZN, Morrison SD, Duscher D, Whittam AJ, Wong VW, Weissman IL, Gurtner GC, Longaker MT (2015) Live fibroblast harvest reveals surface marker shift in vitro. Tissue Eng C Methods 21(3):314–321. https://doi.org/10.1089/ten.TEC.2014.0118
12. Chung MT, Zimmermann AS, Paik KJ, Morrison SD, Hyun JS, Lo DD, McArdle A, Montoro DT, Walmsley GG, Senarath-Yapa K, Sorkin M, Rennert R, Chen HH, Chung AS, Vistnes D, Gurtner GC, Longaker MT, Wan DC (2013) Isolation of human adipose-derived stromal cells using laser-assisted liposuction and their therapeutic potential in regenerative medicine. Stem Cells Transl Med 2(10):808–817. https://doi.org/10.5966/sctm.2012-0183
13. Tevlin R, McArdle A, Brett E, Chung MT, Paik K, Seo EY, Walmsley GG, Duldulao CR, Atashroo D, Zielins E, Vistnes S, Chan CK, Wan DC, Longaker MT (2016) A novel method of human adipose-derived stem cell isolation with resultant increased cell yield. Plast Reconstr Surg 138(6):983e–996e. https://doi.org/10.1097/PRS.0000000000002790
14. Dawson AB (1926) A note on the staining of the skeleton of cleared specimens with alizarin red S. Stain Technol 1(4):123–123
15. Gentile P, Chiono V, Carmagnola I, Hatton PV (2014) An overview of poly(lactic-co-glycolic) acid (PLGA)-based biomaterials for bone tissue engineering. Int J Mol Sci 15(3):3640–3659. https://doi.org/10.3390/ijms15033640

Chapter 3

Cell-Based Gene Therapy System for Delivering BMPs

Austin Dickerson, Eleanor L. Davis, Corinne Sonnet, Alan R. Davis, and Elizabeth A. Olmsted-Davis

Abstract

The use of an adenoviral vector to transduce cells allows for certain secreted proteins or growth factors to be generated in vivo in eukaryotic cells with accurate posttranslational processing. The use of transduced cells eliminates viral toxicity, allows for targeted expression of the secreted factor at a specific site, and ensures that the therapy will be turned off when the cells are cleared by the organism. Here we describe the delivery system which utilizes cells transduced with a non-replicating adenovirus containing bone morphogenetic protein 2 (BMP-2) in the E1 region of the cassette. With this method of delivery, small amounts of the protein can incite de novo bone formation.

Key words Adenoviral transduction, Delivery system, Bone morphogenetic protein, Bone formation, Cell-based gene therapy, Alkaline phosphatase assay, Osteoinduction

1 Introduction

Autologous bone grafts and recombinant human bone morphogenetic protein 2 (rhBMP2) are both systems used currently for bone regeneration and spine fusion [1]. While rhBMP2 has proven to exhibit great osteogenic potential, it causes many complications when delivered directly. These may result from the extremely high doses of protein that must be implanted using a carrier that controls protein release. If the protein is not carefully released over time at high enough concentrations, the bone reaction will not occur reliably. Further potential negative issues associated with this approach are nerve pain and damage, improper wound healing, and antibody formation against the BMP-2 [2]. Early efforts to develop other methods of BMP-2 delivery led to the implementation of gene therapy approaches. BMP delivery via adenovirus or adeno-associated virus directly to the site has been challenged by the efficiency of the virus uptake by the tissues [3] and the risk of free-virus diffusion to other sites [4]. Cell-based delivery systems eliminate many of these complicating factors and ensure a more

Melissa B. Rogers (ed.), *Bone Morphogenetic Proteins: Methods and Protocols*, Methods in Molecular Biology, vol. 1891, https://doi.org/10.1007/978-1-4939-8904-1_3,

reliable expression of the BMP. The cell-based model presented here avoids these common problems as the transgene does not integrate into the chromosome and the injected cells are cleared by the host immune system quickly, removing the risk of delivery of free virus or long-term exposure of the patient to virus integration [5]. Most importantly, this system ensures delivery of sufficient levels of BMP-2 protein delivery in vivo to the desired site without adverse effects resulting from toxic levels of virus and the subsequent long-term expression of BMP-2. Further, the eukaryotic cells continually produce protein processed similarly to the native protein. The release kinetics and activity of cell-delivered recombinant protein are superior to that delivered in a carrier material [6, 7]. Finally, the injection of transduced MC3T3-E1 cells encapsulated in a PEGDA microsphere provides an advantage over direct injection of cells because this biomaterial prevents an immune response against the transduced cells [8]. This chapter details methods designed for the cell-based gene therapy delivery of BMP-2.

2 Materials

When working with adenovirus, care should be taken to follow all local, state, and federal guidelines for working with biological hazardous agents. Cells should be autologous for the species being used (*see* **Note 1**).

2.1 Preparation of Cells

1. Sterile tissue culture dishes or flasks.
2. Cell line: MC3T3-E1 cells are appropriate for a mouse model (*see* **Note 1**).
3. Biologically appropriate media for cell line: For MC3T3 cells, use α-Minimum Essential Medium (α-MEM) + L-glutamine with 10% fetal bovine serum (FBS) and antibiotic-antimycotic containing Fungizone, penicillin, and streptomycin (*see* **Note 2**).
4. Hemocytometer.
5. Inverted light microscope with 10× objective.

2.2 Adenoviral Transduction

1. Biological safety cabinet (BSC) approved for BLS2 work.
2. Cell line compatible with animal model (*see* **Note 1**).
3. Biologically appropriate media for chosen cell line.
4. A standard transfection reagent, such as GeneJammer Transfection Reagent, used according to the manufacturer's instructions.

5. Titered replication-defective E1–E3 deleted adenovirus with a plaque-forming unit (pfu) to particle ratio (pfu:pt) of ≤1:100 that has been shown to be negative for replication-competent adenovirus (*see* **Note 3**).

2.3 Encapsulation of Cells in Microbeads

1. Monoject syringe: 100 μL syringe, 18½G needle.
2. Acrodisc® 13 mm syringe filter with 0.2 μm Supor® Membrane or similar products.
3. 300 mg acetophenone per mL 1-vinyl-2-pyrrolidinone (NVP) (*see* **Note 4**). Vortex solution for 1.5 min, and store in the dark by wrapping foil around microfuge tubes. Filter sterilize in a dark microfuge tube under the hood using a syringe and 0.2 μm filter. Use 3 μL of acetophenone per encapsulation.
4. 0.5 mL microfuge tube with graduated markings.
5. 5 mL glass tubes.
6. 900 mg acetophenone in mineral oil. Add 3 μL of acetophenone to every 1 mL of sterile filtered oil (*see* **Note 5**). Vortex ~30 s until bubbles form. Allow bubbles to dissipate before use (*see* **Note 6**). Use 1 mL of mineral oil per encapsulation.
7. HEPES buffered saline (HBS) pH 7.0–7.4: 10 mM HEPES, 100 mM NaCl.
8. 10 mM eosin Y stock solution.
9. 2× hydrogel (*see* **Note 7**): Mix in the following order: 0.1 g Acrylate-PEG-Acrylate (PEGDA, MW 10 kDa; *see* **Note 9**), 493 μL of HBS, 10 μL eosin Y stock solution, 3.75 μL NVP, 15 μL triethanolamine (TEOA) (*see* **Note 8**), and 10 μL pluronic acid. Vortex 5 s. Make 100 μL of hydrogel per encapsulation. Filter sterilize in a dark, amber microfuge tube under the hood using a syringe and 0.2 μm filter.
10. Biologically appropriate cell culture media.
11. 0.25% Trypsin-1× EDTA.
12. High-intensity light (*see* **Note 9**): Fiber-Lite MH-100 Illuminator with 300–650 nm liquid light guide.
13. Vortex machines (non-illuminated and illuminated vortex).
14. Personal protective equipment (PPE; *see* **Note 10**) and biological safety cabinet.
15. Hemocytometer.
16. 15 mL conical tubes.
17. Phosphate-buffered saline (PBS).

2.4 W20-17 Assay for BMP-2-Induced Alkaline Phosphatase

1. Tissue culture dishes with multiple wells (methods written for 24-well dish).
2. W20-17 mouse bone marrow cell line plated at 50,000 cells/well for a 24-well dish.

3. DMEM with 10% FBS and 1× antibiotic-antimycotic.
4. Conditioned media to be tested (*see* **Note 11**).
5. Phosphate-buffered saline (PBS).
6. Lysis buffer: 25 mM Tris–HCl, pH 8 with 0.5% Triton™ X-100 in PBS.
7. Cell scraper (*see* **Note 12**).
8. 100% ethanol over dry ice for quick freeze bath.
9. Tropix® CSPD® *ready-to-use* or similar chemiluminescent substrate for alkaline phosphatase with Sapphire II chemiluminescence enhancer solution (*see* **Note 13**).
10. A standard luminometer, such as the Glomax® 20/20 Luminometer.

3 Methods

3.1 Preparation of Cells

1. Cell line: MC3T3s are an appropriate cell line for mice (*see* **Note 1**). Allow to proliferate in culture until 80% confluence (Fig. 1) in a sterile incubator at 37 °C in an atmosphere of 5% CO_2 and a relative humidity 95%.
2. Expand into a 150 mm dish to 80% confluence for transduction (Fig. 1).

3.2 Adenovirus Transduction of Cells for Efficient Expression of BMP-2

(Protocol for one 150 mm culture dish.)

1. Prepare 5 mL of media without FBS or antibiotics in a sterile, glass container. Add transfection reagent at the recommended concentration and vortex. Let this mixture incubate at room temperature for 10 min.

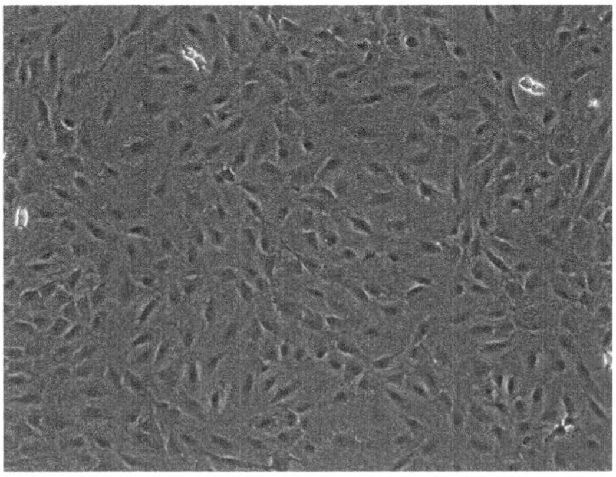

Fig. 1 Photomicrograph of Wistar skin fibroblast monolayer grown to approximately 80% confluence as desired for adenovirus transduction. The cells just stopped active replicating and entered arrest. Transduction will not be efficient in replicating cells that are less than 80% confluent; the cells would potentially lose the virus after initial uptake

2. Add the virus based on the number of cells (*see* **Note 14**) to the transfection reagent/media mixture. Vortex and incubate this mixture at room temperature for 10 min.
3. Prepare cells for transduction: Remove media from cells, and then add 8 mL of fresh media containing 10% FBS and antibiotic-antimycotic to the dish.
4. Add 5 mL of the media/transfection reagent/virus mixture to the cells, and then incubate plate for 4 h at 37 °C.
5. After 4 h, add 12 mL of media with 10% FBS and 1× antibiotic-antimycotic to the plates to bring the media volume up to 25 mL.

3.3 Encapsulation of Cells in Microbeads

One encapsulation is good for 4–6 × 10^6 cells (*see* **Note 15**).

1. Warm media and trypsin and allow light to heat up for 30 min.
2. Harvest cells for encapsulation: Prior to harvesting the cells, collect 1 mL of conditioned media in a microfuge tube, and store at −20 °C. This media should be used to determine BMP-2 expression using an alkaline phosphatase assay. Detach cells from plate using trypsin and count with a hemocytometer. Place cells in a 50 mL centrifuge tube and spin down at 400 rcf for 5 min. Resuspend 4–6 × 10^6 cells in 100 μL of complete media. Transfer the 100 μL cells to a low-binding 0.5 mL microfuge tube. Spin down at 400 × *g* for 5 min in a microfuge.
3. Set up the light (*see* **Note 16**).
4. In a separate 5 mL glass tube, add 1 mL mineral oil.
5. With a 100 μL tip, aspirate supernatant to leave 100 μL of supernatant volume over pellet (*see* **Note 17**). Use this remaining 100 μL supernatant media to resuspend the pellet.
6. Add 100 μL of 2× hydrogel solution to the tube containing the cells. Mix well by pipetting up and down without producing any bubbles.
7. Add the 200 μL cells/hydrogel mix to the bottom of the 1 mL mineral oil, and vortex for two quick pulses. Protect the tube from light for all future steps.
8. Moving quickly and keeping tube covered, pulse test tube briefly on a non-illuminated vortex that is separate from microsphere set-up. Then place the 5 mL tube under the light so that it is angled to touch the light guide and receive maximum light. The tube should be able to vibrate so the contents will be able to emulsify, rather than just swirl around the tube. Vortex for 2 s at high speed (**steps 9** and **10**). Rotate the tube for an additional 20 s under the light to expose all sides to the light, and pulse vortex three times at medium speed (**steps 6** and 7, Fig. 2).

Fig. 2 Schematic of the apparatus used for microencapsulation. The microfuge tube used during encapsulation should be placed on a vortex machine where the light can directly hit the tube. We also recommend putting an opaque cover over the apparatus to block outside light during this procedure. This procedure does not require the use of sterile BSL2 conditions; however, this could be accommodated

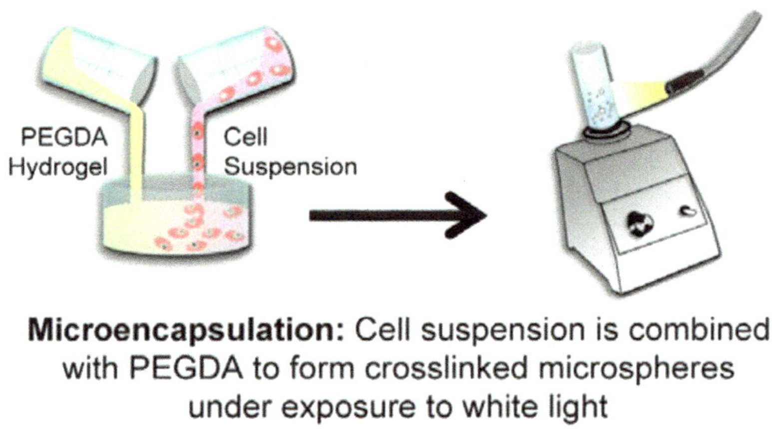

Fig. 3 Graphical depiction of the layering and synthesis of the hydrogel microspheres. Initially after addition of the media, the microencapsulated adenovirus-transduced cells reside between the oil and the aqueous layers. The microspheres are pelleted through centrifugation and placed in fresh media

9. Add 1 mL media with 10% FBS to the tube slowly to pull the microspheres out of the oil. Drop the 5 mL tube into a 15 mL conical tube, and let it incubate for 10 min at room temperature. After repeating process for all desired encapsulations, spin the 15 mL conical tubes at 350–450 rcf.
10. In the biological safety cabinet, slowly aspirate off the supernatant with the excess oil, microbeads, and media, leaving a small amount of media over the pellet (Fig. 3). For injection of encapsulated cells into animals, proceed to **step 11**, and for verification of expression of the transgene, proceed to **step 12**. We recommend that **step 12** is done with at least one of the prepared samples.
11. For injection, resuspend the pellet in 5 mL PBS, and transfer to a new 15 mL conical tube. Rinse the tube with another 5 mL of

PBS, and pool in the 15 mL tube. Spin down at 350–450 rcf for 5 min. Aspirate most of the PBS and repeat this washing step three times. After the last wash, slowly aspirate most of the PBS. Leave a small amount of PBS over the pellet (~100 μL). Prepare 100 μL Monoject syringe with an 18½G needle. Gently mix the microbeads in remaining PBS with the syringe. Add few drops of PBS to gather all beads if necessary (final volume may vary from one syringe to another). Prepare syringe for injection; make sure there are no bubbles. The microspheres can be cryopreserved, or they can be placed into culture. We recommend that the microbeads do not sit but rather are immediately injected for animal studies.

12. For verification of expression of transgene, resuspend the pellet of microbeads with 6 mL of media with 10% FBS. Distribute 2 mL in three distinct transwells containing plated W20-17 cells.

3.4 Alkaline Phosphatase Assay

1. Prepare W20-17 plate: Plate W20-17 cells at 50,000 cells/well in a 24-well plate. After 24 h, aspirate the cell culture media, and place 200 μL fresh media. Add 200 μL of conditioned media to be tested to three different wells of the plate (*see* **Note 18**). Incubate this plate for 72 h at 37 °C/5% CO_2.
2. After 72 h has passed, remove the cell culture media by aspiration. Wash the cells with 1 mL PBS per well. Add 200 μL of lysis buffer to each well. Scrape each well of the dish with cell scraper, and collect the lysate into 1.5 mL microfuge tubes. Store samples at −20 °C until ready to analyze.
3. Subject the samples to three freeze-thaw cycles by thawing the samples in a 37 °C water bath and then freezing the samples in 100% ethanol on dry ice.
4. In 1.5 mL microfuge tubes, prepare 100 μL of CSPD® *ready-to-use* solution in triplicate (*see* **Note 13**). Add 2 μL of the sample to the 100 μL CSPD® solution. Start a 30 s timer (*see* **Note 19**). Vortex 3 s. Read measurement in luminometer at 30 s precisely.

4 Notes

1. If possible, bone marrow mesenchymal stem cells (MSCs) should be used because they do not elicit host immune reactions in genetically matched hosts. These cells can be purchased. Our lab uses the MC3T3 cell line, an osteoblast precursor, for studies involving mice. For rat studies, we use the WSF line, or Wistar skin fibroblast, which we isolate and cultured using Dulbecco's modified Eagle's medium

(DMEM) + L-glutamine with 10% FBS and 1× antibiotic-antimycotic containing Fungizone, penicillin, and streptomycin. Time to confluency ranges based on cell type. Cell passaging should be limited.

2. Cell culture and all associated steps should be done in a Biological Safety Cabinet (BSC) under complete aseptic conditions [9]. The utmost care should be taken to prevent microbial contamination of cell cultures. Our lab uses a separate cell culture room as well as BSCs for all cell culture techniques.
3. Vector Development Laboratory, Houston, TX, USA (www.bcm.edu/vector), provides adenovirus with no less than a pfu:pt ratio of 1:100.
4. The acetophenone solution is best made 1–2 days before use in a limited light enclosure and stored wrapped in foil under argon at 4 °C until use. The NVP should be used within 2 weeks of opening.
5. Mineral oil should be embryo tested by the manufacturer. Mineral oil solution mixed with acetophenone can be stored at −20 °C for up to 1 month. It should only be manipulated with glass pipettes and glass tubes under the hood with no light exposure.
6. This step should be done in the dark. We use a lab bench drawer and keep it closed. The mixture should be homogenous throughout before use.
7. The hydrogel mixture can be stored at −20 °C for 1 month. This mix can be prepared on the benchtop, but it should only be exposed to limited light by using an amber microfuge tube. Prepare this mixture quickly and precisely in the order described, keeping tube covered at all times.
8. TEOA solution is extremely viscous. Take care when drawing it into pipette and mixing with the other hydrogel reagents. The PEGDA must be stored under argon at −20 °C. When ready to use, let the PEGDA stand at room temperature (RT) for 20 min before use.
9. The use of the high-intensity light is one of the most vital steps in the encapsulation process. Our lab selected this combination of light equipment after trial and error and has proven that this setup will work consistently. The liquid light guide cable was purchased separately and has a 0.2 in. core to fit the encapsulation equipment. Keeping a replacement bulb for the MH-100 Illuminator available is recommended.
10. Care must be taken when working with the encapsulation equipment, especially the high-intensity light. We use standard lab coats and gloves as well as Uvex™ Infra-Dura® lens or similar safety glasses. Cover the end of the light cable with

foil when not in use and while powering the equipment on and off to prevent accidental exposure.

11. Media to be tested can be drawn from plates of transduced cells before the cells are harvested and frozen in cryotubes at −20 °C until ready to analyze. The assay requires 200 μL of conditioned media per well for a 24-well plate. We recommend performing this assay in triplicate. We typically reserve 1 mL of conditioned media from the transduced plates to have an ample quantity.
12. We use a commercial cell scraper to scrape the lysate off of the wells. Alternatively, a large micropipette tip may be used.
13. The CSPD® solution is light sensitive. Tightly close the cap on the CSPD® solution bottle to keep light out. Also keep the bottle on ice if needed on the benchtop for an extended period of time.
14. The virus particle (vp) per cell concentration can be adjusted depending on cell type used. Our recommended ratios are 5000 vp/cell for MC3T3 cells and 7500 vp/cell for WSF cells.
15. Do not scale up encapsulation. Only encapsulate the recommended amount at a time. If you have more cells in your culture, then dilute them and do several encapsulations.
16. From this point on, use only glass tubes and process only one tube at a time.
17. It helps to use the graduation markings on the side of the 0.5 mL microfuge tube as a reference.
18. We recommend three experimental replicate assays for each biological sample.
19. Be very consistent with time. When measuring alkaline phosphatase activity, each sample must be allowed to react for exactly the same time period, precisely to the second. If not using an automated luminometer, we recommend starting the timer as soon as you add the lysate sample to the tube containing CSPD® solution and hit "read" on the luminometer as soon as the 30 s is up. Most importantly, be uniform in your technique and timing of each step of this process. Even small variations in the time from sample addition to luminometer reading can affect the data.

Acknowledgments

This work was supported by the Department of Defense grants W81XWH-13-1-0286, W81XWH-12-1-0475, and W81XWH-12-1-0274.

References

1. Lee KB, Taghavi CE, Hsu MS et al (2010) The efficacy of rhBMP-2 versus autograft for posterolateral lumbar spine fusion in elderly patients. Eur Spine J 19(6):924–930
2. Tannoury CA, An HS (2014) Complications with the use of bone morphogenetic protein 2 (BMP-2) in spine surgery. Spine J 14 (3):552–559
3. Olmsted EA, Blum JS, Rill D et al (2001) Adenovirus-mediated BMP2 expression in human bone marrow stromal cells. J Cell Biochem 82(1):11–21
4. Baltzer AW, Lattermann C, Whalen JD et al (2000) Potential role of direct adenoviral gene transfer in enhancing fracture repair. Clin Orthop Relat Res 379(Suppl):S120–S125
5. Lazard ZW, Heggeness MH, Hipp JA et al (2011) Cell-based gene therapy for repair of critical size defects in the rat fibula. J Cell Biochem 112(6):1563–1571
6. Woo EJ (2012) Adverse events reported after the use of recombinant human bone morphogenetic protein 2. J Oral Maxillofac Surg 70 (4):765–767
7. Peel SA, Hu ZM, Clokie CM (2003) In search of the ideal bone morphogenetic protein delivery system: in vitro studies on demineralized bone matrix, purified, and recombinant bone morphogenetic protein. J Craniofac Surg 14(2):284–291
8. Alvarez-Urena P, Davis EL, Sonnet C et al (2016) Encapsulation of adenovirus BMP2-transduced cells with PEG-Da hydrogels allows bone formation in the presence of immune response. Tissue Eng A (in press)
9. Cote RJ (2001) Aseptic techniques for cell culture. Curr Protoc Cell Biol Chapter 1:Unit 1.3

Chapter 4

Generation of Endogenous BMP Transcriptional Reporter Cells Through CRISPR/Cas9 Genome Editing

Luke D. Hutchinson, Polyxeni Bozatzi, Thomas Macartney, and Gopal P. Sapkota

Abstract

Transcriptional reporter systems allow researchers to investigate the function and regulation of transcription factors. Conventional systems employ artificial cDNA overexpression vectors containing either a promoter fragment or specific nucleotide sequence repeats upstream of *firefly* luciferase or fluorescent reporters, such as green fluorescence protein (GFP) cDNA. These systems suffer mainly from the lack of chromatin context. Here, we describe the rapid generation of endogenous transcriptional reporter cells for the bone morphogenetic protein (BMP) pathway using CRISPR/Cas9 genome editing. In principle, our methodology can be applied to any cell line. The endogenous reporters will provide a robust system for the investigation of BMP transcriptional activity in the context of native chromatin landscape and facilitate chemical and genetic screens.

Key words Transcription, Reporter vectors, CRISPR/Cas9, Genome editing

1 Introduction

Bone morphogenetic proteins (BMPs) are members of the transforming growth factor beta (TGFβ) superfamily of cytokines, with crucial roles in embryogenesis and adult tissue homeostasis. Aberrant BMP signaling is thus associated with numerous developmental defects and human pathologies [1]. BMPs signal via binding to and activation of cognate type I and type II BMP receptor serine/threonine kinases. The activated type I BMP receptors subsequently phosphorylate the intracellular SMAD transcription factors 1, 5, and 8 at a C-terminal Ser-X-Ser motif, thereby inducing their interaction with the common mediator SMAD4. The SMAD1/5/8-SMAD4 complex is then able to translocate into the nucleus, where it associates with various transcriptional cofactors to regulate gene expression [2]. Inhibitor of differentiation 1 (ID-1), a dominant negative regulator of basic helix-loop-helix (bHLH) transcription factors, is a direct gene target of BMP signaling

Melissa B. Rogers (ed.), *Bone Morphogenetic Proteins: Methods and Protocols*, Methods in Molecular Biology, vol. 1891, https://doi.org/10.1007/978-1-4939-8904-1_4,

[3]. Therefore, ID-1 mRNA and protein expression are frequently employed as a downstream readout of the BMP signaling pathway.

We have previously reported the generation of endogenous transcriptional reporter cells for the TGFβ pathway using CRISPR/Cas9 genome editing [4]. This system involved the insertion of a transcriptional reporter cassette downstream of the endogenous promoter for the TGFβ target gene plasminogen activator inhibitor 1 (PAI-1). In the following protocol, we describe the generation of an endogenous BMP transcriptional reporter cell line, in which a reporter cassette containing *firefly luciferase* and *GFP* gene has been inserted adjacent to the endogenous *ID-1* promoter sequence. The generation of the cell line thus enables the study of endogenous BMP transcriptional activity as well as chemical and genetic high-throughput screens.

2 Materials

2.1 Mammalian Cell Culture and Cell Transfection Reagents

1. 1× phosphate-buffered saline (PBS) (such as Gibco Dulbecco's phosphate-buffered saline (DPBS) or equivalent).
2. Trypsin solution (such as Gibco™ Trypsin-EDTA (0.05%) or equivalent).
3. Dulbecco's Modified Eagle Medium (DMEM) and 4.5 g L^{-1} D-Glucose (such as Gibco™ DMEM or equivalent).
4. Fetal bovine serum (FBS).
5. 10,000 units mL^{-1} penicillin-streptomycin.
6. 200 mM L-glutamine.
7. Reduced-serum medium (such as Gibco™ Opti-MEM or equivalent).
8. 1 μg mL^{-1} polyethyleneimine (PEI) MAX (MW 25,000).
9. QIAprep Spin Miniprep Kit.

2.2 Normal and Serum-Free Cell Culture Media

The cell lines used in this protocol are grown and maintained in the following culture media: Dulbecco's Modified Eagle Medium (DMEM) supplemented with 10% (v/v) fetal bovine serum (FBS), 2 mM L-glutamine, 100 units mL^{-1} penicillin, and 100 μg mL^{-1} streptomycin (subsequently referred to as D10F medium). For cytokine stimulation, serum-free media are as above however without addition of FBS.

2.3 Cytokines and Inhibitors

1. Recombinant Human BMP-2 Protein
2. SB-505124 (ALK4/5/7 inhibitor)
3. LDN193189 (ALK2/3 inhibitor)
4. Dimethyl sulfoxide (DMSO)

2.4 Luciferase Assay

1. 2× luciferase assay buffer: 50 mM Tris/phosphate (pH 7.8), 16 mM $MgCl_2$, 2 mM DTT (dithiothreitol), 1 mM ATP, 30% (v/v) glycerol, 1% (w/v) bovine serum albumin (BSA), 250 μM D-luciferin, and 8 μM sodium pyrophosphate.
2. 5× luciferase cell Culture lysis reagent (CCLR).
3. Cell culture microplate (96-well), flat-bottom, white, sterile.
4. Plateseal, permanent seal for microplates.
5. Multilabel plate reader (e.g., PerkinElmer 2104-0010A).
6. Coomassie (Bradford) Protein Assay Kit.

3 Methods

3.1 Design and Cloning of Guide RNA (gRNA)

A paired nickase approach for CRISPR was adopted to minimize off-targeting [5]. Guide pairs cutting close to the ATG start codon of ID-1 (ENSG00000125968) were analyzed, and the optimal N-terminal guide pairs (sense 5′-(G)CGGCAAGACAGC-GAGCGGTG and antisense 5′-(G)CGGCCTTCAGCGCG-CAGCTG)) were identified using the Sanger Institute CRISPR web tool (http://www.sanger.ac.uk/htgt/wge/find_crisprs). This pair was chosen on the basis of having the lowest combined off--targeting score while being situated entirely within the coding region to allow silent mutations to be introduced. Complementary oligonucleotides with BbsI-compatible overhangs were designed as described previously [6], and after annealing, the resulting dsDNA guide inserts were ligated into BbsI-digested target vectors. The antisense guide was cloned into the spCas9-D10A expressing vector pX335 (Addgene, #42335) and the sense guide into the puromycin-selectable plasmid pBABED-Puro-U6 (http://mrcppureagents.dundee.ac.uk; #DU33769).

The resulting clones are available from the MRC-PPU Reagent webpage (#DU52298 and #DU52279).

3.2 Construction of Donor Vector Containing the Polycistronic Reporter Cassette

A donor vector containing the following components was designed to be inserted in-frame at the ATG start codon of *ID-1*: complementary DNA (cDNA) sequences encoding for *firefly* luciferase and green fluorescent protein (GFP); two regions (approximately 500 nucleotides) homologous to the *ID-1* gene locus around the start codon, located upstream and downstream of the cassette, to facilitate homologous recombination; silent mutations that were included in the downstream homologous region to prevent cleavage by the gRNA-directed Cas9 nuclease; an internal ribosome entry site (IRES) element that was incorporated immediately downstream of the luciferase gene to enable separate expression of GFP and luciferase enzyme from the same donor vector; and a sequence encoding the 2A self-cleaving peptide, located downstream of the *GFP* gene, which ensures cleavage of the GFP protein from the endogenous ID-1 protein.

This plasmid is available from the MRC-PPU Reagent webpage (#DU52518).

3.3 Transfection of gRNA and Donor Vector

1. Immortalized human keratinocyte HaCaT cells are cultured in 10 cm cell culture plates in 9 mL D10F medium until approximately 60% confluent (*see* **Note 1**).
2. Prepare transfection solution by adding 1 μg of sense gRNA, 1 μg antisense gRNA, 3 μg cDNA donor vector (all prepared using QIAprep Spin Miniprep Kit; #27104), and 20 μL 1 mg mL^{-1} PEI to 1 mL of sterile Opti-MEM (reduced-serum media) in a 15 mL conical centrifuge tube. Vortex for 20 s and subsequently incubate at room temperature for 15 min.
3. Add transfection solution dropwise to the cells (cultured in 9 mL of D10F medium), and swirl plate to mix.
4. 24-h post-transfection, exchange medium for 10 mL fresh D10F medium supplemented with 2 μg mL^{-1} puromycin (*see* **Note 2**).
5. 48-h post-transfection, exchange medium for D10F medium supplemented with 2 μg mL^{-1} puromycin.
6. 72-h post-transfection, exchange medium for D10F without additional antibiotic, and allow the cells to recover for 24 h (*see* **Note 3**).
7. Allow cells to grow until approximately 90% confluent, and prepare them for fluorescence-activated cell sorting (FACS).

3.4 Fluorescence-Activated Cell Sorting (FACS)

1. Aspirate medium and wash cells twice with 1× Dulbecco's phosphate-buffered saline (DPBS).
2. Add trypsin (2 mL per 10 cm plate), and incubate at 37 °C for 5–10 min until the cells detach. Resuspend cells in 10 mL D10F medium.
3. Transfer cell suspension into a 15 mL conical centrifuge tube, and centrifuge at 100 × *g* for 3 min.
4. Aspirate the supernatant, and resuspend the cell pellet in DMEM supplemented with 1% (v/v) FBS at a density of 2.0–5.0 × 10^6 cells mL^{-1}.
5. Viable, single cells are identified on the basis of forward scatter (FSC) vs. side scatter (SSC) and FSC-width vs. FSC-area, respectively. GFP fluorescence is measured with 488 nm excitation and emission detected at 530 ± 30 nm.
6. Collect individual GFP expressing cells into individual wells of a 96-well cell culture plate containing conditioned D10F medium supplemented with 20% (v/v) FBS (*see* **Note 4**).

7. Allow cells to proliferate, transferring into progressively larger surface area cell culture plates until sufficient quantities are obtained for subsequent analysis.

3.5 BMP Signaling Analysis on Reporter Cells

Single-cell clones that were positive for GFP expression and exhibited enhanced luciferase activity compared to control cells will be subsequently analyzed for responsiveness to BMP stimulation.

1. Seed 1.0×10^5 cells per well in a 6-well cell culture plate, and culture until approximately 60% confluent.
2. Replace medium with serum-free culture media for approximately 16 h (*see* **Note 5**).
3. Prepare the required small molecule inhibitors in serum-free culture media, in the presence or absence of 6.25 ng mL^{-1} BMP-2, in 15 mL conical centrifuge tubes.
4. The BMP type I receptor (ALK2/3) inhibitor LDN-193189 can be used at 100 nM concentration as a positive control for BMP pathway inhibition, whereas the TGFβ type I receptor (ALK4/5/7) inhibitor SB-505124 can be used at 1 μM concentration as a negative control for BMP pathway inhibition.
5. Aspirate serum-free culture media, and dispense 2 mL of each cytokine and/or inhibitor combination into the appropriate well. Incubate for 6–8 h at 37 °C (*see* **Note 6**).

3.6 Luciferase Assay

1. Prepare 1× luciferase lysis buffer by adding 1 volume of 5× Cell Culture Lysis Reagent to 4 volumes of ultrapure deionized H_2O (*see* **Note 7**).
2. Aspirate media and wash cells twice with 1× DPBS to remove culture medium (*see* **Note 8**).
3. Dispense sufficient volume of 1× CCLR to cover the cells (approximately 300 μL per well in a 6-well cell culture plate), and incubate cells for 5–10 min at room temperature with gentle agitation in a shaker.
4. Scrape the cells from the plate surface, and transfer lysates into 1.5 mL microcentrifuge tubes on ice.
5. Vortex the cell lysates for 10–15 s, and then centrifuge at $12,000 \times g$ for 2 min at 4 °C. Transfer the supernatant into new 1.5 mL microcentrifuge tubes.
6. Dispense 40 μL of clarified cell lysate per well of a 96-well microplate (in triplicate for each condition), and subsequently add an equal volume of 2× luciferase assay buffer per well.
7. Seal the microplate with adherent sealing tape, and place the microplate on a vibrating platform for 30–60 s. Immediately obtain luminescence values using a microplate reader (*see*

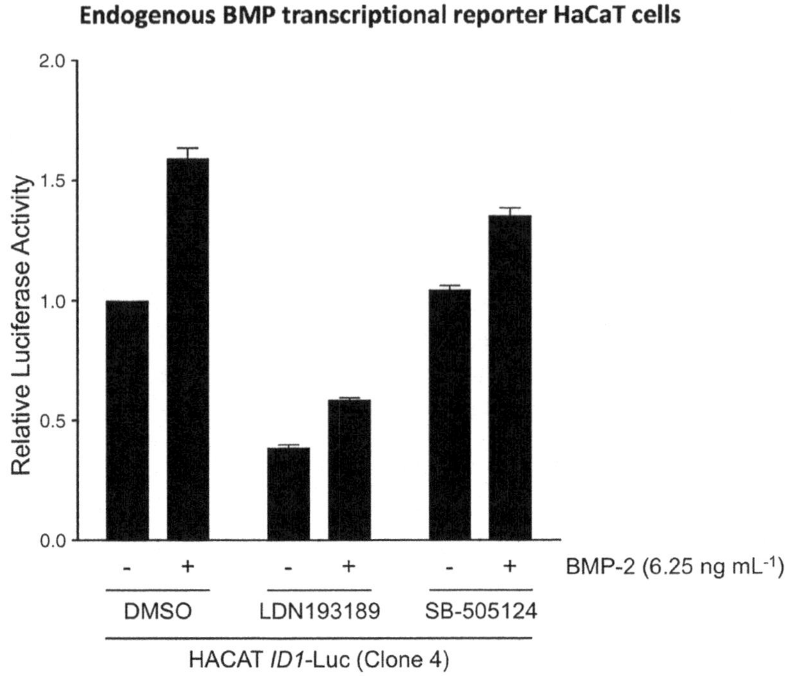

Fig. 1 Validation of the BMP transcriptional reporter HaCaT cells: GFP-positive single-cell isolate (clone 4) was expanded and tested for the insertion of the reporter cassette at the correct locus using BMP induction and BMPR1 (LDN-193189) and TGFBR1 (SB-505124) inhibitors, as indicated

Note 9). An example of BMP-dependent luciferase activity using the endogenous BMP reporter cells is included in Fig. 1.

8. Use residual cell lysate to determine the protein concentration for each sample to enable normalization of luminescence values.

4 Notes

1. Conditioned D10F medium was collected from routine HaCaT cultures and sterile filtered for subsequent use in aiding single cell colony recovery following FACS sorting.
2. Include an untransfected control cell plate to assess effectiveness of antibiotic selection treatment.
3. In order to increase the efficiency of homologous recombination, perform a repeat transfection and antibiotic selection on the same cells following an identical procedure.
4. Centrifuge the 96-well plate at 300 × *g* for approximately 30 s to ensure single cells are situated at the bottom of the wells.

5. Serum deprivation minimizes basal autocrine growth factor/ cytokine signaling, thereby enabling the investigation of a single signaling pathway, in this instance the BMP signaling pathway.
6. 6–8 h incubation with 6.25 ng mL^{-1} BMP-2 is sufficient to enable the expression of endogenous ID-1 transcriptional reporter cassette.
7. Equilibrate 1× CCLR to room temperature prior to use.
8. This cell-washing step can be performed at room temperature.
9. It is very important to minimize the time between the addition of the luciferase assay buffer and obtaining the luminescence values. Therefore, ensure all equipment and reagents are prepared prior to the addition of the buffer.

Acknowledgments

G.P.S. is supported by the UK MRC (MC_UU_12016/3) and the DSTT companies (Boehringer Ingelheim, GlaxoSmithKline, and Merck Serono); L.D.H. and P.B. are funded by the MRC Ph.D. studentships.

References

1. Wang RN, Green J, Wang Z, Deng Y, Qiao M, Peabody M, Zhang Q, Ye J, Yan Z, Denduluri S, Idowu O, Li M, Shen C, Hu A, Haydon RC, Kang R, Mok J, Lee MJ, Luu HL, Shi LL (2014) Bone Morphogenetic Protein (BMP) signaling in development and human diseases. Genes Dis 1:87–105
2. Ross S, Hill CS (2008) How the Smads regulate transcription. Int J Biochem Cell Biol 40:383–408
3. Korchynskyi O, ten Dijke P (2002) Identification and functional characterization of distinct critically important bone morphogenetic protein-specific response elements in the Id1 promoter. J Biol Chem 277:4883–4891
4. Rojas-Fernandez A, Herhaus L, Macartney T, Lachaud C, Hay RT, Sapkota GP (2015) Rapid generation of endogenously driven transcriptional reporters in cells through CRISPR/ Cas9. Sci Rep 5:9811
5. Ran FA, Hsu PD, Lin CY, Gootenberg JS, Konermann S, Trevino AE, Scott DA, Inoue A, Matoba S, Zhang Y, Zhang F (2013) Double nicking by RNA-guided CRISPR Cas9 for enhanced genome editing specificity. Cell 154:1380–1389
6. Ran FA, Hsu PD, Wright J, Agarwala V, Scott DA, Zhang F (2013) Genome engineering using the CRISPR-Cas9 system. Nat Protoc 8:2281–2308

Chapter 5

High-Throughput, Biosensor-Based Approach to Examine Bone Morphogenetic Protein (BMP)–Receptor Interactions

Senem Aykul and Erik Martinez-Hackert

Abstract

Binding of a BMP to its cognate cell surface receptors is the initiating step in the BMP signaling cascade. Thus, knowing which BMP–receptor complexes form is vital for understanding the physiological activities of a particular BMP. Here, we describe a surface plasmon resonance (SPR)-based, high-throughput approach that allows fast identification and evaluation of BMP–receptor complexes. Briefly, the extracellular, BMP-binding domains of receptors are produced as human IgG1-Fc-fusion proteins. The Fc moiety enables simple capture of the Fc-receptor-fusion protein on the sensor chip, supports a highly reproducible, uniform approach of surface regeneration, and ensures full activity of the receptor moiety. BMPs are injected over the captured receptors at one concentration (approximately 60–100 nM), permitting stratification of high-affinity, medium-affinity, and low-affinity binders. Using this concentration range, equilibrium dissociation constants for high-affinity and medium-affinity binders can be estimated with good accuracy and with great precision from the single injection binding curves.

Key words Surface plasmon resonance, Bone morphogenetic protein, Activin, TGF-β, Protein–protein interactions

1 Introduction

The transforming growth factor (TGF)-β family plays a central role in animal cell physiology. Members of the family have vital functions controlling cell growth, differentiation, homeostasis, and cell death [1, 2]. They also play critical roles in many diseases, including inflammation, cancers, and fibrosis [3–5]. In mammals, the family consists of 33 genes encoding TGF-βs, bone morphogenetic proteins (BMPs), growth and differentiation factors (GDFs), activins, and nodal.

The basic mechanism of TGF-β family action is well established: a growth factor (TGF-β, BMP, GDF, activin, or nodal) binds the extracellular domain of its cognate type I and type II receptors, activating an intracellular phosphorylation cascade that leads to transcription of SMAD-regulated target genes [6, 7]. But

Melissa B. Rogers (ed.), *Bone Morphogenetic Proteins: Methods and Protocols*, Methods in Molecular Biology, vol. 1891,
https://doi.org/10.1007/978-1-4939-8904-1_5,

our complete understanding of TGF-β family action remains elusive, as cellular responses depend on specific growth factor–receptor pairings and on co-receptors that regulate the activities of some signaling complexes [8, 9]. Notably, many of these pairings, which combined form the TGF-β family interaction and signal transduction network, remain unknown.

Surface plasmon resonance (SPR) is a powerful and highly valuable tool for analysis of the BMP/TGF-β family interaction network [10]. It can identify specific growth factor–receptor–co-receptor complexes with great accuracy and precision [11–14]. It can also be used to characterize cooperative or inhibitory interactions [14–17]. In a standard SP-binding experiment, one binding partner (the "ligand" in SPR lingo) is immobilized or captured on a sensor chip, while the other binding partner (the "analyte" in SPR lingo) is injected over the sensor chip surface. Note that this protocol follows SPR convention and limits using the word "ligand" to the protein bound to the chip, in this case the BMP receptor. SPR is an optical method that measures the refractive index near the sensor chip surface. Binding is detected as a change in the refractive index. The refractive index increases as mass accumulates due to binding of the SPR "analyte" (the BMP) to the captured SPR "ligand" (the receptor). Changes in the refractive index reveal the association and dissociation of molecules, indicating the kinetic rates of an interaction [10]. Equilibrium binding constants can be derived from these kinetic rates. As SPR is extremely sensitive, it requires relatively small quantities of test material. Importantly, test proteins do not require molecular labels that can compromise their activity. Taken together, these characteristics make SPR especially useful and powerful for studying the TGF-β family and for elucidating the BMP interaction network.

The accurate SPR approach relies on highly reproducible capture methods using receptor-Fc fusion proteins (Fig. 1). The Fc moiety is used to capture the receptor (SPR "ligand") on the sensor chip via ultrahigh-affinity protein A or anti-Fc antibody binding. Capture is simple; sensor chip regeneration is standardized, effective, and reproducible; and receptor activity is not compromised. Injection of BMP solution (SPR "analyte") over the captured receptors provides highly reproducible and informative binding data that enables stratification of high-, medium-, and low-affinity BMP–receptor pairings (Fig. 2). The association and dissociation curves can be used to obtain binding rates and to derive equilibrium binding constants with good accuracy. In addition to mechanistic characterizations, this approach can be used for quality control. Binding curves for benchmarking are easily established, and the activity of a BMP or receptor over time or in different formulations can be monitored quickly and with great precision. The ability to capture any receptor-Fc-fusion protein without compromising

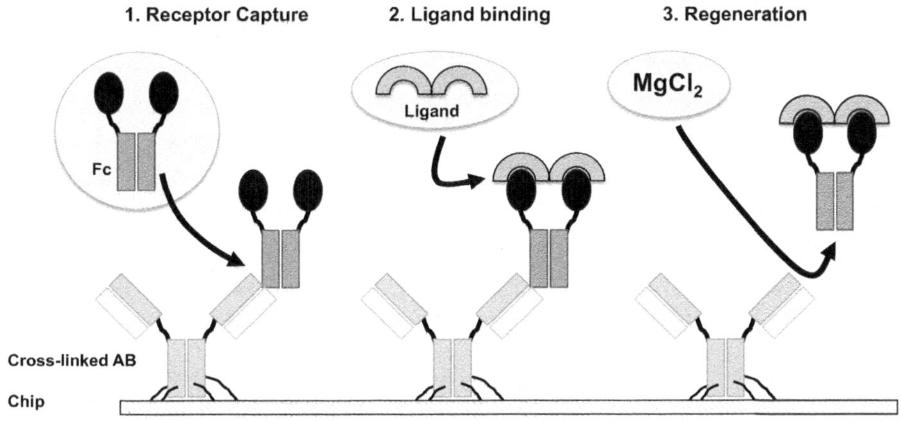

Fig. 1 Schematic representation of the capture, binding, and regeneration cycle. The chip is prepared by cross-linking anti-Fc antibody (AB) on the sensor chip. In the first step of the reaction (1), receptor-Fc fusion protein is captured by the antibody (approximately 150–250 RUs of receptor-Fc fusion protein). In the second step of the reaction (2), 80 nM BMP is injected, and binding to the receptor moiety is detected. In the third step of the reaction (3), the BMP-receptor-Fc fusion complex is eluted with $MgCl_2$ to regenerate the surface. To test binding of a different BMP, new receptor-Fc fusion protein has to be captured (step 1). A protein A chip can also be used to capture receptor-Fc proteins, but regeneration conditions need to be adapted

receptor activity, combined with the ability to rapidly and reproducibly regenerate the sensor chip surface, permits a fast, high-throughput evaluation of any BMP–receptor complex.

2 Materials

2.1 Surface Plasmon Resonance (SPR) Instrument

1. Experiments presented here were carried out on a Biacore 2000 (GE Lifesciences, discontinued) (*see* **Note 1**).

2.2 Surface Preparation Reagents

1. 0.4 M *N*-(3-Dimethylaminopropyl)-*N*-ethylcarbodiimide hydrochloride (EDC) dissolved in water. Solid EDC is hygroscopic and temperature sensitive and must be aliquoted as dry powder under N_2 atmosphere. Sealed EDC aliquots can be stored at −20 °C. EDC solutions must be prepared fresh before use.
2. 0.1 M *N*-Hydroxysuccinimide (NHS) dissolved in water. NHS solutions must be prepared fresh and immediately before use.
3. 1.0 M Ethanolamine diluted in water. The pH is adjusted to 8.2 with concentrated HCl. A pH range between 8.0 and 8.5 is acceptable. Prepare ethanolamine fresh before use (*see* **Note 2**).
4. CM3 or CM5 chip (e.g., Biacore BR-1003-99) stored at 4 °C. Before use, equilibrate at room temperature for at least 30 min. This chemistry is used with the antibody in **item 5**. Sensor chip

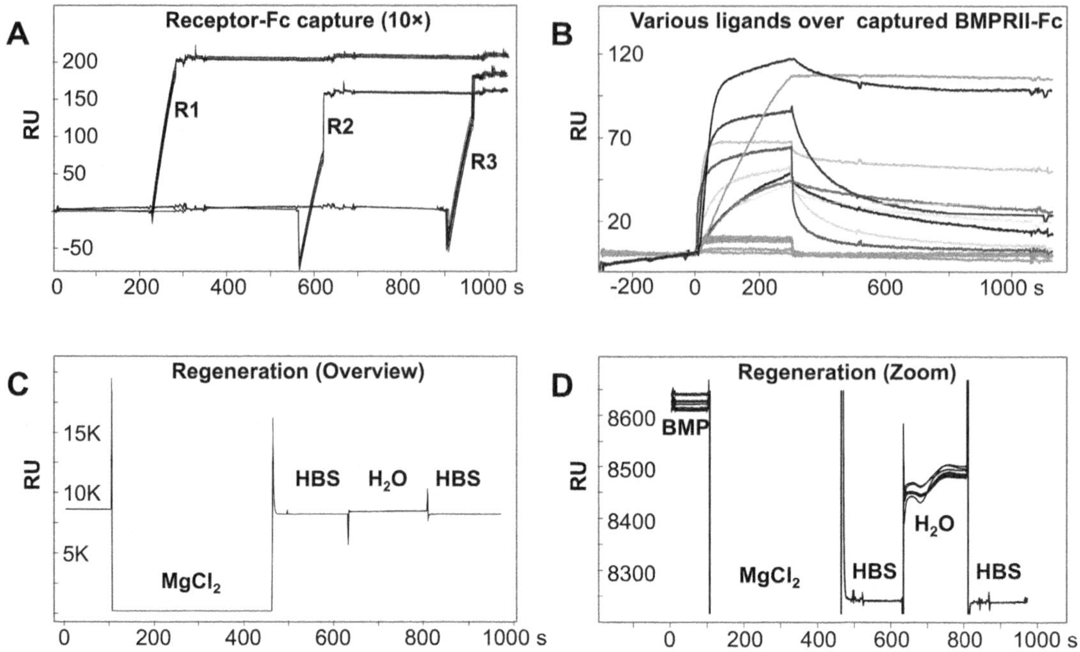

Fig. 2 Example of a multi-reaction cycle. (**a**) Three different receptor-Fc fusion proteins are captured sequentially on a four-channel sensor chip (R1-Fc, R2-Fc, and R3-Fc). Capture is highly reproducible as seen in the exact matching of ten capture cycles for R1-Fc (from approximately 200 to 300 s), R2-Fc (from approximately 550 to 650 s), and R3-Fc (from approximately 900 to 1000 s). (**b**) BMP at a concentration of 80 nM is injected over the sensor chip following receptor capture. In the example shown here, 15 different members of the TGF-β/BMP family were injected to test for binding to BMPRII-Fc. Each curve represents a different molecule. Each protein elicits a specific response as seen in the distinct association rates (from 0 to 300 s) and dissociation rates (from 300 to 1100 s). The strongest BMPRII-Fc binders were activin B (black response curve of approximately 110 RU) and nodal (dark gray response curve of approximately 100 RU). (**c**) Regeneration restores the baseline response ±5 RU (from 900 to 1000 s) by eluting the captured receptor-Fc fusion protein, including the receptor-bound protein. Regeneration conditions are highly reproducible with this approach as seen in this superposition of ten growth factor-binding/regeneration cycles. Regeneration steps are labeled as in the following panel. (**d**) In this close-up view of the sensor chip regeneration, the residual binding response (BMP, from 0 to 100 s) can be observed at the beginning of the regeneration cycle. Magnesium chloride ($MgCl_2$, from 100 to about 450 s) elutes bound receptors and associated ligands. Re-equilibration with buffer (HBS, from about 450 to about 650 s) restores the baseline. A washing step with water (H_2O, from about 650 to about 825 s) is necessary to avoid magnesium chloride accumulation and clogging of the microfluidics unit. A second equilibration with buffer (HBS, from about 825 to about 1000 s) completes the regeneration program, and the sensor chip returns to a constant RU baseline

protein A (e.g., Biacore 29127557) is an alternative option that can be used out of the box.

5. Human Antibody Capture Kit (Biacore, BR-1008-39): stored at 4 °C. This kit contains 0.5 mg/mL antihuman IgG (Fc) antibody, 10 mM sodium acetate pH 5.0, and 3 M $MgCl_2$ (*see* **Notes 3** and **4**).

2.3 SPR Disposables and Running Buffers

1. Capped glass and plastic vials for storing the "analyte," "ligand," and regeneration solutions that are compatible with instrument control, injection needle, and sample racks. For the Biacore 2000 sample racks, use 9 mm glass vials, 7 mm plastic vials, and 8 mm polyethylene dust cover caps.
2. HBS-EPS/BSA running buffer: 0.01 M HEPES, 0.5 M NaCl, 3 mM EDTA, 0.005% (v/v) Tween 20, and 0.1% BSA, pH 7.4 (*see* **Notes 5** and **6**).

2.4 BMPs and Fc-Fusion Receptors

1. 100 μg/mL BMPs dissolved in 4 mM HCl, 0.1 M acetic acid, or 1× PBS. Buffer selection is BMP dependent (follow manufacturer's instructions). 10 μL aliquots are stored at −20 °C for 3 months or at −80 °C for 1 year (*see* **Note 7**).
2. Expression-ready plasmids are not publicly available. The pFUSE-hIgG1-Fc1 vector system (InvivoGen) can be used to create Fc-fusion receptor expression plasmids for transiently and stably transfected mammalian cell cultures. Alternatively, Fc-fusion receptors can be obtained from a number of vendors, including R&D Systems (e.g., BMPRII-Fc, 811-BR) and Prosci (*see* **Note 8**).

3 Methods

3.1 Receptor Preparation and Storage

1. Transfect plasmids encoding the Fc-receptor-fusion gene into mammalian cells. For transient suspension cultures, use the ThermoFisher Scientific FreeStyle 293 or ThermoFisher Expi-CHO Expression systems. For stable suspension cultures, use the ThermoFisher Scientific pcDNA3.3 TOPO, ThermoFisher Scientific Freedom CHO-S, or Lonza's GS Gene Expression systems (*see* **Note 9**).
2. Purify secreted Fc-fusion receptors from conditioned medium using protein A capture chromatography (e.g., HiTrap MabSelect SuRe column, GE11-0034-94). An additional step of size exclusion chromatography (e.g., Superdex 200 16/60, GE 28989335) is recommended to remove aggregated Fc-fusion receptors and degraded components.
3. Combine fractions containing monodisperse, dimeric receptor-Fc fusions, and dialyze in 1× PBS overnight at 4 °C.
4. Some receptor-Fc fusions might require addition of stabilizers such as trehalose after dialysis to maintain activity over time. The need for stabilizers can be determined by monitoring binding affinities over time. Alternatively, the addition of stabilizers is often mentioned in vendor instructions. These instructions could be followed for guidance.

5. Filter sterilize dialyzed Fc-fusion receptors and store aliquots at −20 °C.
6. Purchased Fc-fusion receptors are reconstituted according to the manufacturer's instructions and stored in aliquots at −20 °C.

3.2 Anti-Fc Chip Preparation and Storage

1. The following example is based on the application wizard protocol in the Biacore 2000 control software.
2. The CM5 chip is incubated at room temperature at least for 30 min before it is placed into SPR (Biacore) instrument.
3. For anti-Fc chip preparation, water is used as running buffer, and the surface is primed at least one time before starting the surface activation program.
4. Separately aliquot 120 μL of 0.4 M EDC, 120 μL of 0.1 M NHS, and 100 μL of 1.0 M ethanolamine pH 8.2 into 7 mm plastic vials. Use a separate vial for each solution and each channel that will be activated, i.e., use one vial for EDC, one vial for NHS, one vial for ethanolamine, and one empty vial for EDC/NHS mixing per channel that will be activated.
5. Mix 0.4 M EDC and 0.1 M NHS at a 1:1 ratio.
6. Inject the EDC/NHS mixture separately into each flow channel at 5 μL/min for 400 s.
7. To ensure adequate antibody immobilization, at least a 100 response unit (RU) increase should be achieved following activation with EDC/NHS. Using this protocol, the response increase is between 400 and 600 RUs, which is optimal for this method (*see* **Note 10**).
8. Dilute 13.5 μL 0.5 mg/mL antihuman IgG (Fc) antibody in 876 μL of 10 mM sodium acetate pH 5.0 to a final concentration of 7.7 μg/mL. Immobilize the antibody to both blank and experimental flow channels (i.e., FC 1, 2, 3, and 4) by injecting the antibody containing solution at 10 μL/min for 16 min. 210 μL of diluted antibody is needed for each channel (*see* **Note 11**).
9. To deactivate the surface, inject 1.0 M ethanolamine pH 8.2 to blank and experimental flow channels (i.e., FC 1, 2, 3, and 4) at 5 μL/min for 360 s.
10. Each flow channel (FC) is prepared individually when using the application wizard in the Biacore control software. All steps, from surface activation, antibody cross-linking, to surface deactivation, are carried out for one channel before continuing with the next channel. Solutions must be kept in a separate vial for each tube and channel.
11. If the chip surface is prepared manually, all channels can be prepared simultaneously. EDC/NHS solutions can be premixed and are loaded into all four channels at once.

3.3 Running Conditions

1. Run the SPR instrument at 50 μL/min to minimize mass transport artifacts.

3.4 Fc-Fusion Receptor Capture

1. 150–250 RU Fc-fused receptors are captured separately on each experimental channel. (For the Biacore 2000 used here as example, flow channels 1 and 3 can be used as blanks.) In the following example, channel 1 is used as control, and Fc-fusion receptors are immobilized in experimental flow channels 2–4 (*see* **Note 12**).
2. To get desired receptor levels on the surface, approximately 3.0 μg/mL Fc-fusion receptor in HBS-EPS/BSA running buffer is injected for 20 μL at 20 μL/min per channel.
3. Low surface loading is critical to minimize mass transport artifacts and receptor dissociation from the antibody (*see* **Note 13**).
4. During the receptor capture step, channels are loaded individually. After the receptor capture step, all channels are opened simultaneously. BMP binding and surface regeneration steps are carried out in all four channels at once (*see* **Note 14**).

3.5 BMP Binding

1. To prepare the BMP solution, dilute the BMP aliquot to a final concentration of 80 nM BMP directly from storage conditions into HBS-EPS/BSA running buffer.
2. It is important to prepare at least 320 μL of BMP dilution when using 7 mm plastic tubes and 350 μL of BMP dilution when using 9 mm glass vials in order to prevent injection of air bubbles.
3. For the BMP–receptor association step, inject 250 μL BMP solution at a rate of 50 μL/min into the four channels at once (*see* **Note 15**).
4. For the BMP–receptor dissociation step, flow running buffer for at least 750 s after sample injection is complete. This dissociation time works well when considering the affinity of a typical BMP–receptor complex (*see* **Note 16**).

3.6 Surface Regeneration

1. Regenerate the anti-Fc chip surface by injecting 60 μL of 3.0 M $MgCl_2$ at 10 μL/min (*see* **Note 17**).
2. To prevent $MgCl_2$ accumulation on the surface, inject 30 μL water at 10 μL/min between cycles.
3. To get the same level of Fc-fusion receptor capture in multi-injection experiments, which is critical for high-throughput binding comparisons and for determination of kinetic parameters, the baseline of the anti-Fc cross-linked sensor chip should reach exactly the same level before Fc-fusion receptor immobilization ($\pm$10 RU).

3.7 Evaluating a Binding Experiment

1. Evaluate binding data using software such as BIAevaluation, Scrubber, or Clamp. In the following example, the *y*-axis represents RU (response units), and the *x*-axis represents time (seconds).
2. The following steps are carried out with BMP sample injections to obtain binding information and also with running buffer alone to remove spikes and shifts, which disturb data fitting (*see* **Note 18**).
3. Biacore instruments automatically subtract the blank control channel (i.e., FC 1) from experimental channels (i.e., FC 2, 3, and 4) to obtain a difference between experimental and control channel responses (i.e., 2-1, 3-1, and 4-1).
4. Blank channel subtracted injection curves (i.e., 2-1, 3-1, and 4-1 if FC 1 is the blank channel) are transformed to a zero baseline on the *y*-axis using BIaevaluation software. This step is called "*y*-transformation."
5. Blank channel subtracted binding curves (i.e., 2-1, 3-1, and 4-1) are aligned on the *x*-axis so that the BMP injection time point equals 0 s. This step is called "*x*-transformation."
6. Running buffer injections are subtracted from BMP injections. This process of data analysis (**steps 3–5**) is called "double referencing."

3.8 Calculating Equilibrium Binding Constants

1. Association (k_a) and dissociation rate constants (k_d) are obtained by fitting binding data to the X and Y transformed curves obtained by following the steps described in Subheading 3.5.
2. To obtain k_d and k_a, from a single curve, the separate fitting approach can be used with the high-throughput, single-injection format described in this manuscript (*see* **Note 19**).
3. 1:1 binding models are typically used for data analysis (*see* **Note 20**).
4. BIAevaluation, Scrubber, Clamp, or similar software can be used for data fitting. In this example, we use BIAevaluation.
5. In the separate fitting of a single injection curve, begin by selecting the start and end of the dissociation phase.
6. Fit the selected area to the 1:1 Langmuir model. This step yields the k_d.
7. In the next step, select the start and end of the association phase.
8. By using the k_d obtained in **step 6** and the concentration of the injected BMP, fit the selected association phase to the 1:1 Langmuir model. This step yields the k_a.
9. The equilibrium dissociation constant (K_d) is calculated from the ratio of k_d/k_a. (*see* **Note 21**).

4 Notes

1. The following example can be used with all Biacore instruments that have four flow cells (Biacore 2000, 3000, T100, T200, and S200). Experiments presented here were carried out on a Biacore 2000. Biacore instruments with a different number of flow cells can also be used. Likewise, these methods can be adapted to other SPR instruments.
2. Filtered and vacuum degassed type 1 ultrapure or distilled water should be used for all steps of the anti-Fc chip preparation, especially when preparing the EDC and NHS solutions. EDC, NHS, and ethanolamine should be made freshly before chip preparation. These chemicals are temperature sensitive. In addition, EDC is hygroscopic and labile.
3. Although alternate anti-Fc antibodies could be used, their stability, binding specificity, capture, and regeneration conditions would have to be determined. Using the Antibody Capture Kit is therefore advised.
4. The anti-Fc chips are kept at room temperature in the SPR machine. The lifetime of an anti-Fc chip averages 3 months when HBS-EPS/BSA running buffer is also used as standby buffer. The use of water as standby solution and dry storage outside of the SPR instrument significantly reduces the antibody/chip half-life.
5. The running buffer, reconstitution solutions, and in-house purified protein samples should be sterile filtered (0.2 μm).
6. HBS buffer is preferred over PBS to reduce salt crystal formation after regeneration in the microfluidics unit.
7. Purchased BMPs are reconstituted in sterilized buffers and need not be filtered after reconstitution.
8. We previously found that the linker that connects the ectodomain with the Fc moiety can affect the binding affinity [13], possibly by limiting the range of motion of the receptor domain. A generously sized tag is therefore preferred. We use 20–30 amino acid linkers. Shorter linkers have been used, but this can limit the flexibility and accessibility of the receptor domain and lead to an underestimation of the equilibrium dissociation constant.
9. Adherent and suspension cultures can be used for Fc-fusion receptor expression; however, suspension cultures are recommended for larger-scale production.
10. SPR instruments measure a refractive index change (detected as a RU). The change in refractive index is correlated with mass accumulation on the surface of the chip. A minimum of

100 RU from EDC/NHS activation is required to capture sufficient antibody.

11. This method results in the cross-linking of ~8000 RU of antihuman IgG (Fc) antibody per channel, which is optimal. Excessive antibody cross-linking (>12,000 RU) can lead to Fc-fusion receptor leakage and baseline drift, compromising baseline integrity, exacerbating mass-transport artifacts. Insufficient antibody (<5000 RU) may limit receptor capture and reduce the shelf life of the chip.

12. This approach relies on a simple, standardized capture method of an Fc-tagged protein. Significant advantages are the ability to stably capture the Fc tag with a well-characterized monoclonal antibody or using protein A chemistry and the ability to thoroughly regenerate the surface without compromising its activity. Direct cross-linking of the BMP or receptor domain to the sensor chip via EDC/NHS chemistry may not be desirable, as side chains with primary amines (mainly lysine but also arginine) in these domains can be found at the BMP–receptor interface [11, 18]. Modification of these side chains could therefore restrict BMP–receptor complexation. In addition, regeneration conditions can be harsh and lead to very limited lifetime of the cross-linked receptor or BMP. A particular advantage of this approach is that all captured receptors have the same orientation.

13. A low amount of captured receptor is essential to minimize mass transport limitation artifacts, which frequently leads to overestimation of kinetic rate constants. Limiting receptor levels below 300 RU is therefore advised.

14. As three experimental channels are typically available (Biacore 2000, 3000, T100, T200, S00), three different receptor-Fc fusions or variants can be tested simultaneously.

15. Before starting real binding injections, 2–3 blank injections should be done to equilibrate the Biacore instrument. At least three additional blank injections should be done between binding injections to obtain blanking curves.

16. Importantly, at least a 5% decrease in the SPR response should be observed to obtain an accurate dissociation rate constant.

17. This removes the captured receptors as well as the BMPs that are bound by the receptors. For each new cycle, new receptor-Fc fusions have to be captured.

18. Fc-fusion receptors, BMPs, and blank solutions are prepared and dialyzed in the same running buffer to minimize bulk shift effects, which cause significant problems during the fitting process. In addition, the needle should be dipped in running buffer before injecting BMP to minimize spikes on the

sensograms. Small spikes might be observed when the running buffer injection pump is filled, but these are easy to eliminate in the analysis process. Also, use the "Extraclean" command after each injection to minimize contamination from different samples.

19. High-throughput single injection vs. low-throughput titration. Although multiple curves from an "analyte" titration generally give more accurate kinetic and equilibrium constants, we found that, in a well-executed SPR experiment, a single injection can also yield accurate binding rate constants [12]. In contrast to a low-throughput titration, such an approach can be applied in a high-throughput format, allowing for rapid interrogation of multiple BMP–receptor complexes. In order to get a good agreement with K_d values obtained from titration, the concentration in the single injection should be significantly above K_d. 80 nM is a good starting point for members of the BMP family. Thus, a single-injection approach using the suggested BMP concentrations and captured receptor totals can help rapidly produce a BMP–receptor interaction map, as well as kinetic rates and equilibrium binding constants with good approximation. Notably, any approach that relies on multiple injections is not high-throughput.

20. Many different kinetic models can be used for fitting. In the separate fitting, only the 1:1 Langmuir binding model is available. When simultaneously fitting multiple titration curves, the most commonly used models are "1:1 Langmuir binding," "1:1 binding with mass transfer," "1:1 binding with drifting baseline," "bivalent analyte," "conformational change," "heterogeneous analyte," and "heterogeneous ligand." "Ligand" and "analyte" are used here to reflect the SPR nomenclature. Fitting parameters help identify the appropriate fitting model, i.e., the goodness of the fit, as indicated by the Chi^2 (RU^2) value, shows if a particular model is acceptable. A small Chi^2 value indicates a good fit. For the Biacore 2000, this value should not be above 1.0. Relative residuals should be equally distributed around zero. If the fit is good, the standard error (SE) of the fit becomes negligible. More complex binding models and stoichiometries not described here can also be investigated. For binding analysis of a single injection, separate fitting is used. By contrast, global fitting is preferred when at least five concentrations of one BMP have been sampled.

21. High-throughput BMP binding to Fc-fusion receptors can be studied at a fixed concentration of "analyte"/BMP as suggested in this manuscript. This leads to good K_d estimates. A titration experiment can then be designed based on the

estimated K_d value. At least five protein concentrations of "analyte"/BMP between 0.5 and 20 times the K_d concentration for that binding pair should be used to obtain accurate kinetic parameters. Ideally, binding saturation should be reached at the highest selected analyte concentration.

22. To obtain the SE for the kinetic rates and equilibrium binding constants, the experiment should be repeated at least three times.
23. In order to get steady-state binding kinetics, the flow rate could be reduced, or injection time could be increased. Both approaches give BMPs more time to bind the captured receptors.
24. While BMPs are extremely sensitive to storage conditions, Fc-fusion receptors are stable at 4 °C for at least 2 weeks.
25. For weekly cleaning of Biacore instruments, the anti-Fc chips can be taken out for a few hours. Changes in temperature should be minimized to prevent formation of bubbles on the surface.
26. Water should be avoided as running or standby buffer, as prolonged exposure to water decreases the lifetime of the anti-Fc chips.

References

1. Bragdon B, Moseychuk O, Saldanha S, King D, Julian J, Nohe A (2011) Bone morphogenetic proteins: a critical review. Cell Signal 23 (4):609–620. https://doi.org/10.1016/j.cellsig.2010.10.003
2. Goumans MJ, Mummery C (2000) Functional analysis of the TGFbeta receptor/Smad pathway through gene ablation in mice. Int J Dev Biol 44(3):253–265
3. Werner S, Alzheimer C (2006) Roles of activin in tissue repair, fibrosis, and inflammatory disease. Cytokine Growth Factor Rev 17 (3):157–171. https://doi.org/10.1016/j.cytogfr.2006.01.001
4. Wakefield LM, Hill CS (2013) Beyond TGFbeta: roles of other TGFbeta superfamily members in cancer. Nat Rev Cancer 13(5):328–341. https://doi.org/10.1038/nrc3500
5. Massague J (2008) TGFbeta in cancer. Cell 134(2):215–230. https://doi.org/10.1016/j.cell.2008.07.001
6. Shi Y, Massague J (2003) Mechanisms of TGF-beta signaling from cell membrane to the nucleus. Cell 113(6):685–700
7. Weiss A, Attisano L (2013) The TGFbeta superfamily signaling pathway. Wiley Interdiscip Rev Dev Biol 2(1):47–63. https://doi.org/10.1002/wdev.86
8. Massague J (2012) TGFbeta signalling in context. Nat Rev Mol Cell Biol 13(10):616–630. https://doi.org/10.1038/nrm3434
9. Moustakas A, Heldin CH (2009) The regulation of TGFbeta signal transduction. Development 136(22):3699–3714. https://doi.org/10.1242/dev.030338
10. Olaru A, Bala C, Jaffrezic-Renault N, Aboul-Enein HY (2015) Surface plasmon resonance (SPR) biosensors in pharmaceutical analysis. Crit Rev Anal Chem 45(2):97–105. https://doi.org/10.1080/10408347.2014.881250
11. Townson SA, Martinez-Hackert E, Greppi C, Lowden P, Sako D, Liu J, Ucran JA, Liharska K, Underwood KW, Seehra J, Kumar R, Grinberg AV (2012) Specificity and structure of a high affinity activin receptor-like kinase 1 (ALK1) signaling complex. J Biol Chem 287(33):27313–27325. https://doi.org/10.1074/jbc.M112.377960
12. Aykul S, Martinez-Hackert E (2016) Transforming growth factor-beta family ligands can function as antagonists by competing for type II receptor binding. J Biol Chem 291

(20):10792–10804. https://doi.org/10.1074/jbc.M115.713487

13. Sako D, Grinberg AV, Liu J, Davies MV, Castonguay R, Maniatis S, Andreucci AJ, Pobre EG, Tomkinson KN, Monnell TE, Ucran JA, Martinez-Hackert E, Pearsall RS, Underwood KW, Seehra J, Kumar R (2010) Characterization of the ligand binding functionality of the extracellular domain of activin receptor type IIb. J Biol Chem 285 (27):21037–21048. https://doi.org/10.1074/jbc.M110.114959
14. Aykul S, Ni W, Mutatu W, Martinez-Hackert E (2015) Human cerberus prevents nodal-receptor binding, inhibits nodal signaling, and suppresses nodal-mediated phenotypes. PLoS One 10(1):e0114954. https://doi.org/10.1371/journal.pone.0114954
15. Aykul S, Martinez-Hackert E (2016) New ligand binding function of human cerberus and role of proteolytic processing in regulating ligand-receptor interactions and antagonist activity. J Mol Biol 428(3):590–602. https://doi.org/10.1016/j.jmb.2016.01.011
16. Aykul S, Martinez-Hackert E (2016) Determination of half-maximal inhibitory concentration using biosensor-based protein interaction analysis. Anal Biochem 508:97–103. https://doi.org/10.1016/j.ab.2016.06.025
17. Castonguay R, Werner ED, Matthews RG, Presman E, Mulivor AW, Solban N, Sako D, Pearsall RS, Underwood KW, Seehra J, Kumar R, Grinberg AV (2011) Soluble endoglin specifically binds bone morphogenetic proteins 9 and 10 via its orphan domain, inhibits blood vessel formation, and suppresses tumor growth. J Biol Chem 286 (34):30034–30046. https://doi.org/10.1074/jbc.M111.260133
18. Weber D, Kotzsch A, Nickel J, Harth S, Seher A, Mueller U, Sebald W, Mueller TD (2007) A silent H-bond can be mutationally activated for high-affinity interaction of BMP-2 and activin type IIB receptor. BMC Struct Biol 7:6. https://doi.org/10.1186/1472-6807-7-6

Chapter 6

Mutagenesis and Imaging Studies of BMP Signaling Mechanisms in *C. elegans*

Cathy Savage-Dunn, Ryan J. Gleason, Jun Liu, and Richard W. Padgett

Abstract

C. elegans has played a central role in the elucidation of the TGFβ pathway over the last two decades. This is due to the high conservation of the pathway components and the power of genetic and cell biological approaches applied toward understanding how the pathway signals. In Subheading 3, we detail approaches to study the BMP branch of the TGFβ pathway in *C. elegans*.

Key words *C. elegans*, Body size, Development, Mutagenesis, Genetic screens, Coelomocytes, Mesoderm, Imaging, Trafficking

1 Introduction

1.1 TGFβ Signaling Pathways

The TGFβ pathway is an ancient signaling pathway present in most primitive animals, sponges, and *Trichoplax* but absent from yeast or plants [1, 2]. Members of the vertebrate superfamily number approximately 30 ligands and can be roughly placed into the TGFβ, activin, and BMP families. Evolutionary studies suggest that of the three families, the TGFβ family is a more recent deuterostome invention, with the activin and BMP families being present in the very earliest animals [3]. In mammals, BMPs are associated with many morphological diseases, such as osteogenesis imperfecta and polydactyly, cardiovascular diseases such as pulmonary arterial hypertension, or with the progression of cancer [4]. Essentially, every mammalian tissue has one or more TGFβ ligands expressed, where they play essential roles.

Many aspects of TGFβ signaling are conserved between *C. elegans* and mammals including the structure and function of the ligands and the downstream signaling components. In *C. elegans*, the BMP family is represented by the Sma/Mab pathway, while the activin family is represented by the dauer pathway [5–14]. TGFβ signaling in *C. elegans* is more complicated than two linear pathways as there are five ligands (*dbl-1*, *daf-7*, *unc-129*,

Melissa B. Rogers (ed.), *Bone Morphogenetic Proteins: Methods and Protocols*, Methods in Molecular Biology, vol. 1891, https://doi.org/10.1007/978-1-4939-8904-1_6, © Springer Science+Business Media, LLC, part of Springer Nature 2019

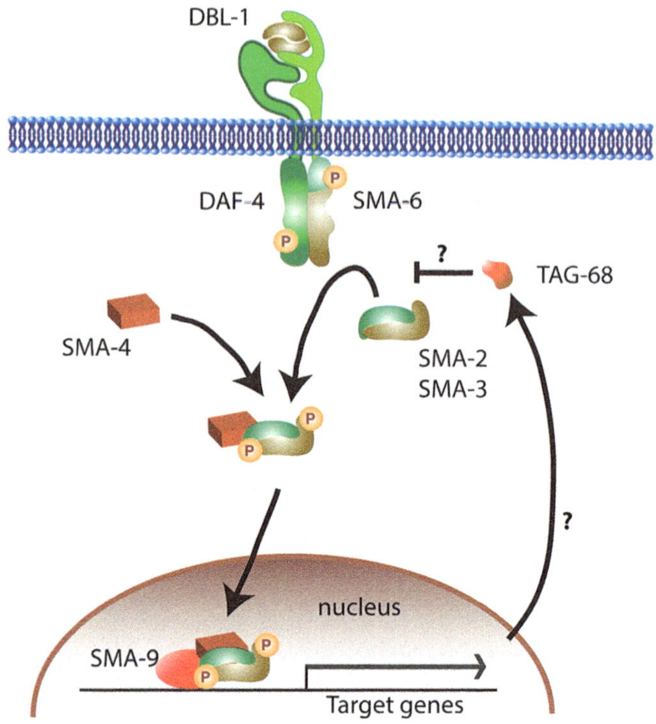

Fig. 1 The *C. elegans* BMP pathway. The ligand (DBL-1) binds to the type I and type II receptors (SMA-6 and DAF-4, respectively), stimulating phosphorylation events on the receptors and on the R-Smads. The two R-Smads of this pathway, SMA-2 and SMA-3, associate with SMA-4, the co-Smad, and translocate to the nucleus. One characterized transcription factor, SMA-9 (SCHNURRI), aids in transcription of downstream target genes. The putative I-Smad, TAG-68, is the only I-Smad homolog in *C. elegans*. It is hypothesized to function similarly to the mammalian I-Smads (Smad6 and Smad7) and would therefore be a target of the pathway and attenuate signaling by interacting with the receptor and/or R-Smads [50]

tig-2, and *tig-3*) but only two complete signaling pathways with possibly multiple ligands in a given pathway (Fig. 1) [15–17]. Signaling begins when the ligand binds to the receptors, which are transmembrane serine-threonine kinases, although they can phosphorylate tyrosine residues as well [18]. Upon binding, the type II receptor phosphorylates the type I receptor on the GS domain, which activates it [19]. The primary target of the type I receptor phosphorylation is the C-terminus of the R-Smads, which allows the R-Smads to bind with the co-Smad. In the Sma/Mab pathway, SMA-2 and SMA-3 are R-Smads, and SMA-4 is the co-Smad. *tag-68* encodes a likely I-Smad, but its function has not been fully determined [16]. This Smad complex moves to the nucleus where it binds DNA and alters transcription with the aid of other transcription factors. One conserved transcription factor used in some developmental contexts is SMA-9/Schnurri, a zinc finger-containing protein [8].

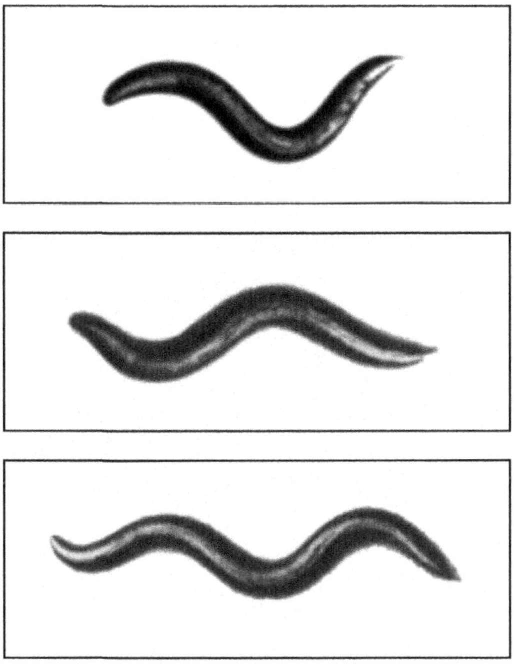

Fig. 2 *C. elegans* body size mutants. Mutations in the *C. elegans* BMP pathway components produce small animals (top panel), while an increase in signaling produces longer animals (bottom panel). Wild-type animals are shown in the middle panel

1.2 Genetic Screens for Components of the Sma/Mab Pathway

The Sma/Mab pathway in *C. elegans* was originally described as a regulator of body size and male tail development: small body size phenotype (Sma) and male tail abnormal phenotype (Mab) caused by mutations in this pathway (Fig. 2). Over time, many additional roles of the pathway have been described, such as its function in innate immunity, reproductive aging, and mesoderm patterning [5, 20–26].

The Mab phenotype is labor-intensive to analyze because it requires a genetic background generating a high incidence of males, and it also requires that animals are screened on slides under a compound microscope, so it has been used in relatively small-scale screens. The small body size and mesoderm development phenotypes have been successfully exploited in genetic screens, as described here. The small body size phenotype is visible under the dissecting microscope and was therefore useful in a large-scale genetic screen for BMP signaling components [11]. From the Sma/Mab screen, all the conserved, known core signaling components (ligand, both receptors, three Smads, and Schnurri) [7, 8, 11, 12] and some important regulatory genes [6, 27–30] were identified. A variation of this screen using genetic suppression of *lon-2*, which encodes an upstream component in the pathway, yielded mutations in many of the same genes [29].

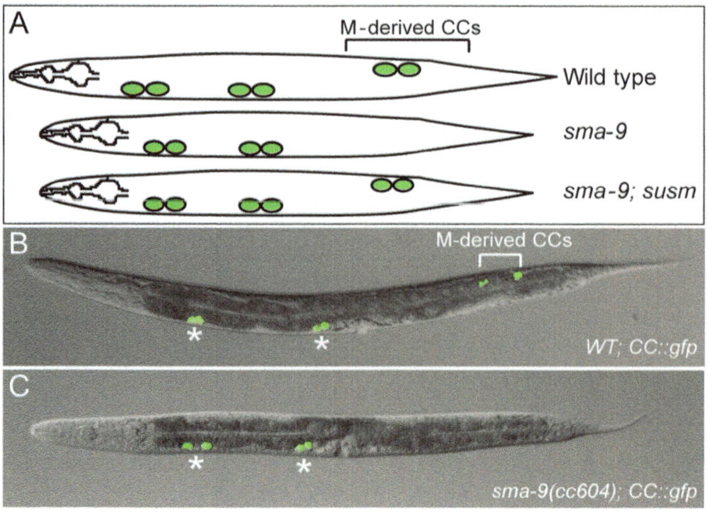

Fig. 3 The coelomocyte (CC) phenotypes of wild-type and *sma-9* mutant worms. (**a**) Diagrams depicting the CC phenotypes of wild-type, *sma-9*, and *sma-9;susm* animals in the LW0040 strain which bears *arIs37* that encodes CC localized GFP. (**b**, **c**) Merged DIC and GFP images of a wild-type worm (**b**) and a *sma-9(cc604)* worm (**c**). *Points to the 4 CCs born during embryogenesis. The two postembryonic M-derived CCs are missing in *sma-9* mutants

Another important role for the Sma/Mab pathway is the specification of mesodermal cell fates. The mesoderm gives rise to muscles and some non-muscle tissues. In *C. elegans*, the postembryonic mesoderm derives from the M precursor cell, which produces a subset of body wall muscles, the sex muscles, and the non-muscle coelomocytes [31]. The Zn finger containing transcription factor SMA-9/Schnurri is critical for specifying the two postembryonically derived coelomocytes (CCs); thus, in *sma-9* mutants, there are 4 CCs instead of the 6 CCs present in wild-type animals (Fig. 3) [5]. This CC defect of *sma-9* mutants can be specifically suppressed by mutations in the Sma/Mab pathway, but not by mutations in any other pathways [5, 9]. This suppression of the *sma-9* CC phenotype, the Susm phenotype, is highly sensitive to altered levels of Sma/Mab signaling: null alleles for a given Sma/Mab pathway gene exhibited higher penetrance of the Susm phenotype compared to partial loss-of-function alleles of the same gene [9]. A *sma-9* suppressor screen has successfully identified the major Sma/Mab signaling components as well as several new, regulatory genes [5, 9, 13, 14]. More is to be learned from both the Sma and the CC screens, as neither have reached saturation for mutants.

2 Materials

2.1 Ethyl Methanesulfonate (EMS) Mutagenesis for Sma Mutants

1. *C. elegans* variety Bristol, strain N2 unless noted otherwise. Note, wild-type, mutant, and transgenic strains (strains containing engineered DNA) can be requested from the *C. elegans* Genetics Center (CGC) (http://www.cbs.umn.edu/cgc/) or from individual laboratories.
2. 60 mm petri dishes with standard worm normal growth medium (NGM) (50 mM NaCl, 15 g/L agar, 20 g/L peptone, 1 mM cholesterol, 1 mM $CaCl_2$, 1 mM $MgSO_4$, 1 mM KP*i* [pH = 7.0]) and seeded with *E. coli* OP50 [32].
3. Pipettors (P20 and P1000 or equivalent) and sterile pipet tips.
4. 15 and 50 mL disposable tubes.
5. 50 mM EMS (*see* **Note 1**).
6. M9 buffer (22 mM KH_2PO_4, 42 mM Na_2HPO_4, 86 mM NaCl, 1 mM $MgSO_4$). Aliquot into 100 mL bottles and autoclave.
7. 1 N NaOH.
8. Glass Pasteur pipettes.
9. Tabletop centrifuge.
10. Stereo dissecting microscope with underneath illumination.

2.2 Body Size Measurements

1. *C. elegans* variety Bristol, strain N2.
2. Control mutant animals: *dbl-1(wk70)*, *sma-3(wk30)*, or *sma-6 (wk7)* (*see* **Note 2**).
3. 60 mm petri dishes with NGM medium and *E. coli* OP50 [32].
4. Pipettors (P1000 or equivalent) and sterile pipet tips.
5. M9 buffer.
6. 1.5 mL microfuge tubes.
7. Microcentrifuge.
8. Bleach solution: Mix 5 mL 1 N NaOH and 2 mL sodium hypochlorite (commercial bleach). Date and use within 10 days.
9. Stereo dissecting microscope with underneath illumination.
10. Sodium azide.
11. Agarose (electrophoresis grade).
12. Microscope slides and coverslips.
13. Compound microscope equipped with camera.
14. Imaging software, such as Fiji [33].

2.3 Double Mutant Analysis

1. *C. elegans* variety Bristol, strain N2.
2. Control mutant animals: *dbl-1(wk70)*, *sma-3(wk30)*, *sma-6 (wk7)* (*see* **Note 2**), *lon-2(e678)*, and *dbl-1* overexpressing strain *ctIs40* (*see* **Note 3**).
3. 60 mm petri dishes with NGM medium and *E. coli* OP50 [32].
4. Stereo dissecting microscope with underneath illumination.

2.4 EMS Mutagenesis for sma-9 Suppressors

1. *C. elegans* strain LW0040[*arIs37(secreted CC::gfp)* I; *cup-5 (ar465)* III; *sma-9(cc604)* X]. This strain has 4 CCs that are labeled by *arIs37(myo-3p::secreted gfp)*, in which secreted GFP made in the bodywall muscles (BWMs) is taken up by the CCs via endocytosis [34, 35]. The *cup-5(ar465)* mutation, which affects the *C. elegans* mucolipin homolog, enhances the ability of the CCs to uptake the secreted GFP, making the CCs easily identifiable under a fluorescence stereo microscope.
2. A standard fluorescence stereo microscope equipped with a GFP filter set, underneath illumination, 10× eyepieces, and a built-in zoom up to 10×.
3. Other reagents as described in Subheading 2.1.

2.5 Molecular Reporters of BMP Signaling

1. *C. elegans* strains:
   (a) LW2433: *jjIs2433[pCXT51(RAD-SMAD) + LiuFD61 (mec-7p::mRFP)]* X.
   (b) LW2301: *dbl-1(wk70)* V; *jjIs2433[pCXT51(RAD-SMAD) + LiuFD61(mec-7p::mRFP)]* X.
   (c) LW2290: *lon-2(e678) jjIs2433[pCXT51(RAD-SMAD) + LiuFD61(mec-7p::mRFP)]* X.
   (d) LW2437: *jjIs2437[pCXT51(RAD-SMAD) + LiuFD61 (mec-7p::mRFP)]* II.
   (e) LW2319: *jjIs2437[pCXT51(RAD-SMAD) + LiuFD61 (mec-7p::mRFP)]* II; *dbl-1(wk70)* V.
   (f) LW2394: *jjIs2437[pCXT51(RAD-SMAD) + LiuFD61 (mec-7p::mRFP)]* II; *lon-2(e678)* X.
2. Fluorescence compound microscope equipped with a dry or oil 40× objective, a GFP filter set, and a CCD camera.
3. 0.1 μM tetramisole in M9 buffer (*see* **Note 4**).

2.6 Materials for In Vivo Imaging of BMP Receptors

1. *C. elegans* strains: Bristol strain N2, *pwIs921[pvha-6::sma-6::GFP]*, *pwIs922[pvha-6::daf-4::GFP]*, *pwIs849[pvha-6::tagRFP::RAB-7]*.
2. M9 buffer.
3. OP50 bacterial strain.
4. M9 buffer with 0.1 μM tetramisole.

5. NGM media.
6. Agarose (gel electrophoresis grade).
7. Dissecting scope.
8. Petroleum jelly.

3 Methods

Grow *C. elegans* strain N2 under standard conditions in 60 mm petri plates with standard NGM media.

3.1 EMS Mutagenesis to Screen for Sma Mutants

The EMS screening protocol is based on techniques developed by Brenner and colleagues [36] as performed [11].

1. Harvest worms from three 60 mm plates containing many fourth larval stage (L4; *see* **Note 5**) and young adult animals by washing off with M9 buffer. Pipette about 1 mL M9 buffer onto each plate, swirl, and pour animals and liquid into a 15 mL disposable tube. Let the worms settle and then adjust the total volume to 3 mL. Gently aspirate off the top liquid if you have too much liquid or add M9 to bring the total to the 3 mL mark.
2. Prepare the EMS solution. Pour about 100 mL of 1 N NaOH into a 500 mL beaker in the hood (*see* **Note 1**). Add 1 mL of M9 buffer to a 50 mL capped disposable tube. In the hood, add 20 μL of EMS to the M9 in the 50 mL tube. Eject the pipet tip into the beaker containing the 1 N NaOH. Cap the 50 mL tube and swirl gently to mix the EMS with the M9 solution.
3. Swirl the 15 mL tube containing the animals in 3 mL of M9 to resuspend them and then pour them into the 50 mL tube that contains the EMS solution. Leave the animals in the solution in the hood for 4 h. Mix animals gently at least once an hour to resuspend them.
4. After 4 h, prepare a 15 mL disposable tube with 12 mL of M9 buffer. With a glass Pasteur pipette, suck up the worms that have settled to the bottom of the 50 mL tube and transfer them to the 15 mL tube that contains M9. Discard the pipette into the 1 N NaOH solution. Pour the EMS solution into the 1 N NaOH to inactivate the mutagen. Use the Pasteur pipette to wash the 50 mL tube with 1 N NaOH.
5. Gently spin the 15 mL tube containing the animals, about 1 K for 5 min in a tabletop centrifuge. Using a Pasteur pipette, remove pelleted animals from the bottom of the tube. Transfer the animals to the edge of a worm plate spread with bacteria. Leave the plate slightly open in the hood. Animals should crawl around for 30 min to recover. These are the P0 animals.

6. Transfer single mutagenized P0 animals to worm plates and allow to segregate self-progeny (F1 animals). After 3–4 days at 20 °C, transfer two F1 animals each to new plates to segregate F2 animals (*see* **Note 6**).
7. Visually screen F2 progeny using a stereo microscope for a clone of small mutants (~1/8 animals are expected on a plate for a recessive mutation). Pick mutants to new plates to segregate and confirm the phenotype in F3 animals. Retain the segregants from only one animal per F2 plate to ensure independent origin of mutants.
8. To calculate mutation frequency, each screened F2 plate generated from two F1 parents is considered to contain four mutagenized haploid genomes.
9. Outcross identified mutations by mating six N2 males with three L4 *sma* mutant hermaphrodites on a plate. In the next generation, observe the phenotype of males to identify X-linked mutations and the phenotype of hermaphrodite cross progeny to identify dominant or semidominant mutations (*see* **Note 7**). Mate heterozygous *sma/+* or hemizygous *sma/0* males with N2 hermaphrodites. In the following generation, pick individual hermaphrodites (half of which should be *sma/+*) to new plates. Look for the segregation of *sma/sma* progeny from these parents and pick them. These are twice outcrossed mutants. Repeat this process 1–2 additional times.

3.2 Body Size Measurements

1. Grow *C. elegans* wild-type strain N2, a known BMP pathway mutant, and newly isolated *sma* mutant strains under standard conditions.
2. For each strain, obtain a plate containing a large number of gravid hermaphrodites. Wash worms off plates with 1 mL M9 buffer. Transfer to eppendorf tube.
3. Pellet worms at 1000 rpm (120g) for 1 min in microcentrifuge. Remove supernatant.
4. Add 500 μL bleach solution. Mix by inverting. Monitor digestion of worms under the dissecting microscope.
5. Pellet eggs and worm debris at 1000 rpm for 1 min in microcentrifuge.
6. Transfer pellet to new plate. Under the dissecting microscope, many eggs should be visible in the drop. Embryos will hatch and crawl onto the food.
7. After 2 days, plates should contain large numbers of L4 animals (*see* **Note 5**). Pick 20–30 L4 animals or animals at different larval stages (*see* **Note 8**) into a drop of 10% sodium azide in M9 buffer on a 5% agar pad on a glass microscope slide. Cover with cover slip. Measurements should be done within 30 min

after adding sodium azide (a nematode anesthetic) since it is harmful to nematodes after extended exposure.

8. Image worms on a compound microscope at low power (100× magnification) or on a dissecting microscope equipped with a camera. Determine the length of the animals by drawing a line down the center of the animal from the head to the tail (*see* **Note 9**) and analyze using imaging software such as Fiji software [33].
9. An unpaired, two-tailed t test is used to compare the p value between the control *wild-type* animal and each genotype. Figure 2 shows typical wild-type and mutant lengths.

3.3 Double Mutant Analysis

To determine whether new *sma* mutants act within the BMP pathway or independently of it, double mutants should be constructed and analyzed [8, 9, 11, 13, 14, 27, 28, 37].

1. Cross N2 males with *sma* mutant hermaphrodites. Pick *sma/+* males and mate with hermaphrodites of the genotypes *dbl-1 (wk70)*, *sma-3(wk30)*, *sma-6(wk7)*, *lon-2(e678)*, and *ctIs40 [dbl-1++]*.
2. Pick individual hermaphrodite progeny from each cross to separate plates. These should have approximately wild-type phenotype if the mutations are recessive. Half of these will be the double heterozygotes (*see* **Note 10**).
3. In the next generation, pick *sma* and/or *lon* animals individually as potential double mutants. Allow to self-fertilize. Confirm double mutants by complementation testing and/or sequence analysis.
4. Determine body size of double mutants as in Subheading 3.2.
5. For double mutants with *dbl-1*, *sma-3*, and *sma-6*, if the new mutation functions in the same pathway, then the double mutant body size should be the same as the single mutants. If the new mutation functions independently, then the double mutant should be more severe (i.e., smaller) than the single mutants (Fig. 4).
6. For double mutants with *lon-2* and *dbl-1(++)*, new mutations that suppress the long body size phenotype likely represent genes encoding components downstream of the ligand. Other outcomes may indicate function at the level of receptor ligand or independent function (Fig. 4).

3.4 EMS Mutagenesis and CC Screening for sma-9 Suppressors

1. Mutagenize LW0040 animals with EMS using the method described in Subheading 3.1.
2. Pick the mutagenized P0 animals onto NGM plates seeded with *E. coli* OP50 by putting five L4s per plate. Allow worms to grow at 20 °C for 5 days.

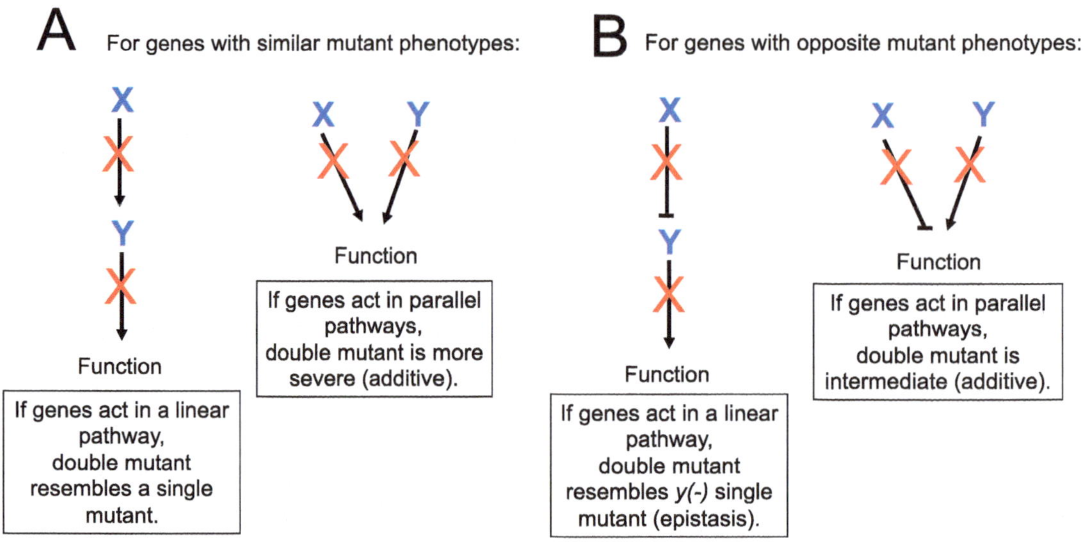

Fig. 4 Double mutant analysis flowchart. Analysis of double mutant phenotypes provides genetic evidence that two gene products, X and Y, act together in a linear pathway or independently in parallel pathways. (**a**) Genes with similar mutant phenotypes. (**b**) Genes with opposite mutant phenotypes

3. Pick the F1 progeny of the mutagenized worms to new plates by plating three animals of similar developmental stages per plate (*see* **Note 11**). Allow worms to grow at 20 °C for 5 days.
4. When the majority of the F2 progeny develop to the late L4 to young adult stage, screen each plate under the fluorescence stereo microscope and save plates that have between 5% and 25% of animals with 6 CCs (*see* **Note 12**).
5. From each of the saved plates described in **step 4** above, pick all the animals with 6 CCs (up to 20 from each plate) and plate them one worm/plate. Those animals that produce high penetrance of progeny with 6 CCs are saved as candidate *sma-9* suppressors (*see* **Note 13**).
6. The putative *sma-9* suppressors can be subjected to further analysis, which includes determining whether they are recessive or dominant, single locus- or multi-loci- mutation. The corresponding genes can be identified via complementation tests, linkage analysis, and SNP-WGS [38].

3.5 Molecular Reporters of the C. elegans *BMP Pathway and RAD-SMAD Expression and Quantification*

As described above, the basic signaling components of the BMP pathway in *C. elegans* are highly conserved with their vertebrate counterparts. However, there are sufficient differences in the protein sequences such that the large pool of available mammalian antibodies often does not work well in nematodes. A critical tool needed in assaying TGFβ signaling strength is the ability to monitor the level of phosphorylated Smad (pSmad) in the nucleus. A reporter tag for the R-Smads suffers from possible interference

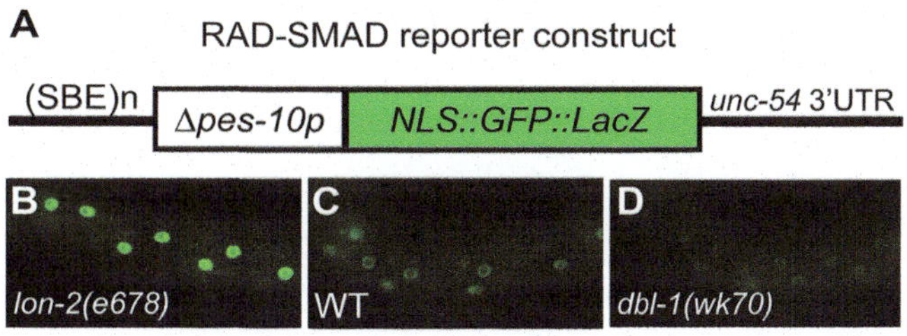

Fig. 5 The RAD-SMAD reporter used to measure the level of Sma/Mab signaling in various mutants. (**a**) A schematic of the RAD-SMAD reporter construct. SBE stands for Smad binding element. (**b–d**) Representative GFP fluorescence images showing the expression of the reporter in *lon-2(e678)* (**b**), wild-type (**c**), and *dbl-1(wk70)* (**d**) animals, respectively

with phosphorylation [39]. The most successful approach was to concatenate Smad binding sites upstream of a minimal promoter, as described below. This reporter allows a readout of signaling strength, with low levels in a null background and higher levels with overexpressed components.

The reporter, named the RAD-SMAD (Reporter acting downstream of Smad) reporter, was generated by placing multiple copies of the Smad boxes (or Smad binding site: GTCT) organized in the RLR orientation (R, rightward and L, leftward) upstream of the minimal *pes-10* promoter and a nuclear localized *gfp::lacZ* fusion protein [13]. The RAD-SMAD reporter showed nuclear GFP expression in intestinal and hypodermal cells, tissues known to be responsive to DBL-1 signaling [15]. The level of RAD-SMAD reporter expression in hypodermal nuclei at the L2 larval stage appears positively correlated with the strength of DBL-1 signaling [13] (Fig. 5). As a control, the BAD-SMAD (Bad reporter of Smad) reporters, which contain mutations in the Smad boxes, showed no GFP expression in any tissues [13]. The RAD-SMAD reporter has been used to quantify the level of BMP signaling in mutants of several newly identified Sma/Mab modulators [6, 9, 13, 14].

1. Bleach gravid adult hermaphrodites of the following genotypes: your mutation of interest, wild-type, *dbl-1(wk70)* and *lon-2(e678)*, all carrying the same RAD-SMAD reporter transgene (*see* **Note 14**), following the protocol described in Subheading 3.2. Wash off the bleach using M9 buffer.
2. Allow embryos to hatch in 1 mL of M9 buffer in a 15 mL conical tube with gentle rotation at 16 °C for 24 h. The embryos will hatch and arrest as synchronized L1 worms.
3. Plate ~200 synchronized L1 worms onto OP50-seeded NGM plates and allow them to grow at 20 °C till the L2 larval stage (*see* **Note 15**).

4. Wash worms off the plates with M9 buffer.
5. Anesthetize worms with 0.1 μM tetramisole in M9 buffer and mount them onto 2% agarose pads.
6. Collect images of a total of 40 worms per genotype, making sure that at least five hypodermal nuclei per worm are within the focal plane imaged using a 40× objective lens on a compound microscope (*see* **Notes 16** and **17**).
7. For each worm, pick five nuclei, measure the fluorescence intensity of each nucleus and the corresponding background signal right outside of that nucleus, using the Fiji software.
8. Calculate the mean GFP signal value for each nucleus by subtracting the nuclear GFP value with the corresponding background value collected in **step 7** above. The resulting values are used to deduce the mean and standard deviation of RAD-SMAD signal for each genotype. Subsequent statistical analysis can be carried out using any standard software package.

3.6 In Vivo Imaging of BMP Receptors

The *C. elegans* BMP pathway is expressed in several tissues, most notably in the hypodermis [7, 10, 12, 28] and in the intestine [22, 24, 40]. The expression levels of many BMP pathway components are sufficiently low in the hypodermis so that tagged receptors are difficult to detect. However, expression levels in the intestine are much more robust, and for this reason, most of the BMP subcellular imaging is done in the intestine.

The intestine houses the nematode innate immunity system, which protects it from environmental pathogens [23, 26, 41–43]. Genetic screens have identified signaling pathways that respond to infection by inducing peptides against bacteria [44]. Some of the induced peptides are regulated by the *C. elegans* BMP pathway making the intestine an appropriate tissue for examination of BMP signaling [22, 24].

The intestine is comprised of 20 large epithelial cells (ideal for imaging) that are positioned as bilaterally symmetric pairs to form a long tube around the lumen with distinct apical and basolateral regions (Fig. 6). It is anchored to the pharyngeal and rectal valves and as an adult has a left-handed 180° twist around the gonad. As a result, the twist causes the displacement of the intestine slightly left of the midline in the anterior half and slightly right of the midline in the posterior half. Care must be taken to find animals positioned correctly to obtain clear images of the intestinal cells (as described below). Imaging in the intestine can be simplified by expressing genes under an intestine-specific promoter so there is less interference from dispersed expression. Several different promoters are available for expression in the intestine, but the *vha-6* promoter works well and is commonly used [6, 40, 45].

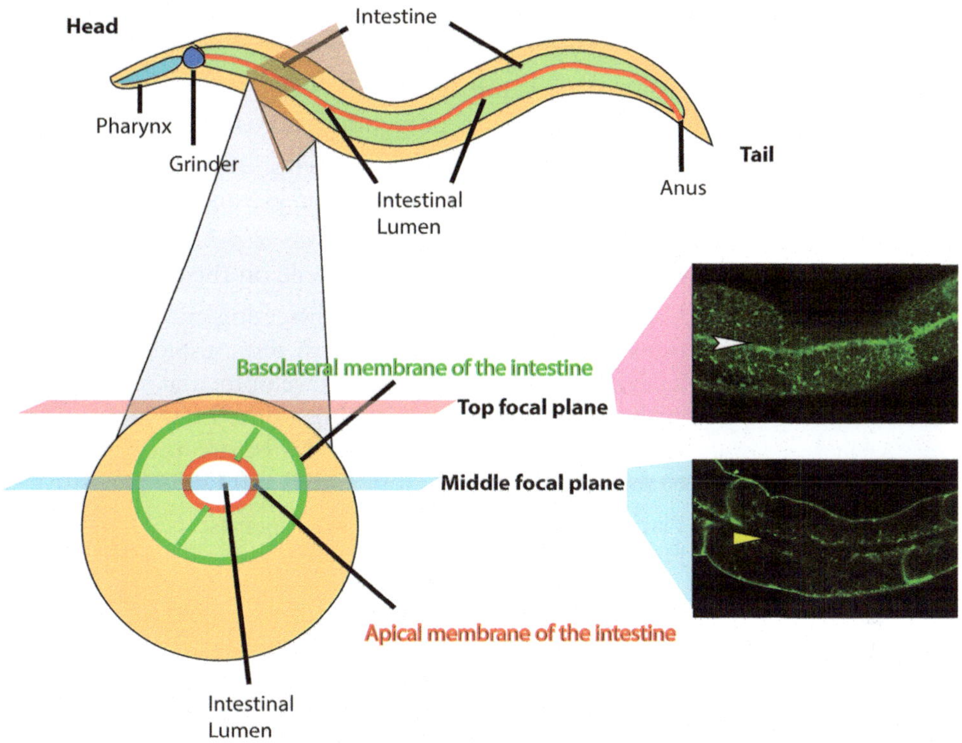

Fig. 6 Schematic depiction of the *C. elegans* intestine to demonstrate focal planes captured to study tagged BMP receptor subcellular localization. An example of DAF-4::GFP localization is depicted in both focal planes. White arrowheads indicate lateral membrane, and yellow arrowheads indicate apical lumen of the intestine. Figure is modified from Gleason et al. [6]

3.6.1 Instruments and Software

1. Leica SP5 confocal microscope system or similar.
2. Image analysis and quantification software iVision (BioVision Technologies), Image J (Fiji), or Metamorph.
3. Spinning-disk confocal microscope (Upright Zeiss Axioimager Z1) with Photometrics Evolve 512 EMCCD camera and C-apochromat water-immersion lens or similar and deconvolution software AutoQuant X3 (Media Cybernetics, Rockville, MD) or similar.

3.6.2 Preparation of an Agarose Pad and Staging Animals for Live Imaging

1. Mix agarose powder in deionized H_2O with a final concentration between 2% and 8% (wt/vol). Heat solution in a microwave oven until agarose is completely dissolved, but do not overboil the solution as it will increase the final concentration of the agarose gel.
2. Using a small pipet bulb and a glass Pasteur pipet place a small drop of heated agarose solution onto a 25 × 75 × 1.0 mm microscope slide. To ensure a flat, uniform surface of the agarose pad place two support slides directly alongside the

first slide one on either side. Also, place a piece of tape along the top of the support slide.

3. Quickly place another 25 × 75 × 1.0 mm microscope slide over the drop of agarose perpendicular to the first slide.
4. Allow the agarose to solidify, then slide the bottom slide or top slide out from under the agarose, leaving the agarose pad stuck to one slide.
5. Place 5 μL of M9 with tetramisole on the agarose pad.
6. Using a wire worm pick and a dissecting microscope, select and transfer animals to the M9 with tetramisole drop. Select and analyze animals at the same developmental stage. To restrict analysis of the samples to young adults 1 day past the L4 stage, L4 stage animals are selected a day before imaging and allowed to grow on NGM medium and *E. coli* OP50 plates overnight.
7. Place a cover slip on top of the agarose pad and sample (*see* **Note 18**).
8. The coverslip can be sealed with petroleum jelly. Heat the petroleum jelly on a 55 °C heat block and apply with a brush.
9. To observe animals on the confocal laser scanning microscope, place your slide with the agarose pad and sample onto the stage of the microscope.
10. Locate and focus on an animal under low magnification (10×). To observe GFP expression in a transgenic animal, switch the fluorescence source on and switch off bright field.
11. Switch to 40×, 63×, or 100× objectives to observe your cells of interest. For most samples, we use a 63× water immersion lens.
12. To monitor subcellular localization of a protein of interest (i.e., the BMP receptors), green fluorescent protein (GFP), or another fluorescent protein, is routinely expressed as a fusion protein with a gene product of interest (*see* **Note 19**). This approach enables the observation of the subcellular localization and dynamics of the tagged protein. This article will focus on GFP excitation and emission but can be adapted to other fluorescent protein fusions of interest. GFP, specifically enhanced GFP (eGFP), has a single excitation peak centered around 488 nm with an emission peak wavelength of 509 nm.
13. To reduce variability of expression of the fusion protein, analysis of protein localization should be restricted to the anterior half of the animal, specifically the second and third pair of cells. As a result, only animals laying with the anterior half of the intestine unobstructed by the gonad are selected for imaging.

14. For imaging the intestinal cells, use the lowest laser power that gives a reasonable signal to avoid damaging the worms. While viewing, be aware of photobleaching.
15. Next, optimize the signal-to-noise ratio in order to establish operating parameters that maximize image quality and minimize specimen damage. Common values to consider include the pinhole, the laser unit or the percent of the laser illumination source, as well as the detector, frequently a photomultiplier tube (PMT). Also, use Leica HyD™ detectors and/or Zeiss GaAsP™ detectors if available (*see* **Note 20**). The size of the confocal pinhole aperture determines the thickness of the optical section. Generally, a pinhole of 1 airy unit (AU) is determined based on the wavelength and numerical aperture, and modern confocal microscopes automatically set the pinhole diameter to 1 AU. To adjust the sensitivity and background of the detectors, fine-tune the gain (also called the high voltage) and the offset (also called the black level). Gain is the relative measure of the application to the photon detection system; higher gains will result in brighter images (*see* **Note 21**). To eliminate background, adjust the offset such that the lowest true values of your sample are collected and background signal is dispensed.
16. A unique concern while imaging in the *C. elegans* intestine is the broad-spectrum autofluorescence of lipofuscin-containing gut granules (lysosome-related organelles). This can be removed through spectral imaging and subsequent linear unmixing allowing the separation of highly overlapping fluorophores into pure signals. Autofluorescence in other tissues is minimal. For example, the emission spectrum of GFP and autofluorescence substantially overlap. To be able to calculate the fluorophore and autofluorescent contributions, linear unmixing requires knowledge of the reference spectra. Predefined and custom fluorescent profiles can be utilized for unmixing. Ideally, a user will collect spectral profiles of the fluorophores as well as the autofluorescent spectra for the tissue being sampled. For accuracy, the reference spectra are taken from samples containing only the fluorophore of interest or in the absence of the fluorophore in the case of autofluorescence. For example, control animals with no GFP expressed are scanned under lambda mode (multispectrum mode) ranging from 480 to 700 nm to collect the broad autofluorescent spectral profile. GFP-expressing control animals are also imaged in lambda mode to collect the GFP emission spectra from 480 to 700 nm. Note, it is preferable to collect the GFP spectra from a tissue with little to no autofluorescence such as oocytes or neuronal expressed GFP. The collected spectra are saved into the spectral database. Intestinal autofluorescence is

particularly strong in the DAPI channel. Therefore, objects that appear in both the GFP and DAPI channels are considered autofluorescence and canceled out, while true GFP signal appears in the GFP channel but does not colocalize with the DAPI channel (*see* **Note 22**).

17. Data sets are collected for 10–12 animals of each genotype or condition being tested (*see* **Note 23**).
18. Quantification of images is performed with open-source Fiji software [33].
19. Regions of interest (ROI) are selected in Fiji software using the "Specify" function. Square pixel boxes are set to a predefined area of 100 × 100 pixels. Three regions are analyzed for GFP intensity per sample using the "Measure function." When comparing intensity, an average of six samples are quantified for a total of 18 ROIs.
20. An unpaired, two-tailed t test is used to compare the p value between the control *wild-type* animal and each genotype.

3.6.3 Colocalization of BMP Receptor Protein Fusions with Subcellular Markers

1. To test the localization of GFP-tagged BMP signal transduction components, subcellular compartments are marked with another fluorophore fusion protein that localizes to the distinct compartment of interest. RFP- or mCherry-tagged compartment markers are typically used to measure colocalization of the GFP-tagged receptors. For example, SMA-6::GFP normally colocalizes to the plasma membrane as well as intracellular endosomal compartments including the late endosome [6]. To label the late endosome, a fusion protein of RAB-7 and RFP is expressed in the intestine and measured for colocalization with SMA-6::GFP.
2. To collect the image, place your slide with the agarose pad and sample onto the stage of the microscope, as described above. Locate and focus on an animal under low magnification (10×). To observe GFP expression in a transgenic animal, switch the fluorescence source on and switch off bright field.
3. Optimize the signal-to-noise ratio for each individual channel as described in the previous section. Typically, settings are adjusted for the control sample for optimal GFP and RFP, signal-to-noise, and also DAPI if using this channel for autofluorescent detection (*see* **Notes 22** and **23**).
4. After the samples have been acquired, reference spectra are selected, and channel and color are assigned for each reference spectrum. Unmixing of the images is performed, and background channels are turned off to obtain the true GFP signal.
5. Regions of interest (ROI) are selected in Fiji software using the polygon selections function. Individual cells outlined by an

ROI are analyzed for GFP colocalization with RFP using the Analyze → Colocalizations → Colocalization thresholds function. Select the image that contains the ROI as Channel 1. Percent of GFP and RFP overlap are determined per volume area, as well as both Mander's overlap coefficient (MOC) and Pearson correlation coefficient (PCC), which are both used to quantify the degree of colocalization between fluorophores. When comparing colocalization of fluorophores, an average of six samples are quantified for a total of 6–12 ROIs (*see* **Note 24**).

6. An unpaired, two-tailed *t* test is used to compare the *p* value between the control *wild-type* animal and each genotype.

4 Notes

1. 1 N NaOH solution is used to inactivate EMS which is a potent mutagen. Always wear gloves when handling EMS. Keep EMS, and anything that comes in contact with it, in the hood. Inactivate anything that comes in contact with it by soaking in 1 N NaOH overnight.
2. The strains *dbl-1(wk70)*, *sma-3(wk30)*, and *sma-6(wk7)* contain premature termination codons generating null alleles in the genes encoding the ligand DBL-1, the Smad SMA-3, and the receptor SMA-6 [7, 12, 39]. Other mutant backgrounds can be analyzed as desired, but it should be kept in mind that for interpretation of double mutant phenotypes, at least one of the mutations must be a null allele of the corresponding gene.
3. *lon-2(e678)* is a loss-of-function mutation in *lon-2*, which encodes a glypican that negatively regulates DBL-1 signaling [29]. *ctIs40* is a transgene carrying extra copies of *dbl-1(+)* that results in overexpression of the ligand [12]. In both genetic backgrounds, the DBL-1 pathway is overactive, leading to a long body size.
4. Sodium azide can quench the GFP fluorescence of the RAD-SMAD reporter. Thus, tetramisole is used to anesthetize the worms.
5. L4 larvae are readily identified due to the process of vulval morphogenesis that is occurring during this stage. In the dissecting microscope, this process is visible as a lighter semicircle seen on the ventral surface midway between the head and the tail. In the compound microscope under Nomarski DIC optics, L4 larvae are characterized by the "Christmas tree structure" formed during vulval morphogenesis [46, 47].

6. Since the Sma/Mab signaling mutants are slow growing compared with controls, it is not possible to screen the F2 progeny on the same plates with other segregants from the P0.
7. For X-linked mutations, male progeny from the cross N2 males X *sma/sma* hermaphrodites will be hemizygous *sma/0* genotype, and all of these males will be small. For dominant and semidominant mutations, all of the progeny, including *sma/+* cross progeny, will have a small body size.
8. To measure young adult animals, pick 30 L4 animals to a new plate. After 24 h, mount the young adults on slides and image. To measure animals at larval stages, harvest worms at 24, 48, and 72 h after bleaching. For larval stage measurements, keep in mind that animals may develop at different rates and that it may be necessary to take this into account when interpreting results. Compare the average length of experimental animals to a wild-type control grown in the same conditions at the same time.
9. Sma/Mab signaling mutants are smaller in both length and width. Measuring body length provides a convenient proxy for growth. Other alternatives can also be used, such as measuring perimeter, cross-sectional area, or calculated volume.
10. This genetic cross assumes that the mutations are unlinked. If the mutations are linked, there are three alternatives: (1) omit the analysis of the particular genetic interaction if other analyses are sufficiently informative; (2) use RNAi instead of mutation to inactivate one of the genes; or (3) use genetic markers to assist in generating the desired double mutant (www.wormbase.org).
11. The 4 CCs phenotype of *sma-9(cc604)* mutants are not 100% penetrant: usually zero to 3% of *sma-9(cc604)* animals have 6 CCs as is found in wild-type animals. Because of this low background and because the *sma-9;susm* double mutant strains are slow growing, it is not possible to screen the F2 progeny on the same plates with other segregants from the P0.
12. Assuming that all the F1 animals on a given plate are fertile and give similar number of F2 progeny, if none of the F1 animals carries a *sma-9* suppressor mutations, near 100% of the F2 animals will have 4 CCs, rather than 6 CCs. If only one of the three F1s carries a *sma-9* suppressor mutation and the mutation is 100% penetrant, then 8.33% of the F2 animals are expected to have 6 CCs. If all three of the F1s carry *sma-9* suppressor mutations and all of the mutations are 100% penetrant, then 25% of the F2 animals are expected to have 6 CCs. To ensure for the recovery of suppressor mutations with a penetrance of at least 50%, plates that have anywhere between 5% and 25% of animals with 6 CCs are kept as candidates.

13. Many of the Sma/Mab pathway mutations exhibit partial dominance in their suppression of the *sma-9* CC phenotype [5, 9]. Thus, multiple 6 CC-bearing animals from the same plate should be single plated. The plates that reproducibly give the highest penetrance of the 6 CCs phenotype (wild-type phenotype) are likely from animals that are homozygous for the suppressor mutation.
14. The RAD-SMAD reporter was integrated into the *C. elegans* genome together with a co-injection marker *mec-7p::mRFP*, which is expressed in the six touch neurons and can be used as a marker in genetic crosses introducing the RAD-SMAD reporter into different mutant background. Two different integrated lines are listed, *jjIs2433* maps to the X chromosome and *jjIs2437* maps to chromosome II. Because the RAD-SMAD reporter exhibits variability in levels of GFP expression between animals, and also between nuclei in the same animal, we recommend the usage of *dbl-1(wk70)* and *lon-2(e678)*, which exhibit reduced or increased RAD-SMAD expression, respectively, as controls prior to the measurement of your mutant of interest.
15. Expression level of the RAD-SMAD reporter is highly dynamic during development, with stronger intestinal expression in embryos and L1s but fainter in adults, strong hypodermal expression in L2 larvae, and even stronger hypodermal expression in L3 larvae. Thus, it is critical that the worms are synchronized when collecting images for the RAD-SMAD reporter.
16. Take images of hypodermal nuclei in the mid-body region of the worm. Avoid the head and tail hypodermal nuclei because they are smaller. Also avoid those hypodermal nuclei that are located adjacent to the most anterior or posterior intestinal nuclei, as those intestinal nuclei tend to exhibit very bright GFP expression.
17. Briefly examine through the slides with WT, *dbl-1(wk70)* and *lon-2(e678)* worms to ensure that an appropriate exposure time is chosen for collecting the RAD-SMAD fluorescence images in all genotypes so that the RAD-SMAD fluorescence signals in *lon-2(e678)* worms are not saturated, while the RAD-SMAD fluorescence signals in *dbl-1(wk70)* worms are slightly above background.
18. An alternative method for nanoparticle immobilization on agar has also been reported using submicron polystyrene beads [48].
19. The ability of the fusion transgene to rescue a mutant phenotype is an excellent test to validate the fusion protein is functional and demonstrates proper subcellular localization. Both

SMA-6::GFP and DAF-4::GFP can rescue *sma-6(wk7)* and *daf-4(m72)* mutant phenotypes [6, 49].

20. Leica HyD (hybrid detectors) or Zeiss GaAsP (Gallium arsenide phosphide) detectors are both optimized for live cell imaging when compared to traditional photomultiplier tubes (PMTs). Sensitive live samples, such as *C. elegans*, need to be imaged under low-light conditions to prevent phototoxicity, as well as photobleaching of the fluorophore. Both the HyD and GaAsP offer much higher sensitivity than a standard PMT detector. Furthermore, the high sensitivity of both these detectors allows reduction of light delivered to sample, further reducing potential background signal due to scattering of the light through tissue slices.
21. Note, as one increases the gain, pixels can be saturated, and noise may become evident in the image.
22. An alternative approach if spectral profiling is not available is to capture a second set of images using a standard commercial DAPI filter and overlay with the GFP image.
23. It is important when comparing samples to image under the same conditions as the control. For example, objectives, pinhole, gain, offset, and laser units should all be kept identical between samples, as well as the age of the animals.
24. Deconvolution software can significantly enhance image quality. Software from Autoquant (X3) or iVision (BioVision Technologies) can be used to process image stacks. Stacks of ten or more images taken at 0.2 mm intervals along the *Z*-axis are ideal for deconvolution in either program.

Acknowledgments

We thank Mehul Vora for images in Fig. 2 and Zhiyu Liu for images in Figs. 3 and 5. Work in the authors' laboratories is supported in part by NIH R15 GM112147 to CSD, NIH R01 GM066953 and R01 GM103869 to J.L., and NIH R01 GM103995 and a Busch Biomedical Grant to R.W.P. R.J.G. was supported in part by a Busch Biomedical Predoctoral Fellowship.

References

1. Srivastava M, Begovic E, Chapman J, Putnam NH, Hellsten U, Kawashima T, Kuo A, Mitros T, Salamov A, Carpenter ML, Signorovitch AY, Moreno MA, Kamm K, Grimwood J, Schmutz J, Shapiro H, Grigoriev IV, Buss LW, Schierwater B, Dellaporta SL, Rokhsar DS (2008) The Trichoplax genome and the nature of placozoans. Nature 454(7207):955–960. https://doi.org/10.1038/nature07191
2. Srivastava M, Simakov O, Chapman J, Fahey B, Gauthier ME, Mitros T, Richards GS, Conaco C, Dacre M, Hellsten U, Larroux C, Putnam NH, Stanke M, Adamska M, Darling A, Degnan SM, Oakley TH, Plachetzki

DC, Zhai Y, Adamski M, Calcino A, Cummins SF, Goodstein DM, Harris C, Jackson DJ, Leys SP, Shu S, Woodcroft BJ, Vervoort M, Kosik KS, Manning G, Degnan BM, Rokhsar DS (2010) The Amphimedon queenslandica genome and the evolution of animal complexity. Nature 466(7307):720–726. https://doi.org/10.1038/nature09201

3. Huminiecki L, Goldovsky L, Freilich S, Moustakas A, Ouzounis C, Heldin C-H (2009) Emergence, development and diversification of the TGF-β signalling pathway within the animal kingdom. BMC Evol Biol 9:28
4. Massagué J (2008) TGFβ in cancer. Cell 134 (2):215–230. https://doi.org/10.1016/j.cell.2008.07.001. pii: S0092-8674(08)00878-7
5. Foehr ML, Lindy AS, Fairbank RC, Amin NM, Xu M, Yanowitz J, Fire AZ, Liu J (2006) An antagonistic role for the *C. elegans* Schnurri homolog SMA-9 in modulating TGFβ signaling during mesodermal patterning. Development 133(15):2887–2896
6. Gleason RJ, Akintobi AM, Grant BD, Padgett RW (2014) BMP signaling requires retromer-dependent recycling of the type I receptor. Proc Natl Acad Sci U S A 111(7):2578–2583
7. Krishna S, Maduzia LL, Padgett RW (1999) Specificity of TGFβ signaling is conferred by distinct type I receptors and their associated SMAD proteins in *Caenorhabditis elegans*. Development 126(2):251–260
8. Liang J, Lints R, Foehr ML, Tokarz R, Yu L, Emmons SW, Liu J, Savage-Dunn C (2003) The *Caenorhabditis elegans schnurri* homolog *sma-9* mediates stage- and cell type-specific responses to DBL-1 BMP-related signaling. Development 130(26):6453–6464
9. Liu Z, Shi H, Szymczak LC, Aydin T, Yun S, Constas K, Schaeffer A, Ranjan S, Kubba S, Alam E, McMahon DE, He J, Shwartz N, Tian C, Plavskin Y, Lindy A, Dad NA, Sheth S, Amin NM, Zimmerman S, Liu D, Schwarz EM, Smith H, Krause MW, Liu J (2015) Promotion of bone morphogenetic protein signaling by tetraspanins and glycosphingolipids. PLoS Genet 11(5):e1005221. https://doi.org/10.1371/journal.pgen.1005221
10. Savage C, Das P, Finelli AL, Townsend SR, Sun CY, Baird SE, Padgett RW (1996) *Caenorhabditis elegans* genes *sma-2, sma-3,* and *sma-4* define a conserved family of transforming growth factor-β pathway components. Proc Natl Acad Sci U S A 93(2):790–794
11. Savage-Dunn C, Maduzia LL, Zimmerman CM, Roberts AF, Cohen S, Tokarz R, Padgett RW (2003) Genetic screen for small body size mutants in *C. elegans* reveals many TGFβ pathway components. Genesis 35(4):239–247
12. Suzuki Y, Yandell MD, Roy PJ, Krishna S, Savage-Dunn C, Ross RM, Padgett RW, Wood WB (1999) A BMP homolog acts as a dose-dependent regulator of body size and male tail patterning in *Caenorhabditis elegans*. Development 126(2):241–250
13. Tian C, Sen D, Shi H, Foehr ML, Plavskin Y, Vatamaniuk OK, Liu J (2010) The RGM protein DRAG-1 positively regulates a BMP-like signaling pathway in *Caenorhabditis elegans*. Development 137(14):2375–2384
14. Tian C, Shi H, Xiong S, Hu F, Xiong WC, Liu J (2013) The neogenin/DCC homolog UNC-40 promotes BMP signaling via the RGM protein DRAG-1 in *C. elegans*. Development 140(19):4070–4080
15. Gumienny TL, Savage-Dunn C (2013) TGF-β signaling in *C. elegans*. WormBook, pp 1–34. https://doi.org/10.1895/wormbook.1.22.2
16. Padgett RW, Patterson GI (2008) TGFβ family signaling in the nematode, *C. elegans*. The TGFβ family. Cold Spring Harbor Press, Cold Spring Harbor, NY
17. Patterson GI, Padgett RW (2000) TGF-β-related pathways. Roles in *Caenorhabditis elegans* development. Trends Genet 16(1):27–33
18. Lawler S, Feng XH, Chen RH, Maruoka EM, Turck CW, Griswold-Prenner I, Derynck R (1997) The type II transforming growth factor-β receptor autophosphorylates not only on serine and threonine but also on tyrosine residues. J Biol Chem 272(23):14850–14859
19. Shi Y, Massagué J (2003) Mechanisms of TGF-β signaling from cell membrane to the nucleus. Cell 113(6):685–700
20. Luo S, Kleemann GA, Ashraf JM, Shaw WM, Murphy CT (2010) TGF-β and insulin signaling regulate reproductive aging via oocyte and germline quality maintenance. Cell 143 (2):299–312
21. Luo S, Shaw WM, Ashraf J, Murphy CT (2009) TGF-β Sma/Mab signaling mutations uncouple reproductive aging from somatic aging. PLoS Genet 5(12):e1000789
22. Mallo GV, Kurz CL, Couillault C, Pujol N, Granjeaud S, Kohara Y, Ewbank JJ (2002) Inducible antibacterial defense system in *C. elegans*. Curr Biol 12(14):1209–1214
23. Nicholas HR, Hodgkin J (2004) Responses to infection and possible recognition strategies in the innate immune system of *Caenorhabditis elegans*. Mol Immunol 41(5):479–493
24. Roberts AF, Gumienny TL, Gleason RJ, Wang H, Padgett RW (2010) Regulation of genes affecting body size and innate immunity

by the DBL-1/BMP-like pathway in *Caenorhabditis elegans*. BMC Dev Biol 10:61

25. Sinha A, Rae R, Iatsenko I, Sommer RJ (2012) System wide analysis of the evolution of innate immunity in the nematode model species *Caenorhabditis elegans* and *Pristionchus pacificus*. PLoS One 7(9):e44255
26. Tan MW, Rahme LG, Sternberg JA, Tompkins RG, Ausubel FM (1999) *Pseudomonas aeruginosa* killing of *Caenorhabditis elegans* used to identify *P. aeruginosa* virulence factors. Proc Natl Acad Sci U S A 96(5):2408–2413
27. Fernando T, Flibotte S, Xiong S, Yin J, Yzeiraj E, Moerman DG, Meléndez A, Savage-Dunn C (2011) *C. elegans* ADAMTS ADT-2 regulates body size by modulating TGFβ signaling and cuticle collagen organization. Dev Biol 352(1):92–103
28. Gumienny TL, Macneil L, Zimmerman CM, Wang H, Chin L, Wrana JL, Padgett RW (2010) *Caenorhabditis elegans* SMA-10/LRIG is a conserved transmembrane protein that enhances bone morphogenetic protein signaling. PLoS Genet 6(5):e1000963. https://doi.org/10.1371/journal.pgen.1000963
29. Gumienny TL, Macneil LT, Wang H, de Bono M, Wrana JL, Padgett RW (2007) Glypican LON-2 is a conserved negative regulator of BMP-like signaling in *Caenorhabditis elegans*. Curr Biol 17(2):159–164
30. Gleason RJ, Vora M, Li Y, Kane NS, Liao K, Padgett RW (2017) *C. elegans* SMA-10 regulates BMP receptor trafficking. PLoS One 12 (7):e0180681. https://doi.org/10.1371/journal.pone.0180681
31. Sulston JE, Horvitz HR (1977) Post-embryonic cell lineages of the nematode, *Caenorhabditis elegans*. Dev Biol 56(1):110–156
32. Stiernagle T (2006) Maintenance of *C. elegans*. WormBook, pp 1–11. https://doi.org/10.1895/wormbook.1.101.1
33. Schindelin J, Arganda-Carreras I, Frise E, Kaynig V, Longair M, Pietzsch T, Preibisch S, Rueden C, Saalfeld S, Schmid B, Tinevez JY, White DJ, Hartenstein V, Eliceiri K, Tomancak P, Cardona A (2012) Fiji: an open-source platform for biological-image analysis. Nat Methods 9(7):676–682. https://doi.org/10.1038/nmeth.2019
34. Fares H, Greenwald I (2001) Genetic analysis of endocytosis in *Caenorhabditis elegans*: coelomocyte uptake defective mutants. Genetics 159(1):133–145
35. Fares H, Greenwald I (2001) Regulation of endocytosis by CUP-5, the *Caenorhabditis elegans* mucolipin-1 homolog. Nat Genet 28 (1):64–68. https://doi.org/10.1038/88281
36. Brenner S (1974) The genetics of *Caenorhabditis elegans*. Genetics 77(1):71–94
37. Maduzia LL, Roberts AF, Wang H, Lin X, Chin LJ, Zimmerman CM, Cohen S, Feng XH, Padgett RW (2005) *C. elegans* serine-threonine kinase KIN-29 modulates TGFβ signaling and regulates body size formation. BMC Dev Biol 5(1):8
38. Doitsidou M, Poole RJ, Sarin S, Bigelow H, Hobert O (2010) *C. elegans* mutant identification with a one-step whole-genome-sequencing and SNP mapping strategy. PLoS One 5 (11):e15435. https://doi.org/10.1371/journal.pone.0015435
39. Savage-Dunn C, Tokarz H, Wang H, Cohen S, Giannikas C, Padgett RW (2000) *SMA-3* Smad has specific and critical functions in DBL-1/SMA-6 TGFβ-related signaling. Dev Biol 223 (1):70–76
40. Wang J, Tokarz R, Savage-Dunn C (2002) The expression of TGFβ signal transducers in the hypodermis regulates body size in *C. elegans*. Development 129(21):4989–4998
41. Ermolaeva MA, Schumacher B (2014) Insights from the worm: the *C. elegans* model for innate immunity. Semin Immunol 26(4):303–309
42. Malone EA, Thomas JH (1994) A screen for nonconditional dauer-constitutive mutations in *Caenorhabditis elegans*. Genetics 136 (3):879–886
43. Thomas JH, Birnby DA, Vowels JJ (1993) Evidence for parallel processing of sensory information controlling dauer formation in *Caenorhabditis elegans*. Genetics 134 (4):1105–1117
44. Mahajan-Miklos S, Tan MW, Rahme LG, Ausubel FM (1999) Molecular mechanisms of bacterial virulence elucidated using a *Pseudomonas aeruginosa-Caenorhabditis elegans* pathogenesis model. Cell 96(1):47–56
45. Shi A, Sun L, Banerjee R, Tobin M, Zhang Y, Grant BD (2009) Regulation of endosomal clathrin and retromer-mediated endosome to Golgi retrograde transport by the J-domain protein RME-8. EMBO J 28(21):3290–3302. https://doi.org/10.1038/emboj.2009.272
46. Eisenmann DM, Kim SK (2000) Protruding vulva mutants identify novel loci and Wnt signaling factors that function during *Caenorhabditis elegans* vulva development. Genetics 156 (3):1097–1116
47. Mok DZ, Sternberg PW, Inoue T (2015) Morphologically defined sub-stages of *C. elegans* vulval development in the fourth larval stage. BMC Dev Biol 15:26. https://doi.org/10.1186/s12861-015-0076-7

48. Kim E, Sun L, Gabel CV, Fang-Yen C (2013) Long-term imaging of *Caenorhabditis elegans* using nanoparticle-mediated immobilization. PLoS One 8(1):e53419. https://doi.org/10.1371/journal.pone.0053419
49. Patterson GI, Koweek A, Wong A, Liu Y, Ruvkun G (1997) The DAF-3 Smad protein antagonizes TGF-β-related receptor signaling in the *Caenorhabditis elegans* dauer pathway. Genes Dev 11(20):2679–2690
50. Massagué J, Blain SW, Lo RS (2000) TGFβ signaling in growth control, cancer, and heritable disorders. Cell 103(2):295–309

Chapter 7

Gene Regulation of BMP Ligands in *Drosophila*

Brian G. Stultz and Deborah A. Hursh

Abstract

Drosophila is a valuable system to study bone morphogenetic proteins (BMPs) due to the high functional conservation of the pathway and the molecular genetic tools available. *Drosophila* has three BMP ligands, *decapentaplegic* (BMP2/4), *screw*, and *glass bottom boat* (BMP5/6/7/8). Of these genes, the transcriptional regulation of *decapentaplegic* has been studied, and some of the enhancers directing its spatially specific gene expression have been described. These analyses have used many of the standard tools of molecular biology, but a valuable method of analysis often used in *Drosophila* is the creation of patches of mutant tissue at any stage and in any location by induced somatic recombination. The ability to create transgenic flies and manipulate the *Drosophila* genome with recombinases is well established. This method can be used to evaluate the requirements for specific transcription factors to act on enhancer elements in vivo, in stage- and tissue-specific manners. The yeast FLP/*FRT* recombination system facilitates experiments to interrogate loss- or gain-of-function for transcription factor activity on known enhancers. This chapter will outline the necessary steps to create the tools needed and conduct somatic cell recombination experiments to interrogate the function of transcription factors on BMP enhancers.

Key words *Drosophila melanogaster*, Bone morphogenetic protein, BMP, Somatic clones, *Decapentaplegic*, *dpp*, *Glass bottom boat*, *gbb*, *screw*, *scw*

1 Introduction

Bone morphogenetic proteins (BMPs) are a functionally well-conserved class of growth factors within the TGF-β superfamily. In *Drosophila*, there are three BMP genes: *decapentaplegic* (*dpp*) is the ortholog of vertebrate BMP 2/4, and *screw* (*scw*) and *glass bottom boat* (*gbb*) are the orthologs of BMP 5/6/7/8. The genetic requirement for *scw* is limited to the specification of dorsal cell fate in the early embryo [1], where it forms heterodimers with *dpp* in order to affect patterning of the early embryo. The *gbb* gene has documented genetic requirements in the ventral nerve cord [2], neuromuscular junctions [3], imaginal wing disc pattern formation [4], midgut formation [5], fat body metabolism [6], and stem cell maintenance in germline and adult stem cells [7, 8]. *dpp* has historically been the most studied *Drosophila* BMP and is required in

Melissa B. Rogers (ed.), *Bone Morphogenetic Proteins: Methods and Protocols*, Methods in Molecular Biology, vol. 1891, https://doi.org/10.1007/978-1-4939-8904-1_7, © Springer Science+Business Media, LLC, part of Springer Nature 2019

virtually every phase of development for a large number of processes including oogenesis [9, 10], embryonic dorsal/ventral polarity [11], patterning of the embryonic mesoderm [12, 13], morphogenesis of the larval midgut [14, 15], larval airway formation [16], imaginal disc patterning and adult appendage development [17], adult ventral head morphogenesis [18], stem cell maintenance [7], and hematopoiesis [19].

Much is known about the expression patterns and signaling pathway of the *Drosophila* BMPs. *scw* expression is ubiquitous but limited to the early *Drosophila* embryo [1]. *gbb* is broadly expressed, starting at gastrulation of the embryo, and continues to be expressed in larval and adult tissue [20]. Based on *Drosophila* genomic data, *scw* and *gbb* have relatively simple gene structures. In contrast, *dpp* has a complex expression pattern that is reflected in the size and complexity of the gene. The *dpp* gene contains five promoters that direct transcription of five distinct mRNAs, all encoding an identical polypeptide [21]. Extensive 5′ and 3′ cis-regulatory regions include a large number of enhancer elements involved in *dpp*'s spatial regulation (Fig. 1). However, given the large number of developmental events that require *dpp*, relatively little is known about the upstream transcriptional activators that regulate these enhancers. The early expression of *dpp* required for embryonic dorsal/ventral patterning is controlled by *dorsal* [22], a *Drosophila* ortholog of rel-NF-κB. *dpp* transcriptional regulation has been extensively studied in the embryonic visceral mesoderm. General activation of the visceral mesoderm enhancer is controlled by the FoxF family member, *biniou* [23], and the Tbx1 ortholog *optomotor-blind-related-gene-1* (*org1*) [24]. Spatially specific visceral mesoderm expression is controlled by the homeotic genes *Ultrabithorax* (*Hox7*) and *abdominal-A* (*Hox8*), in association with the TALE family homeodomain containing proteins *extradenticle* (*Pbx*) and *homothorax* (*Meis*) [25–29]. In addition, the terminal transcription factor of the *wingless* pathway, *pangolin/dTCF*, regulates *dpp* expression in this tissue [30]. An enhancer within the second intron of the Hin coding region (Fig. 1) drives lateral embryonic dorsal ectoderm expression and late embryonic

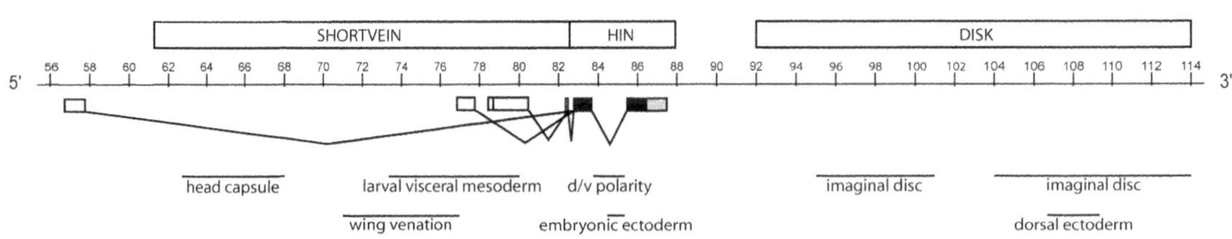

Fig. 1 A map of the complex *dpp* gene including transcripts and known enhancers. Five promotors and transcripts are located just under the numbered coordinates, which encode for a single protein located in the haplo-insufficiency (HIN) region of the gene. The 5′ and 3′ cis-regulatory regions are designated shortvein and disk, respectively. Known enhancers are represented as lines and labeled with their respective functions

expression in larval airway cells, called posterior spiracles. Binding sites for *dTCF*, the *Drosophila* SMAD: *mothers against dpp* (*mad*), and the *dpp* target repressor *brinker* are found within this enhancer [31, 32]. *dpp* expression is also required for post-embryonic development through expression in all the larval imaginal discs. This expression is controlled by the homeodomain transcription factor *engrailed*, and *cubitus interruptus*, a Gli homolog acting as the terminal transcription factor of the *hedgehog* pathway [33, 34]. *dpp* expression related to its role in adult head capsule morphogenesis is driven by the homeotic gene *labial* (*Hox1*), in concert with *extradenticle* and *homothorax* [35], and the zinc-finger transcription factor *odd-paired* [36], which belongs to the *zinc finger in the cerebellum* (*Zic*) family. Finally, *dpp* is controlled through autoregulation in every tissue in which it has been examined [15, 18, 37–40]. This is most likely through activating SMAD sites in the various *dpp* enhancers.

Multiple techniques can be used to study gene regulation in *Drosophila*, including reporter assays in transgenic animals or cell culture to test promoter and enhancer structure, electromobility shift assays (EMSA), and chromatin immunoprecipitation (ChIP) to assess transcription factor binding. One particular method of great utility in *Drosophila* to study transcriptional regulation is the use of mosaic somatic cell clones. Reporter constructs containing enhancer elements can be introduced transgenically, and their dependence on transcription factors can be determined in the tissue of interest by creating patches of tissue that are either mutant for or ectopically expressing a transcription factor or factors. This has been particularly useful for the study of *dpp*, as *dpp* and many of its transcriptional regulators are genetically required during early embryogenesis. In order to identify the function of transcriptional regulators of *dpp* later in development, tools must be used that bypass the requirement for their expression in the early embryo. Mosaic clones can be produced in almost any tissue, during any stage of development through the induction of recombination events. The majority of clonal analyses have been done with the yeast site-specific recombination system: FLP/*FRT* [41–43]. Additional recombinase systems used to modify the genome have been developed for use in *Drosophila*, like Cre/*loxP* [44] or the ϕC31 system [45], but as the principles are largely the same, this chapter will focus on the FLP/*FRT* system. Many tools have been developed in *Drosophila* to take advantage of the FLP/*FRT* system to mark clonal tissue, produce gain or loss of function in a gene of interest, and monitor readout of gene activity or function (reviewed in Blair [46]). Genetic reagents exist to monitor mutant phenotype in the tissue or developmental stage of interest, but to use this method to analyze transcription, a reporter construct containing an enhancer of interest is usually employed. In this chapter we will describe the methods used to make somatic cell clones marked by loss of green fluorescent protein (GFP) expression. The effects on a transcriptional enhancer are monitored using a reporter construct whose expression is detected by immunohistochemistry. We will also describe alterations to the protocol that would allow gain-of-function experiments for a transcription factor of interest. We have used this method with great facility to demonstrate the transcriptional requirements of an enhancer that directs *dpp* expression in the primordia of the adult head: the larval eye/antennal imaginal disc [35, 36].

2 Materials

2.1 Genetics

1. Appropriate *Drosophila* FLP/*FRT* stocks (stocks are available from the Bloomington Stock Center at http://flystocks.bio.indiana.edu).
2. Fly genetics equipment, including stereomicroscope with fiber-optic light source, porous polyethylene fly pad, source of CO_2, and a fly brush (*see* **Note 1**).
3. Fly media: vials, bottles, media formulation, vial and bottle closures.
4. 50 mg/mL Geneticin (G418) stock solution.
5. G418 selection media is made by adding 50 mg/mL stock solution into melted media just before it is poured into bottles and allowed to cool and solidify (*see* **Note 7**).

2.2 Clone Generation

1. BOD (biochemical oxygen demand) incubator set to 25 °C.
2. Water bath set to 38 °C.

2.3 Immunostaining and Analysis

1. 1× phosphate-buffered saline (PBS).
2. Fixative: 4% EM-grade formaldehyde in 1× PBS and 0.01% Triton X-100.
3. Blocking buffer PBTN: 1× PBS plus 0.01% Triton X-100 and 10% normal donkey serum.
4. Wash buffer PBT: 1× PBS plus 0.01% Triton X-100.
5. Baskets: made by cutting the top half from a 2.0 mL screw cap centrifuge tube, punching a hole in the cap and inserting a square of Nitex mesh (120 μm pore size) under the cap (Fig. 2).
6. 3-well spot plate.
7. 24-well culture dish.
8. Dumont #5 forceps with biology tip.
9. Tungsten micro needle (0.125 mm) and pin holder.
10. Nutator or other gentle three dimensional rocking mixer.

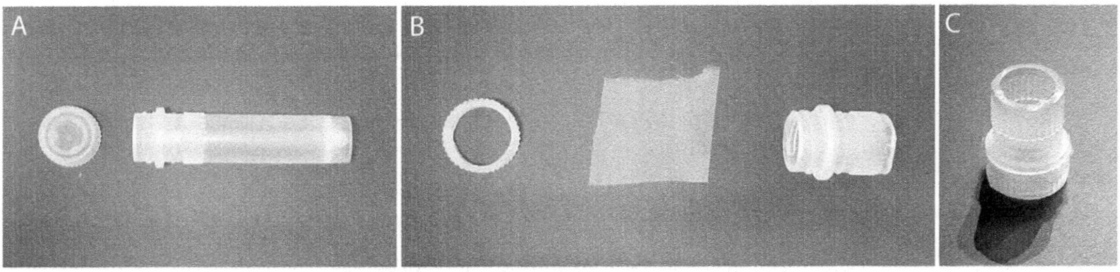

Fig. 2 The construction of baskets for immunohistochemistry on larval tissue. (**a**) One standard 2 mL screw cap microcentrifuge tube is modified by (**b**) cutting the tube in half and cutting the top off the cap. A square of Nitex mesh is inserted between the cap and tube creating a finished (**c**) basket

11. β-Galactosidase primary antibody (*see* **Note 2**).
12. Fluorochrome-conjugated secondary antibody.
13. A 4 mg/mL stock solution of 4′,6-diamidino-2-phenylindole (DAPI) in dimethyl sulfoxide (DMSO).
14. Antifade mounting medium (*see* **Note 3**).
15. Slides and coverslips.

3 Methods

3.1 Genetics

1. Select a gene and identify suitable alleles for somatic cell clone analysis (*see* **Note 4**).
2. Stocks containing chromosomes bearing the yeast *FRT* recombination site associated with the selectable marker neomycin have been created at the base of all *Drosophila* chromosome arms, except chromosome 4 [43]. These chromosomes also carry markers, such as GFP, to identify clones of somatically recombined tissue, and adult markers, such as *white*, to track the chromosome (a full description of available *FRT* chromosomes is located on FlyBase at http://flybase.org). Identify the chromosome arm location of your gene of interest and select a stock that has an *FRT* insertion at the base of the same chromosome arm. The *FRT* will need to be centromeric (proximal) to the gene of interest.
3. A stock containing a mutated gene of interest with a proximal *FRT* insertion on the same chromosome arm must be obtained or generated by meiotic recombination. If you are able to obtain an appropriate previously made mutation/*FRT* stock from a stock center or another lab, proceed to **step 8**.
4. Cross flies bearing the desired gene mutation to an appropriate *P{neoFRT}* stock. In general, mutant strains will be maintained over balancer chromosomes bearing multiple rearrangements to suppress recombination and dominate genetic markers to

allow selection. The goal of this cross is to produce female flies bearing the chromosome with the mutant gene and a homolog bearing the *FRT*, so that meiotic recombination can occur in the ovaries of these females. All genetic crosses in this protocol are maintained in a 25 °C BOD incubator.

5. Collect virgin females containing the desired chromosomes from the previous cross by selecting female flies without any of the visible markers carried on the balancer chromosomes (Fig. 3a). Mate these with males from a balancer stock in bottles containing fly media (*see* **Notes 5** and **6**). The adults are mated for 24 h and then moved to bottles containing G418 media (*see* **Note 7**) to select for the presence of the neomycin gene associated with the *FRT*. The only larvae from this cross that will survive into the next generation will contain non-recombinant *FRT* chromosomes (the vast majority of flies) and recombinant chromosomes possibly containing the mutant gene allele and the *FRT* (Fig. 3b). The second purpose of the cross is to screen for suitable recombinants by using loss of visible markers on the *FRT* chromosome to identify them. For example, in Fig. 3b, a construct able to ubiquitously express GFP (*ubi-GFP*) has been inserted transgenically on the *FRT*-containing chromosome at a locus distal to the *FRT*. This *ubi-GFP* insertion also contains a partial copy of the *white* gene (mini-*white*), which confers an orange/red color to the eye. A suitable meiotic recombination event should have removed *ubi-GFP* in the process of adding the desired mutation to the *FRT* chromosome. Thus the desired recombinant chromosome can be identified by the loss of the visible *white* marker (i.e., look for the rare white-eyed fly among vast majority of orange/red eyed flies).
6. Flies that are recovered from this cross that survive on G418 and have lost the visible marker should immediately be mated to an appropriate balancer chromosome on standard fly media (without G418) to establish a stock (*see* **Note 8**). If lethal alleles of the gene of interest were used to do the initial recombination, then homozygous lethality should be observed in these recombinant lines.
7. Flies from established lines should be retested for G418 resistance and should stably maintain the correct markers, such as eye color. Each stock should be tested against multiple alleles of the gene of interest. This step helps to identify accessory lethal mutations that may exist or that may have been introduced during the recombination (*see* **Note 9**).
8. Once a chromosome has been created that contains the desired mutation and *FRT*, standard genetics is used to create two stocks that contain all the necessary elements for clonal analysis.

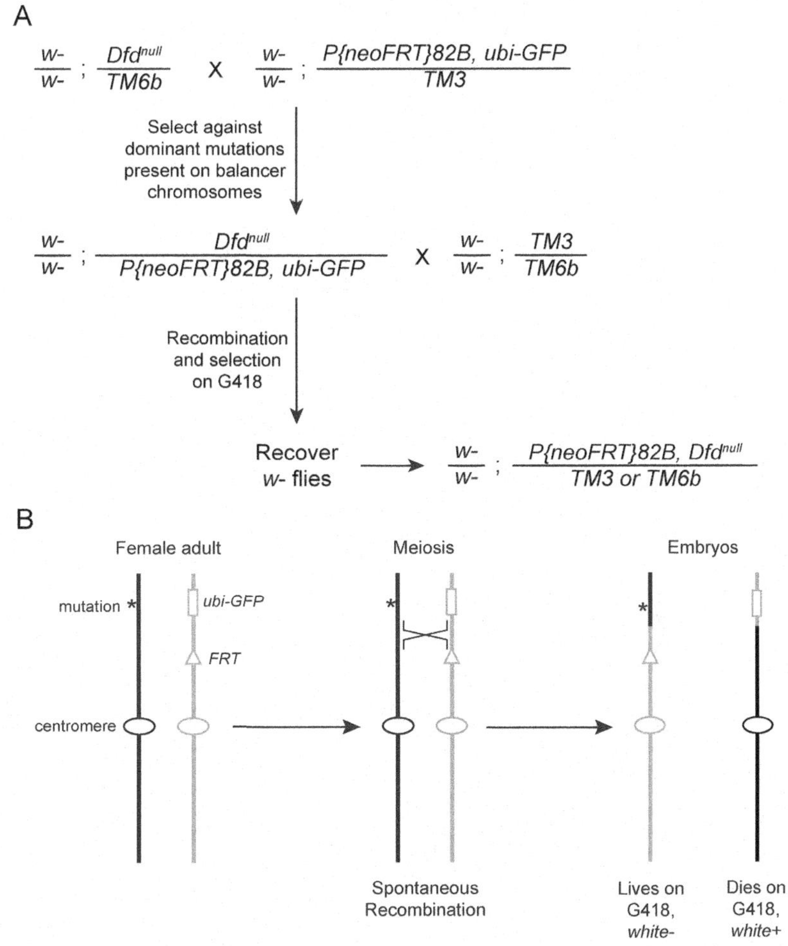

Fig. 3 A diagram of the genetic cross and meiotic recombination event required to create a mutant/FRT chromosome used for somatic cell clone creation. (**a**) A sample genetic cross between a *Deformed* (*Dfd*) null allele and the G418-selectable, *P{neoFRT}82B*, including steps and selection. "w-" indicates a chromosome bearing the *white* gene mutation. TM3 and TM6b indicate the balancer chromosomes, with dominant visible markers for selection. (**b**) A diagram of chromosomes shows the meiotic recombination that must occur to replace the *ubi-GFP* (containing a copy of the *white* gene) with the selected gene mutation

One stock should contain a source of the FLP recombinase (*see* **Note 10**) and a chromosome bearing the same *FRT* used to make the mutant/*FRT* recombinant. The second stock should contain your mutant/*FRT* chromosome and a *LacZ* reporter construct driven by the enhancer that you wish to study (Fig. 4a). Adjustments can be made to the genetics protocol in order to create gain-of-function somatic clones (*see* **Note 11**). The remainder of the protocol will be the same for both types of clonal analysis.

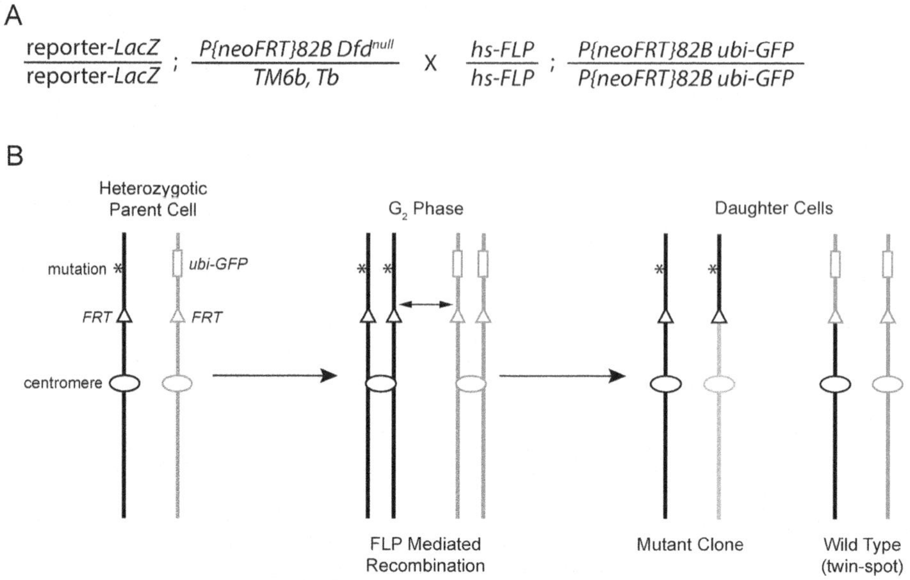

Fig. 4 The genetic cross and diagram of chromosome recombination that occurs during somatic clone creation. (**a**) The genetic cross creating the somatic clones must include two FRT-bearing chromosomes, one with the introduced mutation (here *Dfd*^{null}) and one with a marker to allow clone identification (here *ubi-GFP*). It also contains a gene encoding *flippase* (here hs-FLP) and a reporter *lacZ* gene. In this example, the balancer chromosome for the *Dfd*^{null}/*FRT* chromosome contains the larval marker *Tubby* (*Tb*), so you can select away from this marker during collection of larva for immunohistochemistry. (**b**) During G_2 phase of mitotic cell division, FLP mediates a recombination event between *FRT*s located on sister chromatids. The resulting loss of heterozygosity yields one daughter cell that is homozygous for the mutated gene of interest (mutant clone) and a second daughter cell with two *ubi-GFP* markers (twin spot)

9. Set up a genetic cross of flies containing the appropriate genotypes. Ten to twenty females of one stock are mated to 5–10 males from the second stock in a bottle containing standard fly media.

3.2 Clone Generation

1. Incubate mated adults at 25 °C for 2–3 days (*see* **Notes 12** and **13**).
2. Remove adults and heat shock the bottle by immersing it above the food line in a water bath at 38 °C for 1 h. Heat shocking the larvae will activate the hs-FLP and cause recombination to occur between sister chromatids, resulting in daughter cells with loss of heterozygosity (Fig. 4b). This event results in the loss of GFP expression and the creation of an associated twin spot (*see* **Note 14**).
3. Pick larvae for immunostaining at selected periods after heat shock, typically 2 or 3 days.

3.3 Immunostaining and Analysis

1. With forceps, collect appropriately staged larvae, selecting away from any larval markers on the balancer chromosome, and add these to 3-well spot plate containing 1× PBS (*see* **Note 15**).
2. Dissect imaginal discs by pulling the mouth hooks away from the rest of the body. Attached to the mouth hooks will be the eye/antennal disc, brain, intestine, some fat bodies, and some wing or leg discs (*see* **Note 16**). Transfer 12 sets of mouth hooks to a basket immersed in 1× PBS in a 24-well plate. Discard the remainder of the larval body. For each subsequent step in this protocol, add the appropriate solution to a new well in the plate, and transfer the basket to each well according to the time specified.
3. Fix discs for 25 min in 1 mL fixative at 4 °C while rocking on nutator (*see* **Note 17**).
4. Block tissues for 2 h at 4 °C in 1 mL blocking buffer while rocking on nutator (*see* **Note 18**).
5. Incubate tissues in primary antibodies diluted in 700 μL wash buffer overnight at 4 °C while rocking on nutator (*see* **Note 19**).
6. Wash tissues four times for 20 min each in 1 mL wash buffer while rocking on nutator.
7. Incubate tissues in fluorochrome-conjugated secondary antibodies diluted in 1 mL wash buffer for 2 h at 4 °C while rocking on nutator (*see* **Note 20**). Keep tissues in dark for remainder of protocol by wrapping the plate in aluminum foil.
8. Wash tissues three times for 20 min each in 1 mL wash buffer while rocking on nutator. For counterstaining of tissues, dilute a stock of 4 mg/mL DAPI into the second wash at 1:500.
9. Transfer two or three mouth hooks with imaginal discs attached from baskets to a drop of antifade mounting medium on a microscope slide.
10. Dissect and mount imaginal discs, by carefully removing other tissues attached to the mouth hooks (cuticle, salivary glands, brain, fat bodies, etc.). This is best done by holding the mouth hooks with a forceps and manipulating tissues or imaginal discs with a tungsten needle. Discs can be mounted separately or in the case of the eye/antennal disc can be attached to the mouth hooks.
11. Add a coverslip and visualize tissue on a fluorescent or confocal microscope with the appropriate filters. Store slides at 4 °C for slower fading. For examples of somatic clone experiments as described in this protocol, see de Navascues and Modolell [47], Lee et al. [36], or Stultz et al. [35]. In addition, Blair [46] provides visual examples of multiple types of clonal tissue.

4 Notes

1. For additional information on *Drosophila* husbandry, please refer to *Fly Pushing: The Theory and Practice of Drosophila Genetics* [48].
2. Suitable primary antibodies against β-galactosidase are currently available from the Developmental Studies Hybridoma Bank, catalog #40-1a, or Promega, catalog #Z3781.
3. Different types of antifade mounting media can be purchased from various venders; however, we prefer Vectashield (H-1000 or H-1200, Vector Laboratories).
4. The choice of alleles for clonal analysis is critical, and several alleles of your gene of interest should be assessed for suitability. Usually, a loss of function mutation is most appropriate. However, if total loss of function produces a cellular lethal, no clones will be recovered. In these cases, partial function mutations may prove more suitable. Another consideration is the presence of accessory lethal mutations, other than the gene of interest, on the same chromosome. This can be determined by comparing the lethal phenotype of homozygotes of your chosen allele to transheterozygotes of the allele over other lethal mutations in the same gene. Issues with allele selection are often determined empirically.
5. The goal in this cross is to recover and interrogate the meiotic events occurring in the eggs of the females. This is done by mating them to males carrying appropriately marked balancers that both keep the putative recombinant chromosomes intact, by suppressing further recombination and allowing selection of a visible marker, in our case *white.*
6. The rate of recombination events between two locations on a chromosome arm directly correlates to the distance between them. The closer they are, the less chance a suitable recombination event will occur. Therefore, a gene that lies closer to the *FRT* will require screening more chromosomes in order to acquire one with a suitable recombination event. In the example provided in this chapter (Fig. 3a) between *Deformed* (located at 84A on the cytogenetic map) and *P{neoFRT}82B*, we set up five bottles to recover a fairly rare recombination event (estimate screening 200 flies per bottle).
7. Because different *FRT* chromosomes differ in G418 tolerance and the activity of G418 preparations varies, the optimum G418 concentration for selection must be determined empirically. Prior to beginning the meiotic recombination, find the optimal G418 concentration. This is the concentration which allows maximum

survival of the *neoFRT* chromosome, but kills 100% of a negative control. The range is generally 0.2–0.5 mg/mL.

8. As suitable recombination events emerging from this cross may be rare, any white-eyed flies should be mated immediately to the appropriate males or virgin females from a balancer stock. Have these balancer flies available in advance.
9. It may also be necessary to interrogate the lethal phenotype of the recombinant chromosome by comparing homozygotes to heterozygotes with another allele of the same mutation. Before starting a time-consuming genetic analysis by inducing somatic recombination, ensure the mutant phenotype results from the gene of interest and not another gene on the chromosome.
10. Stocks containing FLP recombinase expressed in a tissue- or organ-specific manner are available from *Drosophila* stock centers. The accuracy of expression, both spatially and temporally, should be verified. The use of a conditional source of FLP recombinase such as hs-FLP (heat-shock-activated FLP recombinase) is often equally effective and is described here. To adjust the protocol for use of non-conditional sources of FLP recombinase, simply eliminate the heat-shocking step.
11. A full analysis of the requirement for transcription factors should include both loss- and gain-of-function experiments. The latter, in particular, can indicate the presence of additional transcriptional regulation that limits expression to certain tissues. Gain-of-function somatic clones can be made using constructs that have ubiquitous expression of the transcriptional activator *Gal4* (i.e., actin promoter driving *Gal4* expression). The promoter and the *Gal4* coding sequence are separated by an element that is flanked by *FRT* elements that can be removed by expression of FLP. A transcription factor driven by the yeast upstream activator sequence (UAS) can then be expressed and the response on the enhancer-driven reporter-*LacZ* construct monitored as outlined in the protocol. A large collection of UAS-driven proteins is available from stock centers. Additionally, a cDNA of any transcription factor can be readily inserted into vectors used to make transgenic *Drosophila* [46].
12. Larvae can be staged within a narrow window of development by collecting embryos from the cross at regular intervals (12 or 24 h is usually suitable), followed by transfer of the adults into fresh bottles of media. The progeny are held until the appropriate stage for heat shock. Multiple rounds of heat shock and clone analysis can be done from the same cross by continuing to transfer adults into fresh bottles of media and collecting embryos for the appropriate amount of time. This could allow for interrogation of multiple developmental stages

simultaneously. In this chapter we describe methods used to create (at first instar) and analyze (at third instar) somatic clones during larval development; however the same techniques could be used for embryonic or pupal development.

13. Preliminary experiments need to be carried out to determine appropriate heat-shock timing. The length of time before heat shock is determined by the stage of development the event of interest occurs in, and to avoid any unrelated, early lethal periods associated with the mutant you are examining. For example, in our experiments, we wished to avoid embryonic lethal phases of the *Deformed* (*Dfd*) gene and focus on events related to patterning of the post-embryonic imaginal discs. The time at which animals are observed after heat shock is determined both by the developmental stage of interest and the size of the clones one wishes to recover. Clones observed soon after heat shock will be small; those observed after longer times will have undergone more cell division and be larger.

14. The creation of a twin spot occurs as the result of one daughter cell getting two copies of *ubi-GFP* (Fig. 3b), in a background of cells that only have a single copy. These cells often have brighter GFP expression and appear juxtaposed to the mutant clonal tissue with GFP-negative cells. The presence of twin spots without mutant clones or clones of drastically reduced size may indicate your mutant clone has a growth disadvantage.

15. In addition to a spot plate well containing PBS, larvae can also be dissected in a drop of PBS placed on a Sylgard (Dow Corning) dissection plate. Place 12 individual drops of PBS on a Sylgard plate, and add a single larva to each for dissection.

16. Depending on the imaginal discs you want to observe, a better method is to cut the larvae in half and with tweezers invert the front half to expose all the discs to the outside. The front half of the larvae will contain all the imaginal discs except the genital discs.

17. 25 min of fixation works fine for most antibodies; however, some antibodies work better with longer fix times.

18. Normal donkey serum and donkey secondary antibodies are often used for their low fluorescent background in the *Drosophila* model system. However, blocking solution with 5% normal donkey serum and 10 μg/mL bovine serum albumin also works well.

19. The primary antibody used in this protocol is a β-galactosidase antibody concentrate from the Developmental Studies Hybridoma Bank (40-1a). It is diluted at a concentration of 1:100 into wash buffer.

20. The secondary fluorochrome-conjugated antibody should have emission spectra that do not overlap with DAPI or GFP. The one used in this protocol is a donkey anti-mouse IgG (H +L) secondary antibody, Alexa Fluor® 555 conjugate (A-31570, Molecular Probes) diluted in wash buffer at 1:500.

Acknowledgments

We thank Drs. John Thomas, Brent McCright, and Mark Mortin for helpful comments that improved the manuscript.

References

1. Arora K, Levine MS, O'Connor MB (1994) The screw gene encodes a ubiquitously expressed member of the TGF-beta family required for specification of dorsal cell fates in the Drosophila embryo. Genes Dev 8 (21):2588–2601
2. Bivik C, Bahrampour S, Ulvklo C, Nilsson P, Angel A, Fransson F, Lundin E, Renhorn J, Thor S (2015) Novel genes involved in controlling specification of Drosophila FMRFamide neuropeptide cells. Genetics 200 (4):1229–1244. https://doi.org/10.1534/genetics.115.178483
3. McCabe BD, Marques G, Haghighi AP, Fetter RD, Crotty ML, Haerry TE, Goodman CS, O'Connor MB (2003) The BMP homolog Gbb provides a retrograde signal that regulates synaptic growth at the Drosophila neuromuscular junction. Neuron 39(2):241–254
4. Khalsa O, Yoon JW, Torres-Schumann S, Wharton KA (1998) TGF-beta/BMP superfamily members, Gbb-60A and Dpp, cooperate to provide pattern information and establish cell identity in the Drosophila wing. Development 125(14):2723–2734
5. Wharton KA, Cook JM, Torres-Schumann S, de Castro K, Borod E, Phillips DA (1999) Genetic analysis of the bone morphogenetic protein-related gene, gbb, identifies multiple requirements during Drosophila development. Genetics 152(2):629–640
6. Ballard SL, Jarolimova J, Wharton KA (2010) Gbb/BMP signaling is required to maintain energy homeostasis in Drosophila. Dev Biol 337(2):375–385. https://doi.org/10.1016/j.ydbio.2009.11.011
7. Kawase E, Wong MD, Ding BC, Xie T (2004) Gbb/Bmp signaling is essential for maintaining germline stem cells and for repressing bam transcription in the Drosophila testis. Development 131(6):1365–1375. https://doi.org/10.1242/dev.01025
8. Tian A, Jiang J (2014) Intestinal epithelium-derived BMP controls stem cell self-renewal in Drosophila adult midgut. Elife 3:e01857. https://doi.org/10.7554/eLife.01857
9. Twombly V, Blackman RK, Jin H, Graff JM, Padgett RW, Gelbart WM (1996) The TGF-beta signaling pathway is essential for Drosophila oogenesis. Development 122 (5):1555–1565
10. Xie T, Spradling AC (1998) Decapentaplegic is essential for the maintenance and division of germline stem cells in the Drosophila ovary. Cell 94(2):251–260
11. Irish VF, Gelbart WM (1987) The decapentaplegic gene is required for dorsal-ventral patterning of the Drosophila embryo. Genes Dev 1(8):868–879
12. Staehling-Hampton K, Hoffmann FM, Baylies MK, Rushton E, Bate M (1994) dpp induces mesodermal gene expression in Drosophila. Nature 372(6508):783–786
13. Frasch M (1995) Induction of visceral and cardiac mesoderm by ectodermal Dpp in the early Drosophila embryo. Nature 374 (6521):464–467
14. Panganiban GE, Reuter R, Scott MP, Hoffmann FM (1990) A Drosophila growth factor homolog, decapentaplegic, regulates homeotic gene expression within and across germ layers during midgut morphogenesis. Development 110(4):1041–1050
15. Hursh DA, Padgett RW, Gelbart WM (1993) Cross regulation of decapentaplegic and Ultrabithorax transcription in the embryonic visceral mesoderm of Drosophila. Development 117 (4):1211–1222
16. Vincent S, Ruberte E, Grieder NC, Chen CK, Haerry T, Schuh R, Affolter M (1997) DPP

controls tracheal cell migration along the dorsoventral body axis of the Drosophila embryo. Development 124(14):2741–2750

17. Spencer FA, Hoffmann FM, Gelbart WM (1982) Decapentaplegic: a gene complex affecting morphogenesis in *Drosophila melanogaster*. Cell 28(3):451–461
18. Stultz BG, Lee HJ, Ramon K, Hursh DA (2006) Decapentaplegic head capsule mutations disrupt novel peripodial expression controlling the morphogenesis of the Drosophila ventral head. Dev Biol 296(2):329–339
19. Pennetier D, Oyallon J, Morin-Poulard I, Dejean S, Vincent A, Crozatier M (2012) Size control of the Drosophila hematopoietic niche by bone morphogenetic protein signaling reveals parallels with mammals. Proc Natl Acad Sci U S A 109(9):3389–3394. https://doi.org/10.1073/pnas.1109407109
20. Wharton KA, Thomsen GH, Gelbart WM (1991) Drosophila 60A gene, another transforming growth factor beta family member, is closely related to human bone morphogenetic proteins. Proc Natl Acad Sci U S A 88 (20):9214–9218
21. St Johnston RD, Hoffmann FM, Blackman RK, Segal D, Grimaila R, Padgett RW, Irick HA, Gelbart WM (1990) Molecular organization of the decapentaplegic gene in *Drosophila melanogaster*. Genes Dev 4(7):1114–1127
22. Huang JD, Schwyter DH, Shirokawa JM, Courey AJ (1993) The interplay between multiple enhancer and silencer elements defines the pattern of decapentaplegic expression. Genes Dev 7(4):694–704
23. Zaffran S, Kuchler A, Lee HH, Frasch M (2001) biniou (FoxF), a central component in a regulatory network controlling visceral mesoderm development and midgut morphogenesis in Drosophila. Genes Dev 15(21):2900–2915
24. Schaub C, Frasch M (2013) Org-1 is required for the diversification of circular visceral muscle founder cells and normal midgut morphogenesis. Dev Biol 376(2):245–259. https://doi.org/10.1016/j.ydbio.2013.01.022
25. Sun B, Hursh DA, Jackson D, Beachy PA (1995) Ultrabithorax protein is necessary but not sufficient for full activation of decapentaplegic expression in the visceral mesoderm. EMBO J 14(3):520–535
26. Stultz BG, Jackson DG, Mortin MA, Yang X, Beachy PA, Hursh DA (2006) Transcriptional activation by extradenticle in the Drosophila visceral mesoderm. Dev Biol 290(2):482–494
27. Capovilla M, Brandt M, Botas J (1994) Direct regulation of decapentaplegic by Ultrabithorax and its role in Drosophila midgut morphogenesis. Cell 76(3):461–475
28. Manak JR, Mathies LD, Scott MP (1994) Regulation of a decapentaplegic midgut enhancer by homeotic proteins. Development 120 (12):3605–3619
29. Capovilla M, Botas J (1998) Functional dominance among Hox genes: repression dominates activation in the regulation of Dpp. Development 125(24):4949–4957
30. Yang X, van Beest M, Clevers H, Jones T, Hursh DA, Mortin MA (2000) Decapentaplegic is a direct target of dTcf repression in the Drosophila visceral mesoderm. Development 127(17):3695–3702
31. Johnson AN, Bergman CM, Kreitman M, Newfeld SJ (2003) Embryonic enhancers in the dpp disk region regulate a second round of Dpp signaling from the dorsal ectoderm to the mesoderm that represses Zfh-1 expression in a subset of pericardial cells. Dev Biol 262 (1):137–151
32. Takaesu NT, Bulanin DS, Johnson AN, Orenic TV, Newfeld SJ (2008) A combinatorial enhancer recognized by Mad, TCF and Brinker first activates then represses dpp expression in the posterior spiracles of Drosophila. Dev Biol 313(2):829–843. https://doi.org/10.1016/j.ydbio.2007.10.021
33. Hepker J, Blackman RK, Holmgren R (1999) Cubitus interruptus is necessary but not sufficient for direct activation of a wing-specific decapentaplegic enhancer. Development 126 (16):3669–3677
34. Muller B, Basler K (2000) The repressor and activator forms of Cubitus interruptus control Hedgehog target genes through common generic gli-binding sites. Development 127 (14):2999–3007
35. Stultz BG, Park SY, Mortin MA, Kennison JA, Hursh DA (2012) Hox proteins coordinate peripodial decapentaplegic expression to direct adult head morphogenesis in Drosophila. Dev Biol 369(2):362–376
36. Lee H, Stultz BG, Hursh DA (2007) The Zic family member, odd-paired, regulates the Drosophila BMP, decapentaplegic, during adult head development. Development 134 (7):1301–1310
37. Biehs B, Francois V, Bier E (1996) The Drosophila short gastrulation gene prevents Dpp from autoactivating and suppressing neurogenesis in the neuroectoderm. Genes Dev 10 (22):2922–2934
38. Chanut F, Heberlein U (1997) Role of decapentaplegic in initiation and progression of the

morphogenetic furrow in the developing Drosophila retina. Development 124(2):559–567

39. Pignoni F, Zipursky SL (1997) Induction of Drosophila eye development by decapentaplegic. Development 124(2):271–278
40. Wiersdorff V, Lecuit T, Cohen SM, Mlodzik M (1996) Mad acts downstream of Dpp receptors, revealing a differential requirement for dpp signaling in initiation and propagation of morphogenesis in the Drosophila eye. Development 122(7):2153–2162
41. Chou TB, Perrimon N (1992) Use of a yeast site-specific recombinase to produce female germline chimeras in Drosophila. Genetics 131(3):643–653
42. Golic KG (1991) Site-specific recombination between homologous chromosomes in Drosophila. Science 252(5008):958–961
43. Xu T, Rubin GM (1993) Analysis of genetic mosaics in developing and adult Drosophila tissues. Development 117:1223–1237
44. Siegal ML, Hartl DL (1996) Transgene Coplacement and high efficiency site-specific recombination with the Cre/loxP system in Drosophila. Genetics 144(2):715–726
45. Groth AC, Fish M, Nusse R, Calos MP (2004) Construction of transgenic Drosophila by using the site-specific integrase from phage phiC31. Genetics 166(4):1775–1782
46. Blair SS (2003) Genetic mosaic techniques for studying Drosophila development. Development 130(21):5065–5072. https://doi.org/10.1242/dev.00774
47. de Navascues J, Modolell J (2010) The pronotum LIM-HD gene tailup is both a positive and a negative regulator of the proneural genes achaete and scute of Drosophila. Mech Dev 127(9–12):393–406. https://doi.org/10.1016/j.mod.2010.05.001
48. Greenspan RJ (2004) Fly pushing: the theory and practice of *Drosophila* genetics, 2nd edn. Cold Spring Harbor Laboratory Press, Cold Spring Harbor, NY

Chapter 8

Using Amphioxus as a Basal Chordate Model to Study BMP Signaling Pathway

Luok Wen Yong, Iryna Kozmikova, and Jr-Kai Yu

Abstract

The BMP signaling pathway has been shown to be involved in different aspects of embryonic development across diverse metazoan phyla. Comparative studies on the roles of the BMP signaling pathway provide crucial insights into the evolution of the animal body plans. In this chapter, we present the general workflow on how to investigate the roles of BMP signaling pathway during amphioxus embryonic development. As amphioxus are basal invertebrate chordates, studies on the BMP signaling pathway in amphioxus could elucidate the functional evolution of BMP pathway in the chordate group. Here, we describe methods for animal husbandry, spawning induction, and manipulation of the BMP signaling pathway during embryonic development through drug inhibitors and recombinant proteins. We also introduce an efficient method of using mesh baskets to handle amphioxus embryos for fluorescence immunostaining and multicolor fluorescence in situ hybridization and to assay the effects of manipulating BMP signaling pathway during amphioxus embryogenesis.

Key words BMP, Amphioxus, Embryo, Dorsomorphin, Immunostaining, In situ hybridization

1 Introduction

Bone morphogenetic proteins (BMP) are evolutionarily conserved signaling molecules in metazoan animals. They play important roles in various processes during embryonic development and exhibit a wide range of biological effects on cell proliferation, differentiation, and morphogenesis [1]. Most notably, BMP pathway regulates major axis formation during early embryonic development in not only bilaterally symmetric animals but also in anthozoans [2–4]. This suggests a deep evolutionary root of utilizing BMP signaling for axis patterning in eumetazoan animals. In most bilaterian animals (with some exceptions like hemichordates and acoels), a dorsoventral BMP gradient specifies neural versus non-neural ectoderm and leads to the formation of the central nervous system (CNS) [5, 6]. This holds true even though the polarity of the BMP signaling gradient and the placement of the

Melissa B. Rogers (ed.), *Bone Morphogenetic Proteins: Methods and Protocols*, Methods in Molecular Biology, vol. 1891, https://doi.org/10.1007/978-1-4939-8904-1_8,

CNS along the dorsoventral axis appear to be inverted in the chordate lineage [7, 8]. In addition, an intermediate level of BMP signaling is necessary for initiating neural crest induction in vertebrates [9–11]. BMP signaling also plays a conserved function during heart development in *Drosophila*, the urochordate *Ciona*, and vertebrates by activating core cardiac transcription factors including orthologs of *GATA* and *Nkx2.5* [12, 13]. Thus, comparative studies on BMP signaling pathway in various metazoan animals can provide important insights for understanding the developmental changes that underlie the evolution of animal body plans.

Cephalochordates (commonly called amphioxus) are invertebrate marine organisms, which together with urochordates (tunicates) and vertebrates constitute the phylum Chordata. Because cephalochordates share an adult body plan highly similar to that of vertebrates, traditionally, they were thought to be the closest relatives of vertebrates. Tunicates were thought to have branched out early at the base of the chordate tree. However, recent molecular analyses have reversed the phylogenetic positions of amphioxus and tunicates, placing amphioxus as the basal chordates and tunicates as the sister group of vertebrates [14–16]. Nevertheless, compared to tunicates, the embryonic development and genomic arrangement of amphioxus and vertebrates are more similar, making amphioxus a suitable model for investigating the basic patterning mechanisms of the chordate body plan and the evolutionary origins of vertebrate features. Through anatomical, genetic, and developmental comparisons, homologous relationships have been suggested between the amphioxus endostyle and vertebrate thyroid, Hatschek's pit and the vertebrate adenohypophysis, and the lamellar organ and vertebrate pineal gland. Furthermore, the vertebrate and amphioxus gill slits, paraxial muscles, dorsal nerve cord, notochord, and kidney appear homologous (reviewed in [17]). In addition, the early development of amphioxus and vertebrate embryos is highly regulative. Both are characterized by the emergence of the dorsal organizer, whose formation and function are mediated by the conserved antagonism between BMP signaling ligands and BMP inhibitors [8, 18–20]. However, compared to vertebrates, amphioxus embryogenesis is certainly much simpler. For example, amphioxus gastrulation is characterized by minimal cell movements, and most organs are formed by one or two cell layers during late embryogenesis [21]. Furthermore, amphioxus genome did not undergo the two rounds of whole-genome duplication that occurred in the vertebrate lineage. In addition, the amphioxus genome apparently preserved extensive micro-synteny between most of the genes that were present in the common ancestor of all chordates [22, 23]. Thus, the amphioxus gene repertoire encoding the components of the Bmp signaling pathway may represent the situation present in the presumptive chordate common ancestor [24].

Currently, among the 35 or so cephalochordate species, mainly four species within the *Branchiostoma* genus have been used for developmental and evolutionary research. They are *Branchiostoma floridae* from the southeastern USA, *B. lanceolatum* from Europe, and *B. belcheri* and *B. japonicum* from the Western Pacific region [25–30]. In regard to being suitable laboratory model organisms amenable to modern molecular experimental approaches, each of these species has its advantages and disadvantages (reviewed in [31, 32]). For example, sequenced genomes are now available for *B. floridae* and *B. belcheri* [23, 33] but not for *B. lanceolatum* and *B. japonicum*, although the genome sequencing project for *B. lanceolatum* is currently underway and its genome assembly should be available soon. Importantly, recent advancements in animal husbandry and spawning induction have made amphioxus laboratory culture feasible. Nowadays, obtaining *B. lanceolatum* embryos on a daily basis during the summer breeding season [25], and even obtaining spawning year-round from tropical species like *B. belcheri*, is possible in the laboratory [34]. Moreover, recent progress in aquaculture techniques now enables long-term amphioxus culture even in inland laboratories [35, 36] and has also enabled complete life cycle to be achieved in the laboratory condition [29, 37].

In this chapter, we describe the techniques of amphioxus husbandry and spawning induction required to study amphioxus embryos in the laboratory. We also describe the use of modulators of BMP signaling pathway in amphioxus embryos and methods to assay the effects of such manipulations.

2 Materials

Unless otherwise indicated, reverse osmosis (RO) purified water, clean distilled water, or millipore-filtered water can be used.

2.1 Amphioxus Husbandry and Spawning Induction

2.1.1 *Branchiostoma lanceolatum*

1. *Branchiostoma lanceolatum* adults are collected in Banyuls-sur-Mer, France, prior to the summer breeding season and raised in the laboratory until spawning.
2. Natural seawater (NSW), filtered natural seawater (FSW).
3. Tropic Marin Sea Salt CLASSIC.
4. 50 mg/mL penicillin stock solution
5. Brown algae *Tisochrysis lutea* obtained from the National Center for Marine Algae and Microbiota, Bigelow Laboratory for Ocean Sciences. Algae is cultured until a titer of 1×10^6 cell/mL before being fed to the animals.
6. 5 L plastic containers, plastic cups 500 mL, air pumps, bottles for algae cultivation.

7. Digital thermometer for testing water temperature, low-temperature incubator.
8. Dark room, lamp with red light, nets.

2.1.2 Branchiostoma floridae

1. *Branchiostoma floridae* adults are collected in Tampa Bay, Florida, USA, during the summer breeding season and subsequently cultured in the laboratory.
2. Plastic boxes as culture tanks (40 cm length × 30 cm width × 20 cm height), tubes, air pump, 1 μm filtered seawater (1 μm FSW: seawater filtered through a 1 μm filter), sand from animal's nature habitat.
3. Algae: *Isochrysis galbana* (Pingtung, Taiwan), *Tetraselmis chuii* (Pingtung, Taiwan), *Rhodomonas lens* (USA), and *Rhodomonas* sp. (Japan). Algae are cultured with 2× Guillard's (F/2) medium [38] in aerated bottles with external lighting. Each type of algae is cultured until they reach a titer of 8–12 × 10^4 cell/mL before being fed to the animals.

2.2 Making Mesh Baskets for Handling Amphioxus Embryos

1. Tools: hot plate, tweezers, scissors, razor blades.
2. Materials: Soft nylon cloth (mesh diameter: 45 μm), 2 mL microcentrifuge, PCR, or conical tubes sized to fit in culture plate wells.

2.3 Manipulating BMP Signaling Levels During Amphioxus Embryogenesis

1. 24-well or 6-well culture plates, microcentrifuge tubes, 1.5 mL screwcap tube (pre-sterilized), mesh baskets.
2. 50 mg/mL penicillin and streptomycin antibiotic stock (P/S). Aliquot and store at −20 °C.
3. Filtered seawater (FSW): filter seawater through a 0.22 μm filter. Note the difference between 1 μm FSW, which is only use to maintain the adult cultures.
4. Antibiotic-filtered seawater (AFSW): filter seawater through a 0.22 μm filter. Add 1 mL of P/S stock to 1000 mL of filtered seawater (FSW) to a final concentration of 50 μg/mL.
5. 5 mM Dorsomorphin (DM, SI-P5499-5MG, Sigma) stock solution: Warm 2.5 mL of dimethyl sulfoxide (DMSO) to 40 °C. Heat a heating block to 40 °C. While waiting, equilibrate the vial of DM to RT. Add the DMSO to the vial and heat the entire vial for 5–10 min. Aliquot the drug in small amounts and seal with parafilm. Store at −20 °C (*see* **Notes 1–2**).
6. 10 μg/mL recombinant zebrafish BMP4 protein (zBMP4; 1125-BM, R&D) stock solution: Add 1 mL of 0.1% BSA in 1× PBS solution to 10 μg of zBMP4 powder. Aliquot and store at −20 °C (*see* **Notes 2** and **3**).

7. MOPS-EGTA (ME) buffer: 0.1 M MOPS, 1 mM EGTA, 2 mM $MgSO_4$, 0.5 M NaCl in diethyl pyrocarbonate-treated H_2O (DEPC-H_2O), pH 7.5. Store at 4 °C.
8. 4% Paraformaldehyde (PFA) in ME buffer, pH 7.5: On the spawning day, add one vial (10 mL) of 16% paraformaldehyde (PFA, 15710, EMS) to 30 mL ME buffer to make 4% PFA in ME. Store the 4% PFA in ME at 4 °C and use within 1 month (*see* **Note 4**).
9. 70% ethanol in DEPC-H_2O: Kept at 4 °C.

2.4 Fluorescence Immunostaining

1. Tools: clean tweezers, mesh baskets, 4- or 24-well culture plate, insect pins.
2. 70% ethanol in DEPC-H_2O: Store in 4 °C.
3. PBSTween20 (PBSTw): 0.1% Tween 20 in 1× PBS. Filter through a 0.22 μm filter (*see* **Note 5**).
4. PBS Triton X-100 (PBSTx): 0.1% of Triton X-100 in 1× PBS. Filter through a 0.22 μm filter.
5. Blocking solutions: (a) 3% BSA and 12.5% goat serum in PBSTx. (b) 3% BSA in PBSTx. Aliquot and store frozen in −20 °C.
6. Primary antibody mix: pSMAD1/5/9 (cell signaling, #13820, Lot No. 0001) in blocking solution with goat serum, dilute 1:150–1:200. Prepare the mix before use (*see* **Notes 6** and 7).
7. Secondary antibody mix: Alexa-flour 555 goat anti-rabbit (Life Technologies, Invitrogen, A21429) in blocking solution without goat serum, dilute 1:400. Prepare the mix before use (*see* **Note 8**).
8. Hoechst stock solution: 10 mg/mL in DEPC-H_2O. Store stock frozen in −20 °C.

2.5 Probe Synthesis

Unless otherwise noted, all solutions are made up with DEPC-treated water. All tools and containers are strictly segregated from RNase containing reagents (*see* **Notes 9** and **10**).

1. RNaseZap (AM9780, Ambion).
2. Probe synthesis template: PCR product of desired genes, including RNA polymerase binding sites at the ends of the desired fragment. Alternatively, linearized plasmid DNA can also be used as template (*see* **Note 11**).
3. Appropriate RNA polymerase that will amplify the antisense fragment of the template: We commonly use T7, SP6, and T3 RNA polymerase. Store at −20 °C.
4. 5× Transcription buffer: consists of 200 mM Tris–HCl (pH 7.9), 30 mM $MgCl_2$, 50 mM NaCl, 10 mM spermidine.

Use buffers which are optimized for a specific polymerase. Store at −20 °C

5. 100 mM Dithiothreitol (DTT), Store at −20 °C.
6. 10× digoxigenin (DIG) or fluorescein (FLUO) labeling kit: consists of 10 mM rATP, rCTP, rGTP, 6.5 mM rUTP, 3.5 mM DIG-11-rUTP. Store at −20 °C.
7. Dinitrophenol (DNP)-labeling kits. Follow manufacturer's protocol for usage and storage. Label IT® DNP Nucleic Acid Labeling kit (MIR3825, Mirus) works well in our hands.
8. Unlabeled nucleotides: consists of 10 mM rATP, rCTP, rGTP, rUTP each. Store in −20 °C.
9. 40 U/μL RNase inhibitor: Store at −20 °C.
10. 1 U/μL RNase-free DNase: Store at −20 °C.
11. TE buffer: 10 mM Tris and 1 mM EDTA in DEPC-H_2O, pH 7.4 in DEPC-H_2O. Filter through a 0.22 μm filter. Aliquot and store at 4 °C.
12. 7.5 M of LiCl solution. Aliquot and store at 4 °C.
13. 100% EtOH. Aliquot and store at 4 °C.
14. HB buffer (as per Subheading 2.6, **item 10**).

2.6 Fluorescence In Situ Hybridization (FISH)

Unless otherwise noted, all solutions are made up with DEPC-treated water, and all tools and containers are strictly segregated from RNase-containing reagents (*see* **Notes 9** and **10**).

1. Tools: clean tweezers, mesh baskets, 4- or 24-well culture plate, insect pins.
2. 70% ethanol in DEPC-H_2O: Store at 4 °C.
3. RNaseZap (AM9780, Ambion).
4. PBSTween20 (PBSTw): *see* Subheading 2.4.
5. 20 mg/mL Proteinase K (pK): Aliquot and store frozen at −20 °C.
6. 100 mg/mL glycine: Aliquot and store frozen at −20 °C.
7. 4% paraformaldehyde in 1× PBS (4% PFA). Store a working aliquot at 4 °C for use within 2–3 weeks, and freeze the rest at −20 °C.
8. 0.1 M triethanolamine (TEA), bring to pH 8.0. Store working aliquot at 4 °C for no longer than 3 months and the rest frozen at −20 °C.
9. Acetic anhydride: Keep in dry cabinet, limit exposure to moisture.
10. Hybridization buffer (HB): 50% deionized formamide, 100 μg/mL of heparin, 5× SSC, 0.1% Tween20, 5 mM EDTA, 1× Denhardt's solution, 1 mg/mL yeast RNA.

Aliquot with microcentrifuge tubes and store at −20 °C (*see* **Note 5**).

11. Pre-hybridization buffer (PreHB): Identical to HB but replace yeast RNA with DEPC-H_2O. Store at −20 °C.
12. Wash solution 1: 50% deionized formamide, 5× SSC, 1% SDS. Store at −20 °C.
13. Wash solution 2: 50% deionized formamide, 2× SSC, 1% SDS. Store at −20 °C.
14. Wash solution 3: 2× SSC, 0.1% Tween-20; Store working aliquot in 4 °C and freeze the rest at −20 °C.
15. Wash solution 4: 0.2× SSC, 0.1% Tween-20; Store working aliquot in 4 °C and freeze the rest at −20 °C.
16. Wash solution 5: 1× PBS, 0.1% Tween-20, 2 mg/mL BSA. Aliquot and store frozen at −20 °C.
17. RNase stocks: Prepare 10 mg/mL RNase A and 10,000 U/mL RNase T1. Both stocks are stored frozen separately from all other ISH solutions at −20 °C.
18. 30% Hydrogen peroxide (H_2O_2).
19. TBS: 0.1 M Tris–HCl, 0.15 M NaCl. Bring to pH 7.5. Filter through 0.22 μm filters.
20. TBSTw: 0.1% Tween20 in 1× TBS. Add 50 μL of Tween20 to 50 mL of TBS. Filter through a 0.22 μm filter.
21. TNB buffer: 0.1% Blocking Reagent (11096176001, Roche) and 0.1% of Tween 20 in TBS. Add 0.25 g of blocking reagent into 40 mL of TBS and heat at 55 °C in a water bath, with frequent mixing. Once all the blocking reagent melts, bring the volume to 50 mL. The solution may appear slightly opaque. Let the solution cool down and add 50 μL of Tween 20. Vortex and aliquot the solution. Store frozen at −20 °C.
22. Antibodies: Anti-DIG-POD (11207733910, Roche), anti-DNP-HRP (FP1129, Perkin Elmer), anti-FLUO-POD (11426346910, Roche).
23. TSA plus fluorescence kit (Perkin Elmer)-containing stock fluorophore fluorescein, Cy3 (NEL752001KT), and Cy5 (NECL745E001KT), with 1× plus amplification diluent included in each kit.
24. Antibody killer solution: 0.1 M glycine, pH 2.2, with 0.1% Tween 20. 0.22 μm filtered.
25. Hoechst stock solution (as in Subheading 2.4, **item 8**).

3 Methods

3.1 Animal husbandry

3.1.1 *Branchiostoma lanceolatum* Husbandry

1. Adults of *Branchiostoma lanceolatum* are transferred into 5 L plastic containers containing 2 L of NSW at a density of 20–25 animals per container at a temperature 17 °C. Animals are subjected to the following light/dark cycle: 14 h light (7 am–9 pm) and 10 h dark (9 pm–7 am). It is very important to aerate water in the box with the air pumps continuously. No sand is present in the box, which simplifies cleaning the water of feces and monitoring animal health.
2. Remove feces from the containers daily, and completely change the NSW twice a week. Check the daily appearance of each animal. Healthy animals have the color of white sand, they are not twisted or bent, and their intestines are full of algae. Healthy animals respond to light by fast swimming. If any are red, pink, or twisted, transfer them into separate box. These animals are not healthy and can endanger the whole container. Add 50 μg/mL penicillin to the box with sick animals (*see* **Note 12**).
3. Animals are fed daily with brown algae *Tisochrysis lutea*. About 200 mL of algae culture is used for each 2 L box per day.
4. Culture the algae in aerated medium containing 33 g/L of Tropic Marin Sea Salt CLASSIC and 1× Guillard's (F/2) Marine Water Enrichment Solution (G0154, Sigma).

3.1.2 *Branchiostoma floridae* Husbandry

1. Amphioxus are cultured in an enclosed room where the temperature and lighting are controlled. The temperature is maintained at 25 °C from April to November and at 20–22 °C from December to March. We culture *B. floridae* with a 1–3 cm thick layer of sand in a culture tank with aeration. 1 μm FSW is added until it is 5 cm above the sand. Each tank has approximately 50 adult animals (*see* **Notes 12** and **13**).
2. Because the amphioxus spawn at night, we invert the day/-night cycle. From 8 am to 6 pm each day, the white lights are turned off, and dim red lights are turned on in the culture room.
3. Animals are fed twice a day at 8 am and 5 pm. Before feeding, the algae cultures are filtered through a plankton net to remove debris. *Isochrysis galbana* and *Tetraselmis chuii* are mixed at a ratio of 2:1. *Rhodomonas lens* and *Rhodomonas* sp. can also be used to replace *Isochrysis galbana*. As a thumb of rule, each animal is fed 15 mL of algae mixture each time.
4. The tank is cleaned weekly. To clean the tank, shake the tank till the amphioxus come out of the sand. Put the animals into a beaker. Wash the sand with water until the water no longer

turns murky. Rinse the sand twice with 1 μm FSW and add sufficient 1 μm FSW. Remove debris from the beaker with animals and check the condition of the animals before putting them back to the tank.

5. Every day, half an hour after lights off, check if the animals spawned spontaneously. Use a sieve (mesh size <100 μm) to scoop the seawater in the tank. If the eggs are present, tiny speckles will be visible on the sieve. The eggs can then be collected into a beaker.

3.2 Spawning Induction (for Branchiostoma lanceolatum) (See Notes 14–15)

1. To find out if the animal gonads are matured, it is necessary to perform spawning induction (try first spawning in the middle of June). To induce spawning, transfer selected animals with full gonads in the evening before spawning to an incubator acclimatized to 23 °C. The light/dark cycle of 14 h light and 10 h dark must be maintained.
2. The following afternoon (the day of spawning), separate animals individually into 500 mL cups with 50 mL FSW. Wash them with FSW to avoid contaminating females by the small amounts of sperm prematurely released by males during the temperature shift.
3. Transfer cups with animals into room allowing to set up the dark cycle. Spawning typically occurs within 1–3 h after commencing the dark cycle. Check animals every 30 min using a lamp with red light.
4. Transfer cups with spawned animals from dark room to the place you are working. Gently collect the eggs by outpouring. Spread about 500–1000 eggs per 10 cm petri dishes. Collect the sperm by outpouring from several males, mix, and keep on ice during the whole period of spawning.
5. Fertilize eggs with several (2–5) drops of sperms. The eggs of *B. lanceolatum* stored at 19 °C may be fertilized during the first 3 h after spawning.

3.3 Making Mesh Baskets for Washing Embryos

See Fig. 1a–c.

1. Cover the surface of a hotplate with clean aluminum foil.
2. Heat the hotplate to approximately 150 °C—sufficient to melt plastic.
3. While the hotplate heats, cut microcentrifuge tubes into the desired height with a razor blade. Larger polypropylene conical tubes or small PCR tubes also can be used for larger single well or smaller multiwell mesh baskets (*see* **Note 16**).
4. Put the nylon mesh on the hotplate. Then, use tweezers to press the cut tubes firmly onto the nylon mesh. The tubes will melt and thus attach to the nylon mesh.

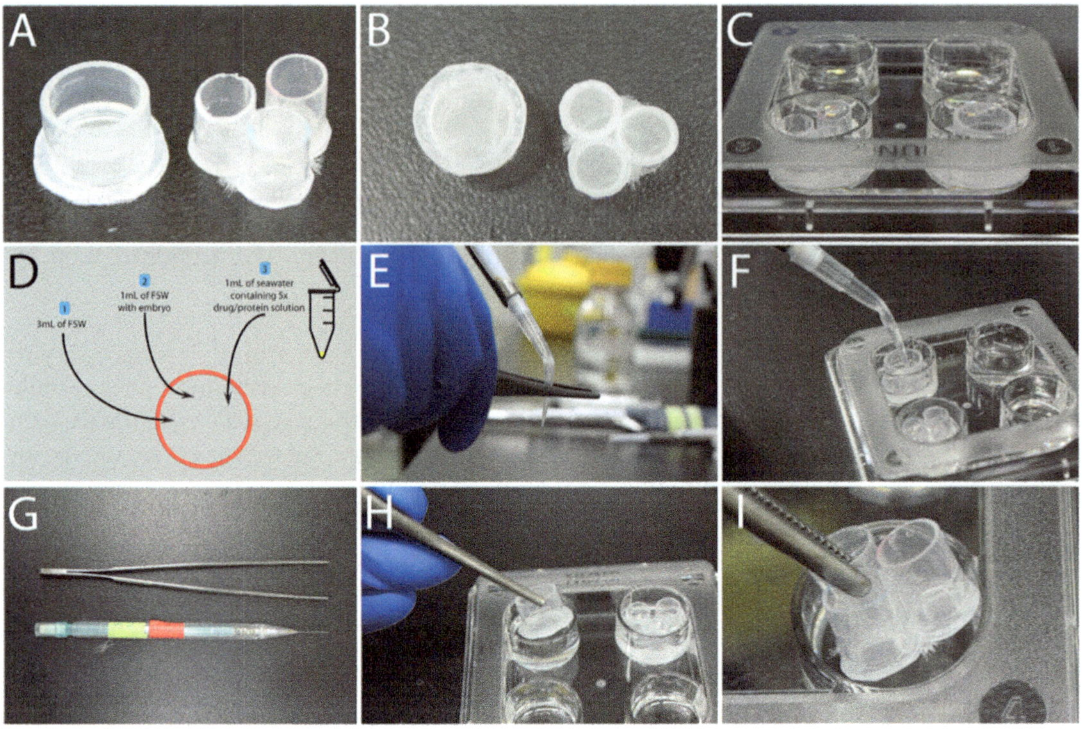

Fig. 1 (**a**) Single well and multiwell mesh baskets. (**b**) Bottom of the well. (**c**) Height of the mesh baskets in the "wells" of the culture plate. (**d**) Embryonic development manipulation steps. (**e**, **f**) Bending pipette tips with a tweezer to ease pipetting from a mesh basket. (**g**) Tweezer (top) and insect pin in a holder (bottom). (**h**, **i**) Transfer mesh by grabbing the top edge of the mesh

5. Remove the mesh basket by lifting the nylon mesh.
6. Trim the mesh outside the basket, so the basket fits into the wells of the culture dish.

3.4 Treatment of Recombinant BMP Protein and BMP Signaling Inhibitor

After animals spawning, the embryos are collected with a sieve into a beaker. The seawater usually contains fecal matter from the adult amphioxus, aggregated algae, and other organic debris. Remove as much debris as possible and replace with clean FSW or AFSW. Culture the embryos in an incubator at (a) 25 or 29 °C for *B. floridae* and (b) 18–22 °C for *B. lanceolatum* (*see* **Note 17**).

3.4.1 Preparation for Drug Treatment in a 6-Well Culture Plate (See Fig. 1d)

1. Add 3 mL of FSW to each well (*see* **Note 18**).
2. Cut the tip of the 1 mL pipette tip, so the embryos are not damaged during pipetting.
3. Add 1 mL of embryos into the plate. Make sure the embryos form no more than one single layer on the bottom of the well (*see* **Note 19**).
4. At the designated time point, add the appropriate volume of drug/protein to the well. Gently pipette up and down to mix

the drug or gently swirl the culture plate. Do not let any liquid spill over to different wells to avoid drug contamination (*see* **Note 20**).

5. Cover the plate with aluminum foil to avoid degradation of the drug or protein.
6. Incubate at the designated temperature.

3.4.2 Washing Out Drug Treatment

1. Put a clean mesh basket in the center of the well.
2. Pipette all embryos into the mesh basket.
3. Holding the edge of the mesh basket with tweezers, swiftly transfer the embryos into a 12-well culture plate with 1 mL of FSW. Transfer the mesh basket into a well with FSW for five times or more.
4. Transfer the mesh basket into a new 6-well culture plate with 5 mL of FSW.
5. In the well, pipette out the embryos and remove the mesh basket. Remember to cut the pipette tip to avoid damaging the embryos (*see* Fig. 1e, f).

3.4.3 Embryo Fixation

1. Pipette 0.5–1 mL of seawater containing the needed number of embryos and place them in a screwcap tube. Remove the same amount of liquid from each well to keep conditions in each well as uniform as possible. Usually, we do not remove more than 3 mL from a well.
2. Remove as much seawater as possible from the screwcap tube, leaving no more than 100 μL of seawater (*see* **Notes 21** and **22**).
3. Add 1 mL of 4% PFA in ME and tilt a few times to mix.
4. Fix the embryo for 1 h at RT or 4 °C overnight.
5. Remove as much PFA as possible and replace with ice cold 70% EtOH (*see* **Note 23**).
6. Tilt the tubes to give the contents a good mix. Leave on ice or in a 4 °C chamber before giving it another change of 70% EtOH. Change for fresh 70% EtOH for a total of three times.
7. Label the tube and store it in −20 °C (*see* **Notes 23** and **24**).

3.5 Fluorescence Immunostaining of pSmad1/5/9

500 μL of solution/buffer is sufficient to immerse the embryos completely in the wells of 4-well or 24-well culture plates (*see* **Note 25**). Unless otherwise stated most steps require rotation on an orbital shaker (55–70 rpm) at room temperature (RT). To change solutions, hold the edge of the mesh basket with tweezers and to swiftly transfer embryos to a new well with fresh buffer/solution (*see* Fig. 1g–i).

3.5.1 Immunostaining Day 1

1. Put mesh baskets into wells containing fresh 70% EtOH. Pipette, with an empty tip, 150 μL into the mesh basket to remove any debris and to check if there are any holes in the mesh.
2. Select embryos and larvae—usually five or more for each stage. Remove fertilization membrane from pre-hatching embryos with sharpened insect pins. Put embryos and larvae into designated mesh baskets.
3. Transfer into fresh PBSTw and rotate for 5 min. Repeat this step twice. For each change, check if there are any leftover embryos in the previous wells.
4. Transfer into fresh PBSTx and rotate for 30 min.
5. While waiting, thaw the blocking solution. Centrifuge the blocking solution very briefly (10,000 × *g*, 20 s), or avoid pipetting any debris on the next step.
6. Change the mesh into the blocking solution and let it rotate at RT for 1 h.
7. While waiting, prepare the primary antibody mix. Tilt tube a few times, mix the primary antibody, and let the tube rotate at 4 °C until the next step.
8. Prior to **step 9**, centrifuge the tube with primary antibody mix very briefly (10,000 × *g*, 20 s), and avoid pipetting any debris into the primary antibody mix (*see* **Notes 26–27**).
9. Transfer embryos into the primary antibody mix. Let it rotate at 4 °C overnight.

3.5.2 Immunostaining Day 2

1. Transfer the embryos into fresh PBSTx to wash away the antibody and rotate for 20 min. Repeat this step for five more times. Use a generous volume of PBSTx (*see* **Note 28**).
2. During the last wash, prepare the secondary antibody mix. Vortex it briefly and centrifuge (10,000 × *g*, 30 s–1 min) before use. Protect the antibody from light from this step onwards.
3. Transfer into secondary antibody mix. Rotate at RT for 2 h.
4. While waiting, prepare the Hoechst working solution (1 μg/mL) with PBSTw. Vortex and then centrifuge the solution (10,000 × *g*) for 3 min.
5. Transfer into Hoechst working solution to stain nuclei. Rotate for 30 min.
6. Transfer into PBSTw and rotate for 10 min. Repeat the PBSTw wash five more times to wash away the dyes.
7. Remove the embryos from the mesh basket for visualization in PBSTw. For prolonged storage, store them in 80% glycerol with 0.1% sodium azide at 4 °C (*see* **Notes 29** and **30**).

3.6 RNA Probe Synthesis

Work carefully to minimize RNase contamination (*see* **Notes 9** and **10**).

1. Clone the gene needed and insert desired fragment into a vector containing different RNA polymerase binding sites on both ends of the cloned fragment. If no such vector is available, design a pair of primers that contain different RNA polymerase binding site on its respective 5′ ends (*see* **Note 11**).
2. Run a PCR using appropriate primers and plasmid DNA to synthesize template.
3. Prepare the probe synthesis reaction mix. A single reaction mix should contain: 500–1000 ng of probe template, 4 μL of 5× Transcription buffer, 1 μL of 100 mM DTT, 2 μL of 10× nucleotide labeling mix or unlabeled rNTPs, 0.5 μL of 40 U/μL RNase inhibitor, 2 μL of 20 U/μL RNA polymerase, and DEPC-H_2O to make a final volume of 20 μL. Only use the polymerase that binds to the antisense strand of the template. Do not vortex the reaction mix, gently flick it, and spin down the content. Incubate the reaction mix at 37 °C for 1 h and 45 min.
4. Add 2 μL of 1 U/μL DNase to the reaction mix to digest the template at 37 °C for 15 min.
5. Precipitate the probe for at least an hour at −20°C by adding 100 μL of TE buffer, 10 μL of 7.5 M LiCl solution, and 300 μL of 100% EtOH to the reaction mix.
6. Centrifuge the solution for 30 min at 4 °C. Remove the supernatant without disrupting the opaque-white pellet. Add 1 mL of ice cold 100% EtOH to the white pellet.
7. Centrifuge at 4 °C for another 15 min. Remove supernatant and let the RNA pellet air-dry.
8. Reconstitute the pellet with 15 μL of DEPC-treated H_2O. Run 1 μL of the probe on an agarose gel to check the integrity of the probe.
9. Take 1 μL of probe to measure the concentration on a UV spectrophotometer if desired. Usually, the concentration should fall around 500–700 ng/μL.
10. Add HB buffer to the probe, so the final volume reaches 50 μL. Store the probe stock in −20 °C.

3.7 Fluorescence In Situ Hybridization (FISH) with Multiplexing Steps

General execution for FISH is similar to immunostaining. All buffers are made or diluted with DEPC-H_2O, and extra precaution is taken to minimize RNase contamination (*see* **Notes 9**, **10**, and **25**).

3.7.1 FISH Day 1

1. Clear bench and tools with 70% EtOH. Wipe work surfaces and tools with RNaseZap. Wear a facemask for extra precaution (*see* **Notes 9** and **10**).
2. Put embryos into mesh baskets as in subheading 3.5.1 (**step** 1 and **step** 2). Clean insect pins with 70% EtOH, wipe with RNaseZap, and then rinse off with DEPC-H_2O to minimize RNase contamination.
3. Transfer into PBSTw and rotate for 5 min to rehydrate the embryos. Repeat this step twice with fresh PBSTw. For each change, check if there are any leftover embryos in the previous wells.
4. For gastrulating and later stage embryos: During the rehydration step, thaw proteinase K (pK) and glycine working solutions. Dilute pK stock 1:4000 in PBSTw to make a working concentration of 5 μg/mL. Dilute glycine stock 1:50 to make a working solution of 2 mg/mL glycine. Skip **steps** 4–7 for embryos during and before blastula stage.
5. Transfer into pK working solution. Let it sit for 10 min. Do not rotate. (*See* **Note 31** for pK times.)
6. Add 4 μL of 10% glycine stock solution directly into the center of the mesh basket. Then, quickly transfer the embryos into glycine working solution (2 mg/mL), and incubate for 5 min. Do not rotate (*see* **Note 32**).
7. Transfer into 4% PFA in PBS for 30 min to 1 h at RT. Do not rotate.
8. During this waiting period, turn on the 60 °C oven and heat sufficient amount of PreHB and HB.
9. Transfer into fresh 0.1 M TEA for a quick rinse, and then change mesh again into fresh 0.1 M TEA. Rotate for 5 min.
10. During the 5-min wait, prepare freshly made 0.25% acetic anhydride in 1 mL 0.1 M TEA (*see* **Note 33**).
11. Transfer into freshly made 0.25% acetic anhydride. Let it sit for 5 min. Do not rotate. While waiting, prepare freshly made 0.5% acetic anhydride in 1 mL 0.1 M TEA.
12. Transfer into freshly made 0.5% acetic anhydride. Let it sit for 5 min.
13. Transfer into fresh PBSTw for a quick rinse, and then change mesh again into fresh PBSTw. Rotate for 5 min.
14. Transfer into pre-warmed PreHB for a quick rinse. Then quickly transfer into warmed fresh HB. Do this step swiftly but carefully to maintain the temperature.
15. Rock the culture plate in the hybridization oven at 60 °C for 2 h or longer. Seal the culture plate in a clean ziplock bag to prevent extensive evaporation.

16. While waiting, prepare probe working solution (probe-HB). Dilute probe stock with HB. Use 100–500 ng of probe per well. For multiplexing, add probes with different end conjugates into the same tube of HB buffer (*see* **Note 34**).
17. Warm probe-HB to 72 °C on a heating block for 5 min. Carefully transfer embryos into probe-HB. Be alert and swift during this step to avoid probe contamination while maintaining the temperature. Grab the mesh basket with the tip of the tweezers, and do not let it contact any liquid.
18. Put culture plate back into the clean ziplock bag, and move to hybridization oven; rock gently at 60 or 65 °C overnight (*see* **Note 35**).

3.7.2 FISH Day 2

1. Prewarm wash 1 and 2 to 60 °C; Thaw wash 3 and 4.
2. Transfer into wash solution 1. Rotate at 60 °C for 5 min. Save the probe-HB for further reuse (*see* **Note 36**).
3. Transfer into fresh wash 1 and rotate at 60 °C for 15 min. Repeat one more time.
4. Transfer into wash solution 2. Rotate at 60 °C for 5 min (time carefully), and then rotate for another 10 min at room temperature (RT).
5. Transfer into fresh RT wash 2 and rotate for 15 min (time carefully).
6. While waiting, thaw RNaseA and RNase T1 stocks. Add 2 μL RNaseA stock and 1 μL RNase T1 stock to 1 mL of wash 3. Turn on a 37 °C incubator, and preheat the RNase mixture for approximately 10 min (*see* **Note 37**).
7. Transfer into wash solution 3 for a quick rinse. Then, quickly change mesh into fresh wash 3 and rotate for 5 min.
8. Transfer into RNase mixture and rotate in 37 °C incubator for 30 min or less.
9. Transfer into wash solution 3 for a quick rinse. Then, quickly transfer again into fresh wash 3 and rotate for 20 min. After this, change mesh into fresh wash 3 once more and rotate for 20 min.
10. Transfer into wash 4 and rotate for 20 min. While waiting, thaw wash 5 and TNB buffer in a 4 °C fridge.
11. Transfer into PBSTw for a quick rinse and then change mesh into wash 5. Rotate for 5 min.
12. Transfer into 1× PBS and rotate for 5 min.
13. Transfer into 1 or 3% H_2O_2 in 1× PBS, and rotate for 30 min to quench endogenous peroxidase (*see* **Note 38**).

Table 1
Recommended antibody dilutions

Antibodies	Dilution
Anti-DNP-HRP	1:1000
Anti-DIG-POD	1:1000–1:2000
Anti-Fluorescein-POD	1:1000–1:2000

14. Transfer into fresh PBSTw to wash away H_2O_2. Rotate for 5 min.
15. Transfer into TBSTw. Rotate for 5 min.
16. Transfer into TNB solution. Rotate for no more than 30 min.
17. While waiting, dilute the antibody with TNB buffer (*see* Table 1). Mix by tilting, and centrifuge briefly before use (10,000–13,000 × *g*) (*see* **Notes 39** and **40**).
18. Transfer into the antibody working solution. Rotate at 4 °C overnight.

3.7.3 FISH Day 3

1. Transfer embryos into fresh TBSTw and rotate for 15 min. Repeat this step twice. (*see* **Note 40**).
2. Meanwhile, thaw the stock fluorophore. Dilute the stock fluorophore with 1× Plus amplification diluent at a ratio of 1:100–1:400. Vortex the fluorophore solution vigorously and briefly centrifuge (x 10,000 g 1 min). Protect from light from this point onwards (*see* **Note 41**).
3. Change the mesh into the fluorophore working solution. Rotate for 15 min. Time this step carefully (*see* **Note 41**).
4. Transfer into fresh TBSTw. Rotate for 5 min. Repeat this step three times to wash away the extra fluorophore.

*3.7.4 For Multiple Color FISH, Skip this subheading, and Proceed to Subheading **3.7.5***

1. To stain nuclei, dilute the Hoechst stock solution to 1 μg/mL in PBSTw. Centrifuge before use.
2. Transfer embryos into Hoechst working solution. Rotate for 30 min.
3. Transfer into fresh PBSTw. Rotate for 10 min to wash away excess dyes. Repeat this step twice.
4. Remove embryos from mesh basket, and put them in PBSTw before transferring to 80% glycerol with 0.1% sodium azide diluted with 1× PBS (for visualization and prolonged storage as in Subheading 3.5.2, **step** 7) (*see* **Notes 29**, **30**, **43**, **44**, and **45**).

*3.7.5 For Multiplex FISH (see **Note 42**)*

1. Transfer embryos into 1–3% H_2O_2 (in 1× TBS). Rotate for 5 min.
2. Transfer into TBST. Rotate for 5 min.
3. Transfer into antibody killer solution. Rotate for 5 min.
4. Transfer into TBSTw. Rotate for 5 min. Repeat three more times to wash away the antibody killer solution thoroughly.
5. Transfer into TNB solution. Rotate for no more than 30 min.
6. Change into a second antibody working solution. Rotate at RT for 1 h.

3.7.6 Mesh Maintenance

1. Rinse the mesh three times with RO H_2O.
2. Sonicate the mesh for 15 min with RO water, *or* wash the mesh with a strong stream of water (attach a 1 mL pipette tip to the tap hose and wash the mesh).
3. Rinse the mesh three times with RO H_2O.
4. Rinse the mesh with diluted neutral detergent for at least 10 min.
5. Rinse off the detergent thoroughly with RO-H_2O.
6. Rinse the mesh with 3% of H_2O_2 in H_2O for 15 min.
7. Rinse the mesh two times with DEPC-H_2O.
8. Rinse the mesh with 70% EtOH in DEPC-H_2O for two times.
9. Store in 70% EtOH in DEPC-H_2O.

4 Notes

1. Dorsomorphin does not last long after reconstitution, so use the reconstituted drug as soon as possible. Older drugs will form crystals in it which require further heating at 40 °C to completely dissolve it before use. The older the drug gets, the longer the heating time will be. We recommend using fresh drugs and avoid freeze thawing the drug.
2. The catalog numbers of Dorsomorphin and zBMP4 protein represent the brand we frequently used. We have not formally tested if the same products from other companies would work or not.
3. In our experiments, treatments with a range 50–250 ng/mL of zBMP4 or 8–20 μM of Dorsomorphin have a concentration-dependent effect.
4. The use of liquid PFA for live embryo fixing is just for convenience. Any form of molecular grade PFA can be reconstituted in ME buffer. The fixed embryos do not show significant difference in terms of morphology or RNA preservation.

5. We prepare PBS, SSC, and Denhardt's buffer through dilution of concentrated solution. These buffers can be purchased or prepared by standard protocols [39]:
   (a) 1× PBS: 137 mM NaCl, 2.7 mM KCl, 10 mM Na_2HPO_4, 2 mM KH_2PO_4 in DEPC-H_2O. Adjust to pH 7.4. Filter through 0.22 μm filter.
   (b) 20× SSC: 3.0 M NaCl, 0.3 M sodium citrate in DEPC-H_2O.
   (c) 50× Denhardt's solution: 1% (w/v) Ficoll 400, 1% (w/v) polyvinylpyrrolidone, 1% (w/v) bovine serum albumin, in DEPC-H_2O.
6. Previous studies have used the pSMAD1/5/8 antibody from cell signaling catalog number #9511. Because the product has been discontinued, we have used pSMAD1/5/9 antibody of catalog number #13820. Do take note of the lot number as the efficacy of the staining may vary from lot to lot.
7. Dilution ratio stated is specific to pSMAD1/5/9.
8. Various secondary antibodies with different fluorophores can be used according to experimental needs and the setup of the confocal microscope. Alexa-flour 488, 594, 635 secondary antibody (from Life Technologies, Invitrogen) works fine in our hands.
9. To minimize RNA contamination, do not speak while handling the samples or handle samples with bare hands. Wear gloves and face mask especially for steps in day 1 of in situ hybridization or probe synthesis.
10. For extra precaution, after cleaning working bench with 70% EtOH, spray working surface and tweezers with RNaseZap.
11. An arrayed cDNA library is available for *Branchiostoma floridae* (http://amphioxus.icob.sinica.edu.tw/) [40]. The representative clone for a gene is picked and the plasmid is extracted. Vector-specific primers are used to amplify the fragments out. The vector-specific primers contain T7 or SP6 RNA polymerase binding sites—antisense fragments are on the direction with the T7 RNA polymerase binding sites. The cDNA clones are available upon request.
   (a) pDONR222-T7-R1:

       5'-TAATACGACTCACTATAGGGAGGGGATATCAG CTGGATG-3'.

   (b) pDONR222-SP6-F1:

       5'-ATTTAGGTGACACTATAGAAGACGGCCAGTCT-TAAGCTC-3'.

(c) PCR condition: 94 °C 2' | (5 cycles) 94 °C 30″, 45 °C 30″, 68 °C 3′ | (25 cycles) 94 °C 30″, 58 °C 30″, 68 °C 3′ | (1 cycle) 68 °C 7′.

12. Pink, red, or twisted animals are unhealthy. Remove them from the tank, and culture them in isolation with penicillin added (final concentration of 50 μg/mL).
13. If animals die in the sand, the sand around the dead animal will become light gray. Clean the tank immediately, or the whole tank of amphioxus may be affected.
14. In contrast to *B. lanceolatum*, success rate of spawning induction in *B. floridae* is low. We mainly depend on spontaneous spawning during breeding season.
15. If *B. lanceolatum* spawn spontaneously, it can also be collected from the sand as per Subheading 3.1.2 (**step 5**). Conversely, if separate gametes can be obtained from *B. floridae*, they are collected in the same way as *B. lanceolatum* (as per Subheading 3.2, **step 4**)
16. Cut the height of the mesh so it is lower than the rim of the well. Do not let the rim of the mesh basket touch the cover of the culture plate even if the cover still seals properly; otherwise, the liquid in the wells might be drawn up along the mesh basket during rotation and spill over to the border of the culture plate.
17. Embryos develop faster at a higher temperature. Choose the temperature of cultivation according to your experimental needs.
18. Use of antibiotics FSW (AFSW) is recommended especially if embryos are to be cultured to larval stages.
19. Occasionally, the embryos may stick to the plastic wells or tips. To avoid this problem, coat the bottom of the well with 1% gelatin in RO-H_2O. Leave the gelatin solution for approximately 30 s, then rinse off with RO water. Then, rinse once with FSW. For tips, pipette the 1% gelatin once and discard the solution. Let it dry, and pipette it a few times with RO water and then with FSW before using the tip.
20. We tend to prepare a 5× drug mix (with 1 mL of FSW) in a microcentrifuge tube, and then add it to the designated well with 4 mL of FSW and animals. This additional step lets us check if the drug has completely dissolved.
21. If it is impossible to leave less than 100 μL of seawater in the tube, add 1 mL of PFA regardless. As the embryos sink, remove as much of the seawater/PFA mixture, and add more fresh 4% PFA in ME. Do this as quickly as possible.

22. At the onset of neurulation, the embryos start to secrete hatching enzymes. Therefore, using a large volume of PFA in ME buffer is imperative to avoid compromising embryo morphology (especially the ectoderm).
23. If subsequent experiments require the embryos to be methanol- or ethanol-free (e.g., for phalloidin staining), store embryos in 1× PBS with 0.1% of sodium azide instead of ethanol. The embryos can be stored in a 4 °C fridge for at least a year.
24. With good practice, animals stored in 70% EtOH are good for ISH for 7 years in our hands. For the naked eye, no obvious loss of signal is seen with the Nodal ISH. However, efficacy for immunostaining may decrease (has not been systematically tested).
25. 500 μL is sufficient to cover one single wells of the 4-well or 24-well dish from Nunclon™ Delta Surface, Thermo Scientific. Different amount of liquid may be needed for culture dish of different sizes or brands.
26. Phalloidin stains filamentous actin as well as the cell boundary. 20 min in RT is sufficient for the muscle actin staining to show, but an overnight 4 °C staining is crucial for the cell boundary staining to appear. For overnight staining, phalloidin can be added to the primary antibody mix at a ratio of 1:40. By itself, block the embryo in 1% BSA in PBST for an hour before proceeding with the staining.
27. If multiplex immunostaining is to be done, both primary antibody can be incubated together as long as they are from different hosts and do not cross react. Secondary antibody that recognizes different host antigens can be added together and incubated at the same time.
28. We do not reuse the antibody.
29. For embryos around blastula stages (>128-cell stage to G2 stage), a slow graduated change into 80% glycerol is recommended to prevent the embryos from collapsing.
30. For confocal microscopy, we recommend adding a bit of Vectashield antifade mounting media (H-1000, Vector laboratories) to the glycerol before putting the coverslip to reduce photobleaching.
31. Proteinase K treatment time may vary depending on the stages of the embryo. For early embryos up to blastula stage, pK is unnecessary. For embryos between gastrula and neurula, incubate in pK for 5–8 min. For embryos from neurula stages onward, incubate for 8–10 min.
32. If a multiwell mesh basket is used, lift the mesh basket, and add the glycine directly to the well. Then, put the mesh back into

the same well before transferring to the glycine working solution.

33. Acetic anhydride is very sensitive to moisture. Prepare mixture 1 and 2 just before use.

34. For multiplex FISH, synthesize and mix probes with different conjugates for different genes. We commonly use digoxigenin (DIG), fluorescein (FLUO), and DNP-conjugated probe. In our experience, DIG and DNP probe work better than FLUO-conjugated probe. For multiplexing, the color for the FLUO-probed gene is frequently weaker with higher background levels as well; thus, FLUO probe is usually reserved for specific, highly expressed genes in multiplex FISH.

35. We usually do probe hybridization at 65 °C, to increase the specificity of probe binding. However, if the signal for some genes is weak, a hybridization temperature of 60 °C can be attempted.

36. DNP-conjugated probes are not reused. DIG- and FLUO-conjugated probes are usually recovered for reuse.

37. Treatment with RNase A and RNase T1 reduces background levels. It is optional but highly recommended.

38. For quenching endogenous peroxidase, 3% of H_2O_2 in 1× PBS (or any amount within recommended range of the kit) may be used when there is higher than expected background levels.

39. For antibody binding, our lab preferentially uses this sequence: anti-DNP-HRP (overnight, 4 °C) → anti-DIG-POD (1 h RT) → anti-FLUO-POD (1 h RT).

40. We normally do not reuse our antibodies for fluorescence in situ hybridization (FISH).

41. The ratio of fluorophore dilution and amplification time needs to be optimized for each specific gene. We frequently use a dilution ratio of 1:100 to 1:400 and staining duration of 15 min up to 30 min.

42. Steps to follow for multiple-color FISH:
    (a) One-color: Subheading 3.7.1 **Steps 1–18** → Subheading 3.7.2 **Steps 1–18** → Subheading 3.7.3 **Steps 1–4** → Subheading 3.7.4 **Steps 1–4**.
    (b) Two-color: Subheading 3.7.1 **Steps 1–18** → Subheading 3.7.2 **Steps 1–18** → Subheading 3.7.3 **Steps 1–4** → Subheading 3.7.5 **Steps 1–6** → Subheading 3.7.3 **Steps 1–4** → Subheading 3.7.4 **Steps 1–4**.
    (c) Three-color: Subheading 3.7.1 **Steps 1–18** → Subheading 3.7.2 **Steps 1–18** → Subheading 3.7.3 **Steps 1–4** → Subheading 3.7.5 **Steps 1–6** → Subheading 3.7.3 **Steps 1–4** → Subheading 3.7.5 **Steps 1–6** → Subheading 3.7.3 **Steps 1–4** → Subheading 3.7.4 **Steps 1–4**.

43. After FISH, immunostaining can also be done on the same embryo. Do not remove the embryos from the mesh basket, and do not put them into 80% glycerol. Proceed to the fluorescence immunostaining protocol (Subheading 3.5.1 **step 4**).
44. Color for FISH usually is not as well-defined as regular chromogenic ISH with the alkaline phosphatase-NBT/BCIP system. We usually prefer to observe the gene expression using chromogenic ISH before doing FISH, which is suitable mainly for multiplex ISH.
45. We never reuse our culture plates for multiple ISH. The used meshes are stored in 70% EtOH before being washed in bulk.

Acknowledgments

This work was supported by grants from the Ministry of Science and Technology (MOST 104-2923-B-001-002-MY3; 105-2628-B-001-003-MY3) and Academia Sinica, Taiwan (AS 98-CDA-L06), and grant GC15-21285J from Grantová Agentura České Republiky. We would like to thank all the past and present members of the Research Group of Development and Evolution at ICOB Academia Sinica, especially Yi-Chih Chen, Tsai-Ming Lu, Yi-Jyun Luo, and Tzu-Kai Huang, for their help on developing these protocols. We would also like to thank Yann le Petillon for the advice regarding the properties of Dorsomorphin.

References

1. Bier E, De Robertis EM (2015) BMP gradients: a paradigm for morphogen-mediated developmental patterning. Science 348 (6242):aaa5838. https://doi.org/10.1126/science.aaa5838
2. Genikhovich G, Fried P, Prunster MM, Schinko JB, Gilles AF, Fredman D, Meier K, Iber D, Technau U (2015) Axis patterning by BMPs: Cnidarian network reveals evolutionary constraints. Cell Rep. https://doi.org/10.1016/j.celrep.2015.02.035
3. Matus DQ, Thomsen GH, Martindale MQ (2006) Dorso/ventral genes are asymmetrically expressed and involved in germ-layer demarcation during cnidarian gastrulation. Curr Biol 16(5):499–505. https://doi.org/10.1016/j.cub.2006.01.052
4. Saina M, Genikhovich G, Renfer E, Technau U (2009) BMPs and chordin regulate patterning of the directive axis in a sea anemone. Proc Natl Acad Sci U S A 106(44):18592–18597. https://doi.org/10.1073/pnas.0900151106
5. Denes AS, Jekely G, Steinmetz PR, Raible F, Snyman H, Prud'homme B, Ferrier DE, Balavoine G, Arendt D (2007) Molecular architecture of annelid nerve cord supports common origin of nervous system centralization in bilateria. Cell 129(2):277–288
6. Mizutani CM, Meyer N, Roelink H, Bier E (2006) Threshold-dependent BMP-mediated repression: a model for a conserved mechanism that patterns the neuroectoderm. PLoS Biol 4 (10):e313. https://doi.org/10.1371/journal.pbio.0040313
7. De Robertis EM (2008) Evo-devo: variations on ancestral themes. Cell 132(2):185–195. https://doi.org/10.1016/j.cell.2008.01.003
8. Yu JK, Satou Y, Holland ND, Shin IT, Kohara Y, Satoh N, Bronner-Fraser M, Holland LZ (2007) Axial patterning in cephalochordates and the evolution of the organizer. Nature 445(7128):613–617
9. Basch ML, Bronner-Fraser M (2006) Neural crest inducing signals. Adv Exp Med Biol

589:24–31. https://doi.org/10.1007/978-0-387-46954-6_2

10. Milet C, Monsoro-Burq AH (2012) Neural crest induction at the neural plate border in vertebrates. Dev Biol 366(1):22–33. https://doi.org/10.1016/j.ydbio.2012.01.013
11. Schumacher JA, Hashiguchi M, Nguyen VH, Mullins MC (2011) An intermediate level of BMP signaling directly specifies cranial neural crest progenitor cells in zebrafish. PLoS One 6 (11):e27403. https://doi.org/10.1371/journal.pone.0027403
12. Christiaen L, Stolfi A, Levine M (2010) BMP signaling coordinates gene expression and cell migration during precardiac mesoderm development. Dev Biol 340(2):179–187. https://doi.org/10.1016/j.ydbio.2009.11.006
13. Lee KH, Evans S, Ruan TY, Lassar AB (2004) SMAD-mediated modulation of YY1 activity regulates the BMP response and cardiac-specific expression of a GATA4/5/6-dependent chick Nkx2.5 enhancer. Development 131(19):4709–4723. https://doi.org/10.1242/dev.01344
14. Bourlat SJ, Juliusdottir T, Lowe CJ, Freeman R, Aronowicz J, Kirschner M, Lander ES, Thorndyke M, Nakano H, Kohn AB, Heyland A, Moroz LL, Copley RR, Telford MJ (2006) Deuterostome phylogeny reveals monophyletic chordates and the new phylum Xenoturbellida. Nature 444(7115):85–88
15. Delsuc F, Brinkmann H, Chourrout D, Philippe H (2006) Tunicates and not cephalochordates are the closest living relatives of vertebrates. Nature 439(7079):965–968
16. Delsuc F, Tsagkogeorga G, Lartillot N, Philippe H (2008) Additional molecular support for the new chordate phylogeny. Genesis 46 (11):592–604. https://doi.org/10.1002/dvg.20450
17. Holland ND, Chen J (2001) Origin and early evolution of the vertebrates: new insights from advances in molecular biology, anatomy, and palaeontology. BioEssays 23(2):142–151. https://doi.org/10.1002/1521-1878(200102)23:2<142::AID-BIES1021>3.0.CO;2-5
18. Holland LZ (2015) Genomics, evolution and development of amphioxus and tunicates: the Goldilocks principle. J Exp Zool B Mol Dev Evol 324(4):342–352. https://doi.org/10.1002/jez.b.22569
19. Kozmikova I, Candiani S, Fabian P, Gurska D, Kozmik Z (2013) Essential role of Bmp signaling and its positive feedback loop in the early cell fate evolution of chordates. Dev Biol 382 (2):538–554. https://doi.org/10.1016/j.ydbio.2013.07.021
20. Onai T, Yu JK, Blitz IL, Cho KW, Holland LZ (2010) Opposing Nodal/Vg1 and BMP signals mediate axial patterning in embryos of the basal chordate amphioxus. Dev Biol 344 (1):377–389. https://doi.org/10.1016/j.ydbio.2010.05.016
21. Holland LZ (2015) Cephalochordata. In: Wanninger A (ed) Evolutionary developmental biology of invertebrates, vol 6. Springer, Vienna, pp 91–133
22. Holland LZ, Albalat R, Azumi K, Benito-Gutierrez E, Blow MJ, Bronner-Fraser M, Brunet F, Butts T, Candiani S, Dishaw LJ, Ferrier DE, Garcia-Fernandez J, Gibson-Brown JJ, Gissi C, Godzik A, Hallbook F, Hirose D, Hosomichi K, Ikuta T, Inoko H, Kasahara M, Kasamatsu J, Kawashima T, Kimura A, Kobayashi M, Kozmik Z, Kubokawa K, Laudet V, Litman GW, McHardy AC, Meulemans D, Nonaka M, Olinski RP, Pancer Z, Pennacchio LA, Pestarino M, Rast JP, Rigoutsos I, Robinson-Rechavi M, Roch G, Saiga H, Sasakura Y, Satake M, Satou Y, Schubert M, Sherwood N, Shiina T, Takatori N, Tello J, Vopalensky P, Wada S, Xu A, Ye Y, Yoshida K, Yoshizaki F, Yu JK, Zhang Q, Zmasek CM, de Jong PJ, Osoegawa K, Putnam NH, Rokhsar DS, Satoh N, Holland PW (2008) The amphioxus genome illuminates vertebrate origins and cephalochordate biology. Genome Res 18 (7):1100–1111. https://doi.org/10.1101/gr.073676.107
23. Putnam NH, Butts T, Ferrier DE, Furlong RF, Hellsten U, Kawashima T, Robinson-Rechavi M, Shoguchi E, Terry A, Yu JK, Benito-Gutierrez EL, Dubchak I, Garcia-Fernandez J, Gibson-Brown JJ, Grigoriev IV, Horton AC, de Jong PJ, Jurka J, Kapitonov VV, Kohara Y, Kuroki Y, Lindquist E, Lucas S, Osoegawa K, Pennacchio LA, Salamov AA, Satou Y, Sauka-Spengler T, Schmutz J, Shin IT, Toyoda A, Bronner-Fraser M, Fujiyama A, Holland LZ, Holland PW, Satoh N, Rokhsar DS (2008) The amphioxus genome and the evolution of the chordate karyotype. Nature 453 (7198):1064–1071. https://doi.org/10.1038/nature06967
24. Satou Y, Wada S, Sasakura Y, Satoh N (2008) Regulatory genes in the ancestral chordate genomes. Dev Genes Evol 218 (11–12):715–721. https://doi.org/10.1007/s00427-008-0219-y
25. Fuentes M, Benito E, Bertrand S, Paris M, Mignardot A, Godoy L, Jimenez-Delgado S,

Oliveri D, Candiani S, Hirsinger E, D'Aniello S, Pascual-Anaya J, Maeso I, Pestarino M, Vernier P, Nicolas JF, Schubert M, Laudet V, Geneviere AM, Albalat R, Garcia Fernandez J, Holland ND, Escriva H (2007) Insights into spawning behavior and development of the European amphioxus (*Branchiostoma lanceolatum*). J Exp Zool B Mol Dev Evol 308(4):484–493. https://doi.org/10.1002/jez.b.21179

26. Fuentes M, Schubert M, Dalfo D, Candiani S, Benito E, Gardenyes J, Godoy L, Moret F, Illas M, Patten I, Permanyer J, Oliveri D, Boeuf G, Falcon J, Pestarino M, Fernandez JG, Albalat R, Laudet V, Vernier P, Escriva H (2004) Preliminary observations on the spawning conditions of the European amphioxus (*Branchiostoma lanceolatum*) in captivity. J Exp Zool B Mol Dev Evol 302(4):384–391. https://doi.org/10.1002/jez.b.20025
27. Holland LZ, Yu JK (2004) Cephalochordate (amphioxus) embryos: procurement, culture, and basic methods. Methods Cell Biol 74:195–215
28. Yasui K, Urata M, Yamaguchi N, Ueda H, Henmi Y (2007) Laboratory culture of the Oriental lancelet *Branchiostoma belcheri*. Zool Sci 24(5):514–520. https://doi.org/10.2108/zsj.24.514
29. Zhang QJ, Sun Y, Zhong J, Li G, Lu XM, Wang YQ (2007) Continuous culture of two lancelets and production of the second filial generations in the laboratory. J Exp Zool B Mol Dev Evol 308(4):464–472. https://doi.org/10.1002/jez.b.21172
30. Zhang QJ, Zhong J, Fang SH, Wang YQ (2006) *Branchiostoma japonicum* and *B. belcheri* are distinct lancelets (Cephalochordata) in Xiamen waters in China. Zool Sci 23(6):573–579. https://doi.org/10.2108/zsj.23.573
31. Bertrand S, Escriva H (2011) Evolutionary crossroads in developmental biology: amphioxus. Development 138(22):4819–4830. https://doi.org/10.1242/dev.066720
32. Yong LW, Yu JK (2016) Tracing the evolutionary origin of vertebrate skeletal tissues: insights from cephalochordate amphioxus. Curr Opin Genet Dev 39:55–62. https://doi.org/10.1016/j.gde.2016.05.022
33. Huang S, Chen Z, Yan X, Yu T, Huang G, Yan Q, Pontarotti PA, Zhao H, Li J, Yang P, Wang R, Li R, Tao X, Deng T, Wang Y, Li G, Zhang Q, Zhou S, You L, Yuan S, Fu Y, Wu F, Dong M, Chen S, Xu A (2014) Decelerated genome evolution in modern vertebrates revealed by analysis of multiple lancelet genomes. Nat Commun 5:5896. https://doi.org/10.1038/ncomms6896
34. Li G, Yang X, Shu Z, Chen X, Wang Y (2012) Consecutive spawnings of Chinese amphioxus, *Branchiostoma belcheri*, in captivity. PLoS One 7(12):e50838. https://doi.org/10.1371/journal.pone.0050838
35. Benito-Gutierrez E, Weber H, Bryant DV, Arendt D (2013) Methods for generating year-round access to amphioxus in the laboratory. PLoS One 8(8):e71599. https://doi.org/10.1371/journal.pone.0071599
36. Theodosiou M, Colin A, Schulz J, Laudet V, Peyrieras N, Nicolas JF, Schubert M, Hirsinger E (2011) Amphioxus spawning behavior in an artificial seawater facility. J Exp Zool B Mol Dev Evol 316(4):263–275. https://doi.org/10.1002/jez.b.21397
37. Yasui K, Igawa T, Kaji T, Henmi Y (2013) Stable aquaculture of the Japanese lancelet *Branchiostoma japonicum* for 7 years. J Exp Zool B Mol Dev Evol 320(8):538–547. https://doi.org/10.1002/jez.b.22540
38. Guillard RRL (1975) Culture of phytoplankton for feeding marine invertebrates. In: Smith WL, Chanley MH (eds) Culture of marine invertebrate animals. Plenum Press, New York, pp 26–60
39. Sambrook J, Russell DW (2001) Appendix. In: Molecular cloning: a laboratory manual, vol 3, 3rd edn. Cold Spring Harbor Laboratory Press, Cold Spring Harbor, NY, pp A1.7–A1.14
40. Yu JK, Wang MC, Shin IT, Kohara Y, Holland LZ, Satoh N, Satou Y (2008) A cDNA resource for the cephalochordate amphioxus *Branchiostoma floridae*. Dev Genes Evol 218 (11–12):723–727

Chapter 9

Proteolytic Activation of Bmps: Analysis of Cleavage in *Xenopus* Oocytes and Embryos

Hyung-Seok Kim, Autumn McKnite, and Jan L. Christian

Abstract

Bone morphogenetic proteins (Bmps) are synthesized as inactive precursors that are cleaved to generate active ligands, along with prodomain fragments that can modulate growth factor activity. Here we provide three protocols that can be used to examine the process of proteolytic activation of Bmps. The first protocol describes how to generate radiolabeled Bmp precursor proteins in *Xenopus* oocytes and then analyze the time course of precursor cleavage by recombinant enzymes in vitro. The second protocol details how to analyze cleavage of radiolabeled precursor proteins in *Xenopus* oocytes over time using pulse-chase analysis and autoradiography. This protocol can also be used to analyze folding and cleavage of radiolabeled precursor proteins at steady state. Finally, the third protocol details methods for isolating Bmp cleavage products from the blastocoele of *Xenopus* embryos and then analyzing them on immunoblots.

Key words *Xenopus laevis*, Oocytes, Embryos, Bone morphogenetic protein, Proteolytic activation, Proprotein convertase, Autoradiography, Immunoblot

1 Introduction

Bone morphogenetic proteins (Bmps) are synthesized as inactive precursor proteins that dimerize and are then cleaved to generate the active, disulfide bonded ligand dimer along with prodomain fragments [1] (Fig. 1). A novel family of enzymes, called proprotein convertases (PCs), are responsible for proteolytic activation of Bmp [2]. There are seven members of the PC family, four of which, furin, PACE4, PC6, and PC7, are good candidates for endogenous Bmp family convertases based on their broad expression patterns, localization within the constitutive secretory pathway, and phenotypes observed in mice lacking these PCs [3, 4]. Bmp precursor proteins dimerize and fold in the endoplasmic reticulum (ER). PCs are first active in the trans-Golgi network (TGN), but most PCs cycle between the TGN and the cell surface and/or are shed from the

Hyung-Seok Kim and Autumn McKnite are contributed equally to this work.

Melissa B. Rogers (ed.), *Bone Morphogenetic Proteins: Methods and Protocols*, Methods in Molecular Biology, vol. 1891, https://doi.org/10.1007/978-1-4939-8904-1_9, © Springer Science+Business Media, LLC, part of Springer Nature 2019

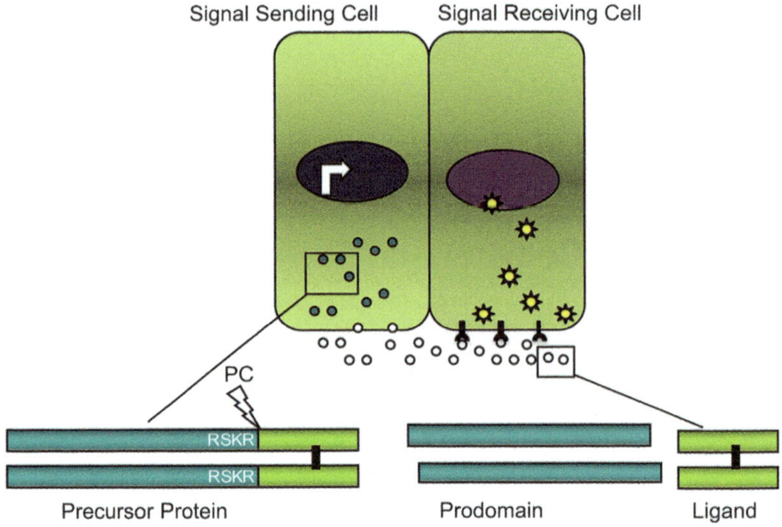

Fig. 1 Illustration of Bmp cleavage. Bmps are synthesized in signal-sending cells as precursor proteins that dimerize via a single disulfide linkage (black bar) in the ligand domain (light green fragment). Bmps are cleaved by proprotein convertases (PCs) during or after secretion to generate prodomain fragments and the mature, dimerized ligand that can activate its receptors on signal receiving cells

cell. Thus, Bmp precursor proteins can potentially be cleaved at any point in the secretory pathway after exiting the ER, or in the extracellular space.

The process of proteolytic maturation varies between different members of the Bmp family in terms of the number and location of cleavage sites that are utilized, the subcellular or extracellular compartment in which cleavage occurs, and whether the prodomain remains associated with the mature ligand following cleavage [3, 5–12]. Although the prodomain fragment that is released during proteolytic maturation lacks any known signaling activity, it aids in dimerization and folding of the Bmp precursor protein, can target the cleaved growth factor to specific extracellular matrix-binding partners, and can regulate ligand stability [3, 13]. The critical role of proteolytic activation and of the prodomain in generating and maintaining an active Bmp ligand is highlighted by the finding that point mutations within the prodomains of different Bmps are associated with syndromes of orofacial clefting, microphthalmia, defects in fertility, and skeletal malformations in humans [3]. For these reasons, it is important to understand how Bmp precursor proteins are proteolytically activated and to identify and decipher the role of all cleavage products that are generated during this process.

In this chapter, we provide detailed protocols for three complementary techniques that can be used to analyze proteolytic activation of Bmps. These include in vitro analysis of cleavage by recombinant PCs and in vivo analysis of cleavage in *Xenopus* oocytes or *Xenopus* embryos. Each protocol has its own strengths and weaknesses, but a general advantage of all three procedures is that they enable one to analyze ectopically expressed, epitope-tagged proteins when antibodies that recognize endogenous Bmps are not available. This is particularly important because, with a few exceptions, antibodies that recognize the cleaved prodomain of Bmps are not commercially available and are difficult to generate.

The first protocol describes how to generate radiolabeled Bmp precursor proteins in *Xenopus* oocytes and to analyze the time course of cleavage by different recombinant PCs. This procedure can be used to determine whether more than one site is cleaved within the prodomain, to examine the timing and order of cleavages if more than one site is used, and to determine which PC family members are capable of cleaving precursors in the absence of any cellular accessory factors. This protocol is especially useful for identifying cryptic upstream cleavage sites within the prodomain. Cleavage of these sites is not readily detected by standard immunoblotting, since only the final cleavage product is observed under steady-state conditions, and since most Bmp antibodies do not recognize the prodomain. By contrast, in vitro cleavage enables one to follow the appearance and disappearance of cleaved fragments as they are generated over time [5, 12]. A potential disadvantage of this protocol is that precursor obtained from oocytes is primarily the unfolded proBmp monomer, rather than the properly folded dimeric precursor that is encountered by PCs in vivo [14]. Despite this shortcoming, even though most Bmps contain multiple consensus PC cleavage motifs (R–X–X–R–) that could potentially be recognized, recombinant PCs faithfully cleave only the site(s) that is/are utilized in vivo [15]. In most cases, if an –R–X–X–R– consensus motif in a substrate is not recognized by purified recombinant PCs in vitro, that enzyme is also not likely to utilize that site in vivo. By contrast, a specific PC family member may cleave a Bmp substrate in vitro, but not in vivo. Thus, the in vitro cleavage assay is useful for identifying PCs that are capable of cleaving a given Bmp family member, but further analysis is required to definitively identify the enzymes responsible for cleavage and to confirm which site(s) is/are cleaved in vivo.

The second protocol describes the use of pulse-chase analysis to examine cleavage of radiolabeled Bmp precursors over time in *Xenopus* oocytes. This procedure enables one to verify that cleavage sites identified in vitro are utilized by endogenous PCs in vivo (e.g., by comparing cleavage of wild-type precursors with that of precursors containing point mutations that disrupt the consensus cleavage motif) [5, 16], to determine the order of sequential cleavage events [6], and to compare the half-life of cleavage products [6]. One can also introduce selective PC inhibitors or antisense oligonucleotides specific for individual PCs into oocytes to knock down expression in order to identify endogenous Bmp convertases [17]. Although the protocol described here includes a pulse-chase procedure, analysis of steady-state levels of radiolabeled Bmp proteins expressed in oocytes is also highly useful and has many advantages over analysis in other in vivo systems. In oocytes, proteins traffic through the secretory system more slowly than they do in cultured mammalian cells. This advantage, coupled with the high sensitivity of autoradiography, makes it possible to detect dimerized precursor protein and intracellular cleavage products in oocyte lysates and to test whether the proteins are properly folded by examining their glycosylation status [14, 18]. By contrast, properly folded, dimerized Bmp precursor proteins are difficult or impossible to detect in cultured cells or *Xenopus* embryos because they very rapidly exit the ER and are cleaved coincident with secretion in these systems [19–21]. The ability to examine precursor protein folding is important, for example, to determine whether prodomain mutations associated with human congenital anomalies exert their hypomorphic effect by inducing misfolding. When expressed in oocytes, misfolded precursor proteins are partially or fully retained in the ER, and cleavage products are absent or are present at lower levels [14]. By contrast, we have found that mutant precursor proteins generated by transient transfection of cultured cells can exceed the capacity for quality control in the ER, allowing misfolded dimers to escape the ER where they can be cleaved by PCs that are first active in the *trans*-Golgi network [14]. This can be partially avoided by transfecting minimal amounts of DNA, but one should be cautious in interpreting the appearance of cleavage products as an indication that a precursor protein is properly folded when expressed in mammalian cells.

The third protocol describes how to aspirate Bmp cleavage products from the blastocoele of *Xenopus* embryos. The major advantage to this protocol is that it provides a very rapid and inexpensive method to obtained highly concentrated Bmp cleavage products in vivo. The entire protocol can be completed in under 24 h and does not require expensive tissue culture reagents or sterile conditions. In *Xenopus*, as in other ectopic expression systems, uncleaved, monomeric precursor protein accumulates at much higher levels than cleavage products, and this, together with abundant yolk proteins present in early embryos, makes it difficult to detect cleavage products in *Xenopus* whole-embryo lysates by immunoblotting. The recognition by Drs. P. Huw Williams and James Smith that the embryo secretes highly concentrated growth factors into the blastocoele [22] has greatly simplified in vivo analysis of Bmp cleavage in *Xenopus*. This protocol can be used to verify that cleavage sites identified in vitro are utilized by endogenous PCs in vivo (e.g., by comparing cleavage of wild-type precursors with that of precursors containing point mutations that disrupt the consensus cleavage motif) [17, 22], to analyze the effects of point mutations in Bmps on their ability to form functional dimers, and to compare formation of Bmp homodimers versus heterodimers [20].

2 Materials

2.1 Generation and In Vitro Cleavage of Radiolabeled Bmps

1. Low temperature (14–25 °C) bioincubator.
2. 0.2% tricaine: Dissolve 2 g tricaine (ethyl 3-aminobenzoate methanesulfonate salt) in 900 mL distilled water, adjust pH to 7.0 by addition of sodium bicarbonate. Add distilled water to a final volume of 1 L. Store at 4 °C.
3. Collagenase I and II stock solution (28 U/mL): Dissolve 1 vial (70 mg) Liberase Blendzyme 3 (Roche), or similar product, in 10 mL sterile distilled water. Reconstitute 30 min on ice. Pipette 200 μL aliquots into microcentrifuge tubes and store at −20 °C. Do not refreeze aliquots after thawing.
4. OR-2: 82.5 mM NaCl, 2 mM KCl, 1 mM $MgCl_2$, 5 mM HEPES, pH 7.5. Make a 20× base stock by dissolving 96.4 g NaCl, 2.98 g KCl, 203.3 g $MgCl_2 \cdot 6H_2O$ in 800 mL distilled water. Add distilled water to a final volume of 1 L. To make OR-2, add 50 mL of 20× base stock and 1.25 g HEPES hemisodium salt to 900 mL distilled water, pH to 7.5 with NaOH. Add distilled water to a final volume of 1 L. Filter sterilize with a 0.2 μM filter. Store at 4 °C.
5. Oocyte dissociation buffer: Add 200 μL of collagenase I and II stock to 20 mL of OR-2. Make fresh for each use.
6. Modified Barth solution (MBSH): 88.0 mM NaCl, 1.0 mM KCl, 0.33 mM $Ca(NO_3)_2$, 0.41 mM $CaCl_2$, 0.82 mM $MgSO_4$, 2.4 mM $NaHCO_3$, 10.0 mM HEPES, pH 7.5–7.55. Make a 20× base stock by dissolving 102.8 g NaCl, 1.49 g KCl, and 1.56 g $Ca(NO_3)_2 \cdot 4H_2O$ in 800 mL distilled water. Add distilled water to a final volume of 1 L. To make MBSH, add 50 mL of 20× base stock, 0.82 mL 1 M $MgSO_4$, 4.8 mL 0.5 M $NaHCO_3$, 2.49 g HEPES-Hemi Na and 250 mg Amikacin to 900 mL distilled water, pH to 7.5 with NaOH. Add distilled water to a final volume of 1 L. Filter sterilize with a 0.2 μM filter. Store at 4 °C.
7. [^{35}S]Met/Cys protein labeling mix: Dry 14.3 mCi [^{35}S]Met/Cys protein labeling mix under vacuum and resuspend in diethylpyrocarbonate (DEPC)-treated water at a concentration of 110 μCi/μL. Store aliquots at −80 °C.
8. In vitro synthesized capped RNAs: Transcribe linearized template DNAs using an appropriate RNA Transcription Kit. Template DNA must contain a polyadenylate (polyA) tail or be in an

appropriate vector (e.g., pCS2+) to add a polyA tail in vivo. Most laboratories use template cDNAs A detailed protocol can be found in Mimoto and Christian [23].

9. Microinjector for oocyte injections and pulled glass capillaries used for injection needles (described in more detail in Mimoto and Christian [23]).
10. TxSWB buffer: 1% Triton X-100, 0.1 M NaCl, 0.1 M Tris–HCl pH 8, 10 mM EDTA. Store at room temperature.
11. Protease inhibitor cocktail: Make a 7× stock of a commercial broad-spectrum protease inhibitor cocktail in TxSWB. Add 1/7 volume to appropriate buffer immediately before use. Store at −20 °C.
12. Homogenization buffer: 0.25 M sucrose, 50 mM Tris–HCl, pH 7.5, 5 mM potassium acetate, 5 mM magnesium acetate, 1 mM DTT. Add DTT and protease inhibitors immediately before use.
13. SDS solution: 1% SDS, 0.1 M Tris–HCl pH 8.9.
14. TxSWB buffer plus phenylmethylsulfonyl fluoride (PMSF): Add 5 μL 200 mM PMSF per 10 mL TxSWB immediately before use.
15. Recombinant protein G-Sepharose 4B or protein A-Sepharose 4B bead solution: One to one suspension of beads in TxSWB, mix immediately before use.
16. Wash solution: 0.1 M NaCl, 0.1 M Tris–HCl pH 8. Store at room temperature.
17. Scintillation counter to determine radioactivity counts.
18. 4× Loading dye: 200 mM Tris–HCl pH 6.8, 8% SDS, 0.4% bromophenol blue, 40% glycerol. To make nonreducing loading dye, mix 2 mL 1 M Tris–HCl, pH 6.8, 0.04 g bromophenol blue, 4 mL glycerol, and 0.8 g SDS. Add distilled water to a final volume of 10 mL. Aliquot and store at −20 °C. To analyze proteins under reducing conditions, add 1/20 volume ß-mercaptoethanol (final concentration 14.4 M) before use.
19. 5× Furin buffer: 0.5 M HEPES pH 7.5, 2.5% Triton X-100, 5 mM $CaCl_2$. Mix 500 μL 1 M HEPES, pH 7.5, 125 μL 20% Triton X-100, 5 μL 1 M $CaCl_2$, and 370 μL dH_2O. Store at −20 °C.
20. Furin reaction mix: For 300 μL: 60 μL 5× Furin buffer, 6 U Furin, 0.3 μL ß-mercaptoethanol, 236.7 μL distilled water. Mix immediately before use.
21. Equipment for polyacrylamide gel electrophoresis.
22. Gel dryer.

2.2 Pulse-Chase Analysis of Bmp Cleavage in Xenopus *Oocytes*

Most of the materials and solutions used for this protocol are the same as those used for protocol 2.1.

1. Methionine/cysteine solution: 5 M methionine, 5 M cysteine in sterile distilled water. Store at −20 °C.
2. 10× media buffer: 10% Triton X-100, 1 M NaCl. Store at room temperature.
3. Peptide *N*-glycosidase F (PNGase) and/or endoglycosidase H (EndoH): if glycosylation required (*see* **Note 20**).

2.3 Analysis of Bmp Cleavage Products in Blastocoele Fluid from Xenopus *Embryos*

1. Low temperature (14–25 °C) bioincubator.
2. Human chorionic gonadotropin (HCG): 10,000 IU dissolved in 2.5 mL distilled water, final concentration is 400 IU/ 0.1 mL.
3. 1 mL syringe and 27 gauge needles.
4. 100 and 35 mm glass or plastic petri dishes.
5. 3 mL plastic transfer pipettes.
6. 10× modified Barth's saline (MBS): 880 mM NaCl, 10 mM KCl, 25 mM $NaHCO_3$, 100 mM HEPES (pH 7.5), 10 mM $MgSO_4$, 0.14 mM $Ca(NO_3)_2$, 0.41 mM $CaCl_2$. Adjust pH to 7.5 with NaOH. Filter sterilize with a 0.2 μM filter. Store at 4 °C.
7. 1× and 0.1× MBS: dilute 10× MBS stock 1:10 and 1:100, respectively, with distilled water.
8. Aged tap water: (used for embryo incubation, where deionized water is not suitable) allow tap water to sit for at least 24 h for chlorine to evaporate. Alternatively, chlorine can be removed from tap water using a carbon filter.
9. Cysteine dejellying solution: 2% (w/v) cysteine dissolved in distilled water. Adjust to pH 7.8–8.0 with NaOH. Prepare fresh the day of use.
10. 20× DeBoer's solution: 100 mM NaCl, 1.3 mM KCl, 0.44 mM $CaCl_2$. Adjust pH to 7.4 with $NaHCO_3$. Store at 4 °C. Dilute 20-fold with distilled water to make working solution.
11. 5% Ficoll solution: 5% (w/v) Ficoll (type 400 DL) dissolved in 0.1× MBS. Adjust pH to 7.5 with NaOH. Store at 4 °C.
12. In vitro synthesized capped RNAs: transcribe linearized template DNAs using an appropriate RNA Transcription Kit. Template DNA must contain a polyadenylate (polyA) tail or be in an appropriate vector (e.g., pCS2+) to add a polyA tail in vivo. A detailed protocol can be found in Mimoto and Christian [23].

13. Microinjector for embryo injections and pulled glass capillaries used for injection needles (described in more detail in Mimoto and Christian [23]).
14. Embryo lysate buffer: 50 mM Tris–HCl, pH 7.5, 150 mM NaCl, 1 mM EDTA, 1% Triton X-100, 0.1% SDS, 2.5%IGEPAL CA-630 (NP-40). Store at room temperature.
15. PNGase: If deglycosylation is required (*see* **Note 37**).
16. 4× Loading dye: 200 mM Tris–HCl pH 6.8, 8% SDS, 0.4% bromophenol blue, 40% glycerol. To make nonreducing loading dye, mix 2 mL 1 M Tris–HCl, pH 6.8, 0.04 g bromophenol blue, 4 mL glycerol, and 0.8 g SDS. Add distilled water to a final volume of 10 mL. Aliquot and store at −20 °C. To analyze proteins under reducing conditions, add 1/20 volume ß-mercaptoethanol (final concentration 14.4 M) before use.
17. Equipment for polyacrylamide gel electrophoresis and electrophoretic transfer of proteins.

3 Methods

3.1 Generation and In Vitro Cleavage of Radiolabeled Bmps

1. Submerge a female *Xenopus laevis* in 0.02% tricaine until fully anesthetized (about 15–30 min) (*see* **Note 1**).
2. Remove frog from tricaine and place back side down on flat surface for dissection. Using a scalpel or a pair of surgical scissors, make a horizontal incision in the skin across the lower abdomen and a second vertical incision running from the midline of the first incision up to the thorax. Cut through the underlying body wall beneath each incision. Locate the beating heart in the thorax and clip the ventricle with a pair of scissors to ensure death by exsanguination (*see* **Note 2**). Remove a piece of ovary from beneath the muscle layer in the lower abdomen (*see* **Note 3**) and examine under a dissecting microscope to ensure that the oocytes are healthy.
3. Tear ovarian tissue into smaller pieces using a sharp pair of forceps and place in a 50 mL screw top centrifuge tube containing 20 mL oocyte dissociation buffer. Incubate with gentle rocking or shaking at 25 °C for 1–3 h to dissociate oocytes from follicle cells (defolliculation) (*see* **Note 4**).
4. Once dissociated, wash oocytes two to three times for 5 min each in OR-2, followed by two to three 5 min washes in MBSH. Transfer to a tissue culture dish containing MBSH (*see* **Note 5**) and remove unhealthy oocytes or immature oocytes that have not reached stage VI [24].
5. Incubate remaining oocytes in MBSH overnight at 18 °C.

6. Select healthy, stage VI oocytes [25] and place them in an injection dish or tray [23] oriented so that the pigmented animal pole is facing upward. Locate the equator between pigmented and nonpigmented poles. Inject oocytes near the equator with a mixture of in vitro-synthesized capped RNAs encoding the desired Bmp precursor (0.5–50 ng) (*see* **Note 6**) and α1-PDX (5 ng; to inhibit endogenous proprotein convertases [26]) together with 700 nCi of [^{35}S]Met/Cys in a total volume of 10–50 nL per oocyte.
7. Transfer injected oocytes into a 12 well tissue culture dish (~10–15 oocytes per well) and culture in MBSH at 18 °C with gentle shaking. The incubation time necessary to achieve maximal protein expression can vary from 20 to 72 h and must be determined empirically for each substrate.
8. After incubation, transfer 6–10 oocytes to a microcentrifuge tube, remove media, and add 30 μL homogenization buffer plus protease inhibitor cocktail. Homogenize oocytes by pipetting up and down using a 200 μL pipette tip and then fully homogenize using a 20 μL pipette tip. Add 100 μL SDS solution and incubate for 30 min at 37 °C (*see* **Note 7**).
9. Add 500 μL of TxSWB buffer plus PMSF and protease inhibitors (*see* **Note 8**) and incubate for 45 min on ice.
10. Microfuge for 10 min at 13523 × *g* at 4 °C. Without disturbing the pellet, transfer the supernatant to a new microcentrifuge tube.
11. Add the appropriate antibody to the supernatant at a concentration of 1–10 μg/mL (determined empirically) and incubate for 2–12 h at 4 °C, with gentle shaking to allow antigen-antibody complexes to form.
12. Once the incubation is complete, add 20 μL of recombinant protein G-Sepharose 4B or Protein A-Sepharose 4B bead solution (*see* **Notes 9** and **10**) and incubate overnight at 4 °C on a rocking platform to allow for binding of complexes to beads.
13. Centrifuge for 30 s at 13523 × *g* at 4 °C. Remove supernatant and place in new tube. Store at −80 °C for up to 1 week (*see* **Note 11**).
14. Add 750 μL ice-cold TxSWB buffer plus PMSF, shake vigorously, centrifuge for 1 min at 6010 × *g* at 4 °C, and remove supernatant. Repeat twice.
15. Wash beads twice in ice cold wash solution as in the previous step.
16. Resuspend pellet in 500 μL wash solution.

17. Determine radioactivity (counts per minute; cpm) in a 5 μL aliquot to ensure protein synthesis before continuing with protocol (*see* **Note 12**).
18. Label a centrifuge tube for each time point (*see* **Note 13**) and add 8 μL of 4× loading dye to each tube. Set aside until **step 19**.
19. Aliquot the appropriate volume (at least 3000 counts per time point, e.g., 30,000 counts for 10 time points; *see* **Note 14**) into a new microcentrifuge tube, spin the beads, and remove the buffer.
20. Resuspend the beads containing the bound substrate in 24 μL of Furin reaction mix (*see* **Note 15**) per time point (e.g., 240 μL for ten time points). Immediately remove a 24 μL aliquot to the zero time point tube using a cut pipette tip (to prevent clogging of the tip). Heat for 5 min at 90–100 °C in order to stop the reaction.
21. Incubate the remaining reaction at 25–30 °C and remove aliquots at various time intervals (minutes to hours, determined empirically), stopping the reaction as described above. Store samples on ice until all reactions are complete, or at −20 °C for up to 4 weeks.
22. Heat samples to 100 °C for 5 min and place on ice.
23. Resolve the cleavage products by polyacrylamide gel electrophoresis and visualize by autoradiography.

3.2 Pulse-Chase Analysis of Bmp Cleavage in Xenopus Oocytes

1. Isolate and culture oocytes as described in protocol 3.1, **steps 1–5**.
2. Select healthy, stage VI oocytes [25] and place them in an injection dish or tray [23] oriented so that the pigmented animal pole is facing upward. Locate the equator between pigmented and nonpigmented poles. Inject oocytes with 0.5–8 ng of in vitro synthesized capped RNA encoding the Bmp of interest (*see* **Notes 16** and **17**), along with 700 nCi of [^{35}S]Met/Cys in a total volume of 10–50 nL per oocyte.
3. Transfer injected oocytes into a 24 well tissue culture plate containing 800 μL of MBSH per well (10–15 oocytes per well per time point to be analyzed) and culture for 3 h at 18 °C.
4. After 3 h, collect the media from one well into a microcentrifuge tube for the zero hour time point. Rinse oocytes with 200 μL MBSH and add to the same tube. Collect 6–10 healthy oocytes from this well into a second microcentrifuge tube for the zero hour time point (*see* **Note 18**). Any remaining oocytes in this well can be discarded. Samples can be frozen in liquid nitrogen and stored at −80 °C until remaining samples are collected and ready for analysis.

5. Add 8 μL 5 M nonradiolabeled methionine/cysteine solution to each of the remaining wells (final concentration of 5 mM) to initiate the chase. Culture oocytes at 18 °C with gentle shaking.
6. At increasing time intervals after the chase (*see* **Note 19**), collect media and oocytes from remaining wells as described in **step 4**.
7. Prepare oocyte lysates as described in protocol 3.1, **steps 8–10**. Add 110 μL of 10× media buffer to all media samples.
8. Immunoprecipitate Bmps from oocyte lysates and from the conditioned media, as described in protocol 3.1, **steps 11–15**. Proteins can be deglycosylated with PNGase and/or EndoH at this point if desired (*see* **Note 20**).
9. Resuspend beads in 24 μL 1× loading dye (with or without ß-mercaptoethanol for nonreducing or reducing conditions, respectively). Heat to 100 °C for 5 min and place on ice.
10. Resolve the cleavage products by polyacrylamide gel electrophoresis and visualize by autoradiography.

3.3 Analysis of Bmp Cleavage Products in Blastocoele Fluid from Xenopus Embryos

1. The evening prior to spawning, inject 0.3 mL (1200 IU) of HCG into the dorsal lymph sac of each of two mature *Xenopus laevis* females using a 1 mL syringe fitted with a 27 gauge needle. After injection, house the females in plastic drawers or buckets in a 15 °C bioincubator. Spawning generally begins 12–14 h later (*see* **Notes 21** and **22**).
2. Collect eggs in a 100 mm petri dish containing 1× MBS by applying gentle pressure to the frogs abdomen (*see* **Note 23**).
3. Remove as much of the 1× MBS as possible using a plastic transfer pipette and apply 100 μL of sperm suspension (*see* **Note 24**) and ~1 mL of 0.1× MBS. Allow 5 min for fertilization and then fill the petri dish with 0.1× MBS or aged tap water and let stand for about 30 min. During this time, a jelly coat will form around each egg such that the eggs adhere to each other and to the dish. Eggs that have been fertilized will rotate within the fertilization membrane such that the pigmented animal pole faces upward.
4. The jelly coat must be removed to prevent the eggs from sticking to the needle during injections. To do so, pour off the 0.1× MBS or water and fill the petri dish with freshly prepared 2% cysteine solution. Swirl the dish for 3–5 min until the jelly coat is dissolved. Remove the cysteine solution by tilting the dish and decanting the solution carefully or use a plastic transfer pipette to remove the cysteine solution. Once the jelly coat is removed, the eggs are fragile and free floating, and from this point on, it is important to tilt the dish when changing solutions to ensure that the eggs are covered with a

small amount of fluid. Fill the dish with 1× DeBoer's solution, swirl embryos briefly to rinse, and then remove solution with a plastic transfer pipette or by decanting. Repeat with two more rinses in 1× DeBoer's solution and then two rinses in 0.1× MBS (*see* **Note 25**).

5. When embryos reach the 2–4 cell stage, transfer them to an injection dish or tray filled with 5% Ficoll/0.1× MBS (*see* **Note 26**). Inject 50–200 pg of in vitro synthesized RNA encoding the desired Bmp in a total volume of 4–10 nL into one cell of each embryo near the animal pole (*see* **Notes 27** and **28**).
6. Once injections are complete, transfer the embryos to 35 mm petri dishes filled with 5% Ficoll/0.1× MBS and culture the embryos overnight at 15–16 °C (*see* **Note 29**).
7. The following morning, remove the Ficoll solution, rinse twice with 0.1× MBS, and culture the embryos in 0.1× MBS at room temperature.
8. When embryos reach stage 10–11, return them to an injection tray or dish filled with 0.1× MBS. Attach an empty injection needle to a micromanipulator connected to a microinjector. Insert the needle (*see* **Note 30**) just beneath the surface of the animal pole so that it enters the blastocoele cavity. Press the fill button while continuously watching the needle and the embryo under a dissecting microscope. The needle should fill with clear fluid over a period of a few seconds. The embryo will visibly collapse, and the animal pole will become concave as the blastocoele is emptied (Fig. 2) (*see* **Notes 31** and **32**). Stop pressing fill if white cloudy matter (cell debris) enters the needle. This indicates that the blastocoele is completely collapsed. Press the inject button one or more times to get rid of

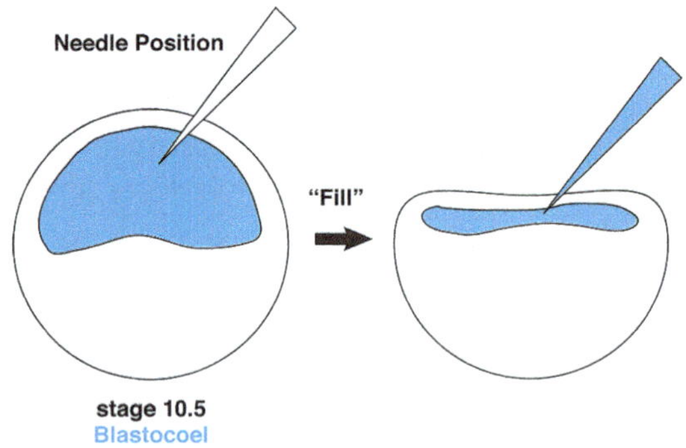

Fig. 2 Illustration of blastocoele aspiration. Schematic illustration of a sagittal section through a stage 10.5 *Xenopus* embryo showing proper positioning of the needle just under the blastocoele roof. When fluid is aspirated into the needle by pushing the fill button, the blastocoele collapses, and the animal pole flattens or becomes concave

any cloudy matter as this will clog your needle and is likely to contain proteases.

9. Repeat **step 8** until fluid has been aspirated from the blastocoele of 20 embryos (*see* **Note 33**).
10. Place a piece of paraffin under the injection needle and pipette 1 μL of sterile distilled water onto it. Lower the needle under the water and press the eject button until the needle is empty (*see* **Note 34**). Transfer the blastocoele fluid to a microcentrifuge tube on ice using a micropipette. You can expect to harvest ~0.5 μL of blastocoele fluid per embryo.
11. Transfer the depleted embryos to a separate tube on ice (*see* **Note 35**). Remove excess 0.1× MBS and add 200 μL of embryo lysate buffer (10 μL per embryo). Homogenize by pipetting up and down 10–20 times with a micropipette.
12. Microfuge for 10 min at 13523 × *g* at 4 °C.
13. Transfer 160 μL (4/5 volume) of supernatant to a new tube on ice, avoiding the yolk, and microfuge for 10 min at 13523 × *g* at 4 °C.
14. Transfer 128 μL (4/5 volume) of supernatant to a new tube on ice, avoiding the yolk (*see* **Note 36**).
15. Proteins present in embryo lysates or blastocoele fluid can be deglycosylated with PNGase F at this point, if desired, by following the manufacturers' instructions (*see* **Note 37**).
16. Add 5 μL of reducing or nonreducing 4× sample buffer to15 μL of blastocoele fluid and to 15 μL of clarified embryo lysate (*see* **Note 38**). Heat to 100 °C for 5 min and place on ice.
17. Resolve precursor proteins and cleavage products by polyacrylamide gel electrophoresis and visualize by immunoblotting using appropriate antibodies.

4 Notes

4.1 Generation and In Vitro Cleavage of Radiolabeled Bmps

1. Frogs are fully anesthetized when a response to a toe pinch is no longer seen.
2. It is also possible to do a survival surgery by making a single small incision in the lower abdomen to remove a piece of ovary, suturing the muscle/fascia layer and skin layer separately using sterile technique and allowing the female to recover in shallow water [27]. If allowed by institutional animal care policies, the same female can then be used multiple (four to six) times to obtain ovary, allowing several months for recovery between each surgery.

3. A single lobe of a healthy ovary yields hundreds of mature oocytes.
4. Enzymatic dissociation can potentially result in reduced viability and impaired ability of oocytes translate RNAs [25]. We have found that using a blend of highly purified collagenase I and II, such as that found in Liberase Blendzyme or similar products, minimizes this problem. In addition, the time of incubation in dissociation medium should be minimized by stopping the reaction once most oocytes have been released from the ovary. If necessary, oocytes can be defolliculated manually.
5. Once defolliculated, oocytes should be cultured in glass or tissue culture treated polystyrene dishes or multiwell plates to prevent sticking.
6. With a few exceptions, we have found that commercial antibodies that specifically and accurately recognize Bmp cleavage products are not available. In this case, sequence encoding different epitope tags for which commercial antibodies are available (e.g., Flag, myc, V5, HA) can be inserted in frame into the prodomain and the mature domain. Tagged proteins should be assayed in vivo to ensure that the epitope does not interfere with proper folding, cleavage, or activity of the mature protein. In general, epitope tags inserted into the prodomain are better tolerated than those in the mature domain.
7. Alternatively, radiolabeled RNAs can be synthesized in vitro in the presence of [^{35}S]Met/Cys using commercially prepared rabbit reticulocyte lysates and following the manufacturers' directions. In this case, the translation reactions should be diluted to 500 μL with TxSwB supplemented with PMSF and protease inhibitors and substrate immunoprecipitated following **steps 11–14** in Subheading 3.1 of the protocol.
8. Add PMSF to TxSWB buffer immediately before use as it loses activity within 20 min.
9. Protein A and Protein G bind antibodies of different subtypes and from different species with varying affinity. It is important to choose the immobilized antibody-binding protein that binds most effectively to your antibody.
10. It is advisable to use a pipette tip cutoff at the end to transfer the beads to prevent the clogging of the pipette tip.
11. Supernatant can be used for additional immunoprecipitations.
12. The counts obtained will vary for different proteins, depending on the number of Cys/Met residues in the molecule. A range of 30,000–200,000 cpm/reaction is recommended. Inability to detect counts could be due to inefficient protein synthesis

(possibly due to degradation of the RNA used for injection or poor quality oocytes) or inefficient immunoprecipitation.

13. Time required for cleavage varies between substrates and must be determined empirically. A good starting point is 0, 5, 10, 30 min, 1 and 3 h.
14. When comparing the cleavage of wild-type and mutant proteins, or when comparing cleavage by multiple PCs, it is important to take equal numbers of counts for each cleavage reaction.
15. Recombinant furin and PC7 are commercially available, and furin is likely to cleave most, if not all Bmps in vitro, but other PCs must be generated de novo. High-level expression of PCs has been achieved using standard techniques such as transient or stable transfection of cultured cells or infection with recombinant vaccinia virus [28] or baculovirus [29]. For in vitro cleavage reactions, PCs can be concentrated by ultrafiltration from the media of expressing cells, and no further purification is required. When comparing cleavage of a substrate by different PCs, the concentration of each PC can be determined as described in Cui et al. [15].

4.2 Pulse-Chase Analysis of Bmp Cleavage in *Xenopus* Oocytes

1. For pulse-chase analysis, inject the minimal amount of RNA necessary to visualize the precursor at the earliest time point. This enhances the ability to see the appearance of cleavage products, and corresponding disappearance of precursor, over a reasonable period of time.
2. With a few exceptions, we have found that commercial antibodies that specifically and accurately recognize Bmp cleavage products are not available. In this case, sequence encoding different epitope tags for which commercial antibodies are available (e.g., Flag, myc, V5, HA) can be inserted in frame into the prodomain and the mature domain. Tagged proteins should be assayed in vivo to ensure that the epitope does not interfere with proper folding, cleavage, or activity of the mature protein. In general, epitope tags inserted into the prodomain are better tolerated than those in the mature domain.
3. Whereas Bmp precursor proteins are retained in the oocyte, cleaved prodomain and mature domain fragments accumulate in both the oocyte and in the culture media, and thus it is important to analyze both fractions.
4. The length of time required for complete cleavage must be determined empirically for each substrate. In general, time points separated by 4–8 h over several days are appropriate.
5. Deglycosylating samples with PNGase F and/or Endo H prior to electrophoresis is not strictly required but can aid in visualizing, quantitating, and accurately identifying precursor proteins and cleavage products. Furthermore, as discussed in the

introduction, analysis of glycosylation status (sensitivity to PNGase F versus Endo H) at late time points enables one to determine whether precursor proteins are properly folded and able to exit the ER. Deglycosylation can be performed immediately after immunoprecipitation by resuspending beads in denaturation buffer and following the manufacturer's instructions.

4.3 Analysis of Bmp Cleavage Products in Blastocoele Fluid from Xenopus Embryos

1. Frogs may be handled with clean hands or with vinyl gloves. Gloves made from other materials (e.g., latex) will damage the frogs' fragile skin. If gloves are not used, hands should be washed thoroughly before and after contact with the frogs to ensure that they are free from lotions and trace detergent, which can be damaging to the frog.
2. It may be necessary to exchange the Frog Bath to bring the frogs up to room temperature more quickly if spawning does not begin. Alternatively, females can be injected the day eggs will be used, housed at room temperature, and will begin spawning within approximately 6 h. We generally spawn 2–3 frogs per experiment to ensure that adequate numbers of healthy eggs are produced.
3. A more detailed description of spawning, culture, and microinjection of RNAs into *Xenopus* embryos can be found in Mimoto and Christian [23].
4. Testes are isolated from euthanized males as described [23]. Just prior to fertilization, cut a small piece of testis (~1/5) and put it into a 1.5 mL tube containing ~500 μL 1× MBS. Crush with a conical tissue grinder to make a testis slurry and store on ice for subsequent fertilizations on the same day. Intact testes can be stored at 4 °C for 2–3 weeks, but crushed testes must be used within 4–5 h.
5. From this point on, embryos may be cultured on the bench or in a bioincubator at varying temperatures ranging from 14 to 25 °C. This allows for control of the rate of development to facilitate injections at a particular stage or to produce embryos at varying stages simultaneously. Incubation at 15–16 °C is recommended prior to injection because it slows the rate of cleavage and enables more embryos to be injected from a given spawn. An approximation of the changes in developmental timing based on temperature are as follows: 25 °C = 100%, 20 °C ≈ 75%, 16 °C ≈ 50%, and 14 °C ≈ 25% normal developmental rate. Timing of normal development is based on staging by Nieuwkoop and Faber [30].
6. Embryos can be injected easily at any time between the 1 and 8 cell stage. We generally wait until embryos have cleaved once

to ensure that they are healthy and dividing normally before injecting them.

7. With a few exceptions, commercial antibodies that specifically and accurately recognize Bmp cleavage products are not available. To get around this problem, sequence encoding epitope tags for which commercial antibodies are available (e.g., Flag, myc, V5, HA) can be inserted in frame into both the prodomain and the mature domain. Tagged proteins should be assayed in vivo to ensure that the epitope does not interfere with proper folding, cleavage, or activity of the mature protein. In general, epitope tags inserted into the prodomain are better tolerated than those in the mature domain.
8. The dose of RNA that will give a detectable signal on an immunoblot must be determined empirically. Most epitope-tagged Bmp proteins can be readily detected following injection of 50–200 pg of RNA, but this will vary depending on how well the RNA is translated in vivo and on how good the antibody is. Many Bmps interfere with gastrulation when ectopically expressed, which can make it more difficult to aspirate fluid from the blastocoele. Thus, one should inject the minimal amount of Bmp RNA that can be detected on immunoblots.
9. Addition of Ficoll to the culture media aids in healing the injection wound but should be removed once embryos begin to gastrulate. Embryos are transferred to small culture dishes for overnight incubation in Ficoll to minimize the cost of this expensive reagent.
10. A good needle is sharp with a small bore at an angle of approximately 45 degrees. Under a dissecting microscope, clip the tip of the needle using a pair of sharp forceps or a razor blade. The diameter of the bore is not critical but will generally be larger than the bore of a needle used for microinjection. If it is too small, fluid will be aspirated very slowly, and the needle is likely to clog, while if it is too large, it will be difficult to control the speed of aspiration. This can be determined empirically by trial and error.
11. The needle should be inserted quite shallowly such that it just penetrates the thin roof of the blastocoele. If the needle is inserted too deeply such that it pierces the vegetal mass on the floor of the blastocoele, white cellular debris will be aspirated, and the needle is likely to clog. If this happens, one can usually clear the needle by briefly pulsing the inject button.
12. If desired, one can better control the rate of aspiration by pulsing the fill button, rather than pressing it continuously.
13. The number of embryos that need to be aspirated to collect sufficient material for analysis must be determined empirically,

but generally 10–20 embryos are sufficient if a good antibody is available.

14. Emptying the needle into a minimal volume of water prevents the fluid from spreading over the Parafilm under the high pressure applied when using the eject feature. Alternatively, one can hit the inject button repeatedly to expel the fluid under lower pressure.
15. Bmp precursor proteins are not secreted into the blastocoele and thus can only be detected in embryo lysates, while cleavage products will be detected primarily in the blastocoele fluid. Depending on the purpose of the experiment, you may choose not to analyze the precursor present in embryo lysates. In initial experiments, however, it is recommended that you harvest and analyze embryo lysates as a control to ensure that the precursor protein is robustly expressed in the event that you are unable to detect cleavage products.
16. Clarified embryo lysates and blastocoele fluid can be stored at −80 °C at this point if desired.
17. Deglycosylation is not strictly required but can aid in visualizing, quantitating, and accurately identifying precursor proteins and cleavage products.
18. Analyzing proteins under both reducing and nonreducing conditions can provide useful information about the mature ligand, which is secreted as a disulfide-linked dimer. By contrast, the prodomain fragment is secreted as a monomer, and most of the precursor protein that accumulates in embryo lysates following ectopic expression tends to be unfolded monomers.

References

1. Bragdon B, Moseychuk O, Saldanha S, King D, Julian J, Nohe A (2011) Bone morphogenetic proteins: a critical review. Cell Signal 23:609–620
2. Seidah NG, Prat A (2012) The biology and therapeutic targeting of the proprotein convertases. Nat Rev Drug Discov 11:367–383
3. Constam DB (2014) Regulation of TGFbeta and related signals by precursor processing. Semin Cell Dev Biol 32:85–97
4. Scamuffa N, Calvo F, Chretien M, Seidah NG, Khatib AM (2006) Proprotein convertases: lessons from knockouts. FASEB J 20:1954–1963
5. Cui Y, Hackenmiller R, Berg L, Jean F, Nakayama T, Thomas G, Christian JL (2001) The activity and signaling range of mature BMP-4 is regulated by sequential cleavage at two sites within the prodomain of the precursor. Genes Dev 15:2797–2802
6. Degnin C, Jean F, Thomas G, Christian JL (2004) Cleavages within the prodomain direct intracellular trafficking and degradation of mature bone morphogenetic protein-4. Mol Biol Cell 15:5012–5020
7. Gregory KE, Ono RN, Charbonneau NL, Kuo CL, Keene DR, Bachinger HP, Sakai LY (2005) The prodomain of BMP-7 targets the BMP-7 complex to the extracellular matrix. J Biol Chem 280:27970–27980
8. Jones WK, Richmond EA, White K, Sasak H, Kusmik W, Smart J, Oppermann H, Rueger DC, Tucker RF (1994) Osteogenic protein-1 (OP-1) expression and processing in Chinese hamster ovary cells: isolation of a soluble complex containing the mature and pro-domains of OP-1. Growth Factors 11:215–225
9. Kunnapuu J, Bjorkgren I, Shimmi O (2009) The Drosophila DPP signal is produced by

cleavage of its proprotein at evolutionary diversified furin-recognition sites. Proc Natl Acad Sci U S A 106:8501–8506

10. Kunnapuu J, Tauscher PM, Tiusanen N, Nguyen M, Loytynoja A, Arora K, Shimmi O (2014) Cleavage of the Drosophila screw prodomain is critical for a dynamic BMP morphogen gradient in embryogenesis. Dev Biol 389:149–159
11. Mi LZ, Brown CT, Gao Y, Tian Y, Le VQ, Walz T, Springer TA (2015) Structure of bone morphogenetic protein 9 procomplex. Proc Natl Acad Sci U S A 112:3710–3715
12. Sopory S, Kwon S, Wehrli M, Christian JL (2010) Regulation of Dpp activity by tissue-specific cleavage of an upstream site within the prodomain. Dev Biol 346:102–112
13. Harrison CA, Al-Musawi SL, Walton KL (2011) Prodomains regulate the synthesis, extracellular localisation and activity of TGF-beta superfamily ligands. Growth Factors 29:174–186
14. Sopory S, Nelsen SM, Degnin C, Wong C, Christian JL (2006) Regulation of bone morphogenetic protein-4 activity by sequence elements within the prodomain. J Biol Chem 281:34021–34031
15. Cui Y, Jean F, Thomas G, Christian JL (1998) BMP-4 is proteolytically activated by furin and/or PC6 during vertebrate embryonic development. EMBO J 17:4735–4743
16. Hawley SH, Wunnenberg-Stapleton K, Hashimoto C, Laurent MN, Watabe T, Blumberg BW, Cho KW (1995) Disruption of BMP signals in embryonic Xenopus ectoderm leads to direct neural induction. Genes Dev 9:2923–2935
17. Nelsen SM, Christian JL (2009) Site-specific cleavage of BMP4 by furin, PC6, and PC7. J Biol Chem 284:27157–27166
18. Freeze HH, Kranz C (2010) Endoglycosidase and glycoamidase release of N-linked glycans. Curr Protoc Mol Biol Chapter 17: Unit 17.13A
19. Kwon S, Christian JL (2011) Sortilin associates with transforming growth factor-beta family proteins to enhance lysosome-mediated degradation. J Biol Chem 286:21876–21885
20. Neugebauer JM, Kwon S, Kim HS, Donley N, Tilak A, Sopory S, Christian JL (2015) The prodomain of BMP4 is necessary and sufficient to generate stable BMP4/7 heterodimers with enhanced bioactivity in vivo. Proc Natl Acad Sci U S A 112:E2307–E2316
21. Tilak A, Nelsen SM, Kim HS, Donley N, McKnite A, Lee H, Christian JL (2014) Simultaneous rather than ordered cleavage of two sites within the BMP4 prodomain leads to loss of ligand in mice. Development 141:3062–3071
22. Birsoy B, Berg L, Williams PH, Smith JC, Wylie CC, Christian JL, Heasman J (2005) XPACE4 is a localized pro-protein convertase required for mesoderm induction and the cleavage of specific TGFbeta proteins in Xenopus development. Development 132:591–602
23. Mimoto MS, Christian JL (2011) Manipulation of gene function in *Xenopus laevis*. Methods Mol Biol 770:55–75
24. Dumont JN (1972) Oogenesis in *Xenopus laevis* (Daudin). I. Stages of oocyte development in laboratory maintained animals. J Morphol 136:153–179
25. Smith LD, Xu WL, Varnold RL (1991) Oogenesis and oocyte isolation. Methods Cell Biol 36:45–60
26. Jean F, Stella K, Thomas L, Liu G, Xiang Y, Reason AJ, Thomas G (1998) alpha1-Antitrypsin Portland, a bioengineered serpin highly selective for furin: application as an antipathogenic agent. Proc Natl Acad Sci U S A 95:7293–7298
27. Sopory S, Christian J (2006) Regulation of TGF-ß family activity by proprotein processing. In: Whitman M, Sater A (eds) Analysis of growth factor signaling in embryos. CRC Press LLC, Boca Raton, FL, pp 38–55
28. Seidah NG, Day R, Hamelin J, Gaspar A, Collard MW, Chretien M (1992) Testicular expression of PC4 in the rat: molecular diversity of a novel germ cell-specific Kex2/subtilisin-like proprotein convertase. Mol Endocrinol 6:1559–1570
29. Denault JB, Leduc R (1996) Furin/PACE/SPC1: a convertase involved in exocytic and endocytic processing of precursor proteins. FEBS Lett 379:113–116
30. Nieuwkoop PD, Faber J (1967) Normal table of *Xenopus laevis*. North Holland Publishing Co., Amsterdam

Chapter 10

Imaging and Quantification of P-Smad1/5 in Zebrafish Blastula and Gastrula Embryos

Joseph Zinski, Francesca Tuazon, Yan Huang, Mary Mullins, and David Umulis

Abstract

Spatiotemporal patterns of morphogen activity drive differential gene expression with a high degree of precision within a developing embryo and reproducibly between embryos. Understanding the formation and function of a morphogen gradient during development requires quantitative measurement of morphogen activity throughout an individual embryo and also between embryos within a population. Quantification of morphogen gradients in to presents unique challenges in imaging and image processing to minimize error and maximize the quality of the data so it may be used in computational models of development and in statistically testing hypotheses. Here we present methods for the preparation, immunostaining, imaging, and quantification of a bone morphogenetic protein (BMP) activity gradient in individual zebrafish embryos as well as methods for acquiring population-level statistics after embryo grouping and alignment. This quantitative approach can be extended to other morphogen systems, and the computational codes can be adapted to other imaging contexts in zebrafish and other organisms.

Key words Morphogen, Gradient, Zebrafish, Quantitative imaging, BMP signaling

1 Introduction

The quantification of morphogen gradient activity through fluorescent imaging has grown increasingly important to studies delineating how gradients form in space and time to drive cellular differentiation during development. In *Drosophila*, where morphogen gradients form along the anterior-posterior (AP) and dorsal-ventral (DV) axes, quantification of morphogen gradients by fluorescent methods has provided insight into how gradients form [1–3], how information is transduced from the gradient to regulate gene expression [4, 5], and how feedback between regulatory motifs can drive pattern formation [6, 7].

The approaches for quantifying morphogen gradients or their activity in zebrafish share many similarities to those used in *Drosophila* morphogen gradients. Morphogens in *Drosophila* can

Melissa B. Rogers (ed.), *Bone Morphogenetic Proteins: Methods and Protocols*, Methods in Molecular Biology, vol. 1891, https://doi.org/10.1007/978-1-4939-8904-1_10,

be visualized directly by in situ methods detecting endogenous proteins, live imaging of fluorescent fusion proteins, or antibody staining of epitope-tagged proteins. These approaches have proven effective for quantifying the AP gradient of the Bicoid (Bcd) transcription factor and the DV gradient of nuclear Dorsal (Dl) in *Drosophila* [6, 8].

In the zebrafish embryo, nodal and retinoic acid (RA) morphogen gradients have also been quantified; nodal signaling activity induces mesendoderm along the animal-vegetal (AV) axis, and RA patterns tissues along the AP axis (reviewed in [9]). Nodal signaling results in the phosphorylation of Smad2, which subsequently accumulates in the nucleus. Multiple methods have leveraged nuclear Smad2 to quantify nodal signaling activity by measuring nuclear Venus-Smad2 intensity, bimolecular fluorescent complementation of Venus between Smad2 and Smad4, and the nucleocytoplasmic ratio of GFP-Smad2 [10–12]. Further, levels of nuclear Smad2 and the distributions of the GFP-Smad2 nucleocytoplasmic ratio have both been related to the expression of nodal direct targets [10, 11]. Additional measurements in the zebrafish embryo include phosphorylated Smad2 levels along the AV axis [13] and nodal-GFP fusion proteins emanating in a gradient from transplants that express nodal-GFP [11]. Finally, the RA gradient has been quantified through the development of fluorescence resonance energy transfer (FRET) probes [14] and, more recently, by using fluorescence lifetime imaging (FLIM) and relying on the inherent fluorescence of RA [15].

However, in both *Drosophila* and zebrafish, the gradient of bone morphogenetic protein (BMP) signaling that directs DV patterning is not easily measured by in situ or fusion protein approaches. While there has been some success in *Drosophila* visualizing Dpp (*Drosophila* BMP2/4 homologue) through a Dpp-GFP fusion protein and an HA epitope-tagged Dpp, neither approach was sufficient for quantification [8, 16]. Instead, quantification of the BMP signaling gradient in both *Drosophila* and zebrafish relies on using an antibody that specifically recognizes the phosphorylated form of Smad1/5/8 (P-Smad5 in zebrafish, P-Mad in *Drosophila*) and the intracellular transcription factor that is directly phosphorylated by BMP receptors and transduces the BMP signal. Though it is difficult to measure Dpp-GFP, a previous study in the *Drosophila* wing imaginal disc overexpressed Dpp-GFP and established a linear relationship between the level of P-Mad and the level of Dpp-GFP, supporting the use of P-Mad as a direct readout of extracellular BMP in this context [17].

In *Drosophila*, the earliest quantitative imaging of BMP signaling and computational models of patterning focused on P-Mad dynamics along the DV axis of the blastoderm embryo. Through nuclear division cycle 14 of *Drosophila* development, the embryo is a syncytium wherein the nuclei are located in a single layer at the embryo periphery and quantification of the gradient relied on measurement of P-Mad levels from extended focus images, where the maximum intensities from the z-sections are extended to a single 2D image of the dorsal half of the embryo [7]. Since these 2D images of the dorsal surface intensity of P-Mad immunostaining proved reliable and the embryo axes are easily identifiable, this methodology was sufficient to quantify differences in P-Mad distribution between wild-type and BMP pathway mutant embryos and develop mathematical models of BMP signaling in *Drosophila* [7]. Equivalent methods were later used to identify mutant BMP components in *Drosophila* that increase embryo-to-embryo variability [18].

However, the methods for quantifying P-Mad in *Drosophila* were not sufficient to quantify BMP patterning in zebrafish for a number of reasons. First, at each stage of zebrafish development, cells are distributed in thick layers, as opposed to a single-layer syncytium, and BMP signals throughout the deep cells. Second, BMP signaling dynamically specifies cell fates along the AP axis [19, 20] coupling the AP and DV axes and thus requiring future studies aimed at modeling 3D spatiotemporal BMP patterns to have 3D data for model testing (reviewed in [9]). Third, acquiring the data IN TOTO gives the most complete view of BMP signaling and enables examination of signaling dynamics in all regions of the embryo. These requirements for high-quality, 3D images of the zebrafish embryo that can be used in quantification introduce new challenges for acquiring data, particularly due to the potential for refractive index (RI) mismatch between cells, media, yolk, immersion oil, and coverslip that all systematically interfere with quantification [21, 22]. RI mismatch systematically introduces error in quantification through signal drop-off with increasing depth, spherical aberration (an issue that increases with the numerical aperture, NA, of the objective), and axial scaling. Furthermore, unlike in *Drosophila*, the DV axis of the zebrafish embryo is not readily identifiable prior to gastrulation, since the embryo morphology and cell densities are largely symmetrical. Thus, new methods were needed for embryo signal quantification, alignment, and registration.

We developed new approaches and tools to quantify BMP signaling activity in all cells of the relatively large zebrafish embryo. First, we established a fluorescent image acquisition method that minimizes imaging artifacts by careful RI matching and embryo clearing. Following image acquisition of fluorescent-antibody stained P-Smad5 embryos, the high-dimensional image data are converted into an array of nuclei positions similar to that done by others [23], where the average channel intensities are calculated for each nucleus position. This dramatically reduces data storage and, subsequently, the time it takes to calculate the model-data fitness. Consider that in acquiring an image stack that is $1024 \times 1024 \times 100$ (x, y, z) pixels, multiplied by the number of channels (N), there are $1.05N \times 10^8$ total points. However, we only need to store the spatial position of each nucleus (x, y, z) and the average channel intensity at each nucleus. Effectively, if an embryo has 10,000 cells, then only $10{,}000 \times 3 \times N$ or $3N \times 10^4$ points are stored, reducing the file size from 1.05×10^8 to 3×10^4 and resulting in a 3500× smaller file.

Following the conversion of the image data, individual embryo point clouds of nuclear positions and associated P-Smad5 levels are registered using an adaptation of coherent point drift (CPD), a robust point alignment method that allows for variability in the total number of points, as well as overall size and orientation of the alignment set (a particular embryo) to the reference set (a reference embryo) [24, 25]. Finally, data from individuals and populations can then be extracted for statistical comparison, including measures of gradient shape and embryo-to-embryo variability of P-Smad5 signaling patterns. In the remainder of this paper, we outline the approach to acquire and process the images to quantify the formation of the BMP signaling gradient in zebrafish embryos.

2 Materials

1. PBS: 137 mM NaCl, 2.7 mM KCl, 8 mM Na_2HPO_4, 1.46 mM KH_2PO_4.
2. 4% PFA: 9.25% of a 37% PFA (formaldehyde) solution (Sigma 252549), 91.75% PBS by volume.
3. PBSTrit: 99.9% PBS, 0.1% Triton X-100 by volume.
4. NCS-PBSTrit: 10% heat-inactivated newborn calf serum, 0.1% Triton X-100, 1% dimethyl sulfoxide, 88.9% PBS.
5. Methanol.
6. BABB: two parts benzyl benzoate, one part benzyl alcohol mixed fresh just prior to use.
7. Long coverslip and a 25 mm × 25 mm silicon wafer with a 10 mm diameter circle cut out of the center and a thickness of 0.7 μm (Grace Biolabs 475271).
8. Nail polish.
9. Vacuum grease.
10. P-Smad1/5/9 Rabbit antibody (Cell Signaling Technology #13820) or Histone H3 Mouse Antibody (Active Motif #39763).
11. Goat Anti-Rabbit Alexa Fluor 647 (Invitrogen A-21244/5) or Alexa Fluor 546 (Invitrogen A-11030), if the Histone H3 antibody is used.
12. 1:2000 dilution of Sytox Green (Invitrogen S7020). Store diluted Sytox Green at −80 °C in small (2–5 μL) aliquots.

3 Methods

3.1 Embryo Fixation

1. Collect zebrafish embryos of the desired stage between 1 and 10 h post fertilization in 1.5 mL tubes, with approximately 25 embryos per tube. Remove as much media as possible from each tube and add 1 mL of 4% PFA solution to each.
2. Place embryos on a rocker for 4 h at room temperature or overnight at 4 °C.
3. Wash embryos four times for 30 min on a rocker with 1 mL of PBS or PBSTrit.

3.2 Removal of the Chorion and Yolk

1. Transfer embryos to a cell culture dish with PBS or PBSTrit using a plastic transfer pipette.
2. Manually remove the chorion from each embryo under a manual dissecting microscope. Use two microdissection forceps or one micro-dissecting forcep and one dissecting needle (Fig. 1a, b). *Do not use proteinase* (*see* **Note 1**).
3. Manually remove the yolk. Use forceps to gently squeeze the embryo at its equator while nudging/pulling the yolky vegetal pole away with another pair of forceps or needle (Fig. 1c, c′).

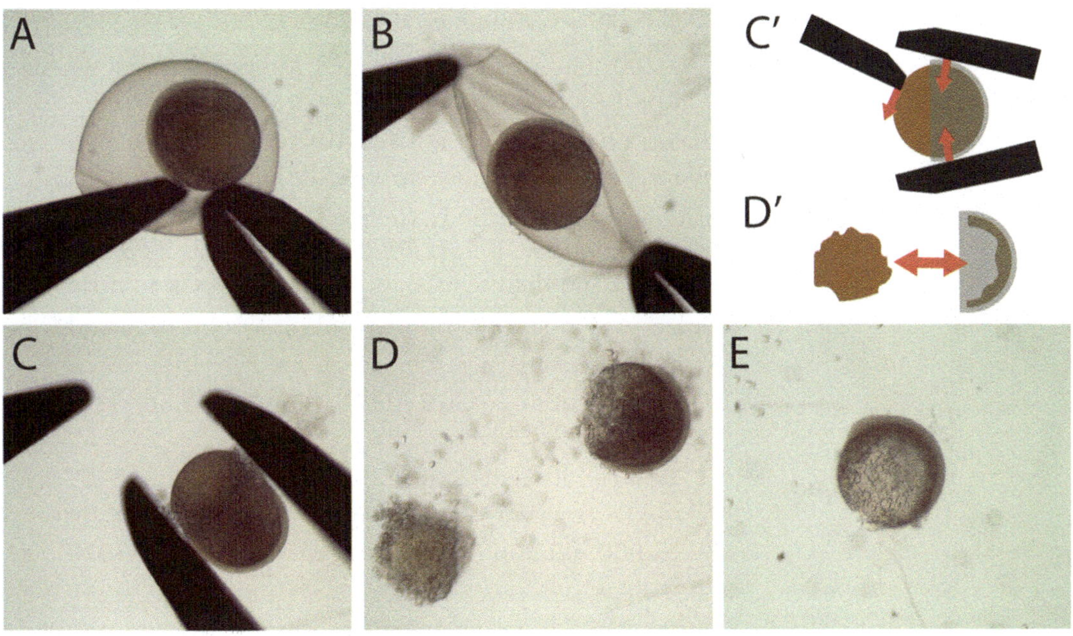

Fig. 1 Removing the chorion and yolk from the fixed embryo. (**a**, **b**) Chorion removal by grabbing the chorion with forceps while holding another part of the chorion against the dish with a probe (**a**) and then ripping a hole and pulling the chorion apart to release the embryo (**b**). (**c**–**e**) Removal of the yolk. (**c**, **c′**) First, gently squeeze the animal cap with forceps while picking at the yolk with a probe. (**d**, **d′**) Most of the yolk should pop out of the animal cap. (**e**) The embryo after gentle vortexing, which removes much of the remaining yolk particles

Part or all of the yolk will separate from the cap of cells (Fig. 1d, d′) (*see* **Note 2**).

4. Remove the remaining yolk granules by carefully returning the embryos to the 1.5 mL tube and gently vortexing. This will remove the yolk (Fig. 1e). Since the embryos are delicate, ensure the 1.5 mL tube is filled with no less than 1 mL of liquid to prevent them from sticking to the sides of the tube.
5. Confirm that the yolk has been fully removed. Carefully return the embryos to the dish of PBS and check that very little yolk is attached to the blastoderm (Fig. 1e).
6. Return the embryos to a new 1.5 mL tube filled with fresh PBS, leaving the yolk fragments in the dish.
7. Wash for 5 min with 1 mL of PBSTrit. Embryos can be stored for up to 2 weeks in PBS. Dehydrate in methanol for longer storage.

3.3 Fluorescent P-Smad5 Staining

1. Block embryos in NCS-PBSTrit. Remove PBSTrit and replace with 1 mL of NCS-PBSTrit. Put on a rocker at room temperature for at least 2 h.
2. Prepare primary antibody solution. In 0.5 mL NCS-PBSTrit, add:
   (a) 1:200 dilution of P-Smad1/5/9 Rabbit Antibody (Cell Signaling Technology #13820)
   (b) 1:200 dilution of Histone H3 Mouse Antibody (Active Motif #39763) (*see* **Note 3**)
3. Remove NCS-PBSTrit and incubate the embryos with the primary antibody solution on a rocker at 4 °C overnight.
4. Wash off primary antibody. Remove primary antibody solution and replace with 1 mL PBSTrit. Place on a rocker for 15–30 min and then replace the PBSTrit. Repeat four times.
5. Prepare secondary antibody solution. In 0.5 mL of NCS-PBSTrit, add:
   (a) 1:500 dilution of Goat Anti-Rabbit Alexa Fluor 647 (Invitrogen A-21244/5).
   (b) 1:500 dilution of Goat Anti-Mouse Alexa Fluor 546 (Invitrogen A-11030), if the Histone H3 antibody is used.
   (c) 1:2000 dilution of Sytox Green (Invitrogen S7020) (*see* **Note 4**).
6. Remove PBSTrit and incubate embryos with secondary antibody solution on a rocker in the dark at 4 °C overnight OR at room temperature for at least 2 h. Avoid exposing samples to light for extended periods in all subsequent steps.

7. Wash off secondary antibody. Remove secondary antibody solution and replace with 1 mL PBSTrit. Place on a rocker for 15–30 min and then replace the PBSTrit. Repeat four times.
8. Remove as much PBSTrit as possible and replace with 1 mL of PBS. Embryos can be stored in PBS at 4 °C for a few weeks, or they can be dehydrated in methanol for indefinite storage at −20 °C.

3.4 Clearing and Mounting for Imaging

1. Remove as much PBS as possible and replace with 1 mL of 50% methanol/50% PBS. Let sit for 5 min.
2. Remove as much 50% methanol/50% PBS as possible and replace with 1 mL of 100% methanol. Let sit for 5 min (*see* **Note 5**).
3. Remove as much methanol as possible and replace with 1 mL of 50% methanol/50% BABB (two parts benzyl benzoate, one part benzyl alcohol). Let sit for 5 min (*see* **Note 6**).
4. Remove as much 50% methanol/50% BABB as possible and replace with 1 mL of 100% BABB. Let sit for 10 min.
5. Remove as much BABB as possible and replace with 1 mL of 100% BABB. Let sit for 5 min (*see* **Note** 7).
6. *Method 1* (Fig. 2a–d): Construct an imaging well from a long coverslip and a 25 mm × 25 mm silicon wafer with a 10 mm diameter circle cut out of the center and a thickness of 0.7 μm (Grace Biolabs 475271) (Fig. 2a). Use nail polish to seal the bottom side down to the coverslip; leave the other side open and uncovered (Fig. 2a blue).
   Method 2 (Fig. 2e–h): Construct an imaging well from a long coverslip and a ring of silicone vacuum grease (Fig. 2e, f). For easy application, backfill a 1.25 mL plastic syringe with the vacuum grease and carefully push through (*see* **Note 8**).
7. Under a fluorescent dissecting microscope with a green light fluorescent filter, remove a single embryo from the 1.5 mL tube and place it into the imaging well.
8. Fill the imaging well with BABB until the meniscus is flat across the imaging well (Fig. 2b).
9. Under the green light of the fluorescent dissecting microscope, adjust the orientation of the embryo. Orient embryos *without* the yolk (stages prior to 70% epiboly) so that the animal pole is down. Orient embryos *with* the yolk (stages after 70% epiboly) so that the vegetal pole is down. If the embryo is damaged, discard. Approximately 25% of embryos usually become damaged during the staining and mounting process depending on how gently the washes are performed.
10. *Method 1* (Fig. 2a–d): Gently lay another long coverslip on top of the imaging well, laying it down from one side to the other

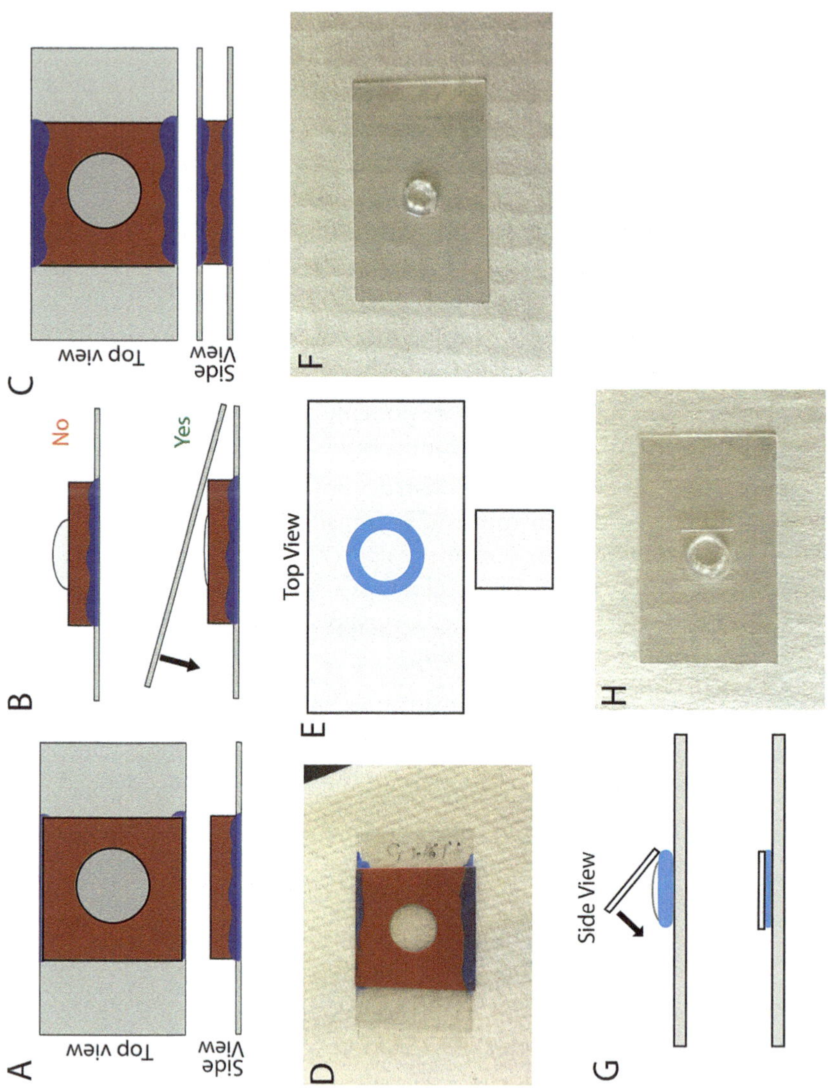

Fig. 2 Mounting the embryo for imaging. A single embryo can be mounted between two coverslips separated by either a silicon wafer (**a–d**) or silicone vacuum grease (**e–h**). (**a**) First, place the silicon wafer onto the coverslip and secure the edges with nail polish (blue). (**b**) Place the embryo in BABB into the newly formed well, taking care to remove enough BABB so that the meniscus is flat before placing a second coverslip on top. (**c**, **d**) After dabbing away excess BABB, seal the edges with nail polish (blue). (**e**, **f**) Make a circle of petroleum jelly on a coverslip. (**g**, **h**) Place the sample in BABB in the well and seal with a smaller coverslip

(Fig. 2b), which reduces air bubbles being trapped below the coverslip. Carefully dab the edges of the coverslip with a Kimwipe to remove any BABB. Dab repeatedly and apply gentle force until very little BABB comes out. Seal the edges with nail polish (Fig. 2c, d) (*see* **Note 9**).

Method 2 (Fig. 2e–h): Gently lay a short coverslip, as above, on top of the imaging well and lightly press down to seal (Fig. 2g, h). The short coverslip may be repositioned if necessary.

11. Image within 8 h, as prolonged time in BABB (~24 h) reduces fluorescence intensity.

3.5 Confocal Imaging

1. Mount the imaging well on a confocal microscope and image. Calibrate the gain, laser power, and all other settings before imaging the first embryo. Keep all settings the same for the remainder of the set (*see* **Notes 10–15** and Fig. 3).
2. After imaging, embryos may be recovered from the imaging wells for genotyping. We extracted DNA using the HotShot technique [26] and genotyped single embryos using KASPar [27] or dCAPs genotyping [28] (*see* **Note 16**).

3.6 Nuclei Identification and P-Smad5 Intensity Extraction

We developed a software program to carry out the procedures that are described in the following section. The files and instructions are available for download at: https://elifesciences.org/download/aHR0cHM6Ly9jZG4uZWxpZmVzY2llbmNlcy5vcmcvYXJ0aWNsZXMvMjIxOTkvZWxpZmUtMjIxOTktc3VwcDEtdjIuemlw/elife-22199-supp1-v2.zip?_hash=Zxk8TVpw5NGza2MNi0jq4pD5GuZ%2FVeluqS15rUkVbUo%3D.

The attached link includes the Matlab codes, sample image files, and a brief tutorial to run the software. Here we go into greater depth on the processes and potential modification points for the image analysis codes to meet the requirements of diverse embryo quantification needs.

1. The first step is completed before starting the MATLAB® script. Convert the ".lsm" confocal stack files to ".tif" files using ImageJ and then import into Matlab as 1024 × 1024 × *Z* multidimensional arrays. *XY* pixels are 0.55 μm, *Z* pixels were 2.2 μm (Fig. 4a).
2. Locate the center points of all nuclei using the Histone H3 (or Sytox) stain. This is done by:
   (a) Smoothing the histone H3 (or Sytox) stain with a 3-D Gaussian kernel with a size of approximately 9 pixels (4.95 μm) by 9 pixels (4.95 μm) by 3 pixels (6.9 μm) and a standard deviation of 1.91 (Fig. 4a′). This kernel is approximately the size of most nuclei, which are about 6 μm in diameter. To eliminate false positives, remove local minima and maxima H-maxima and H-minima

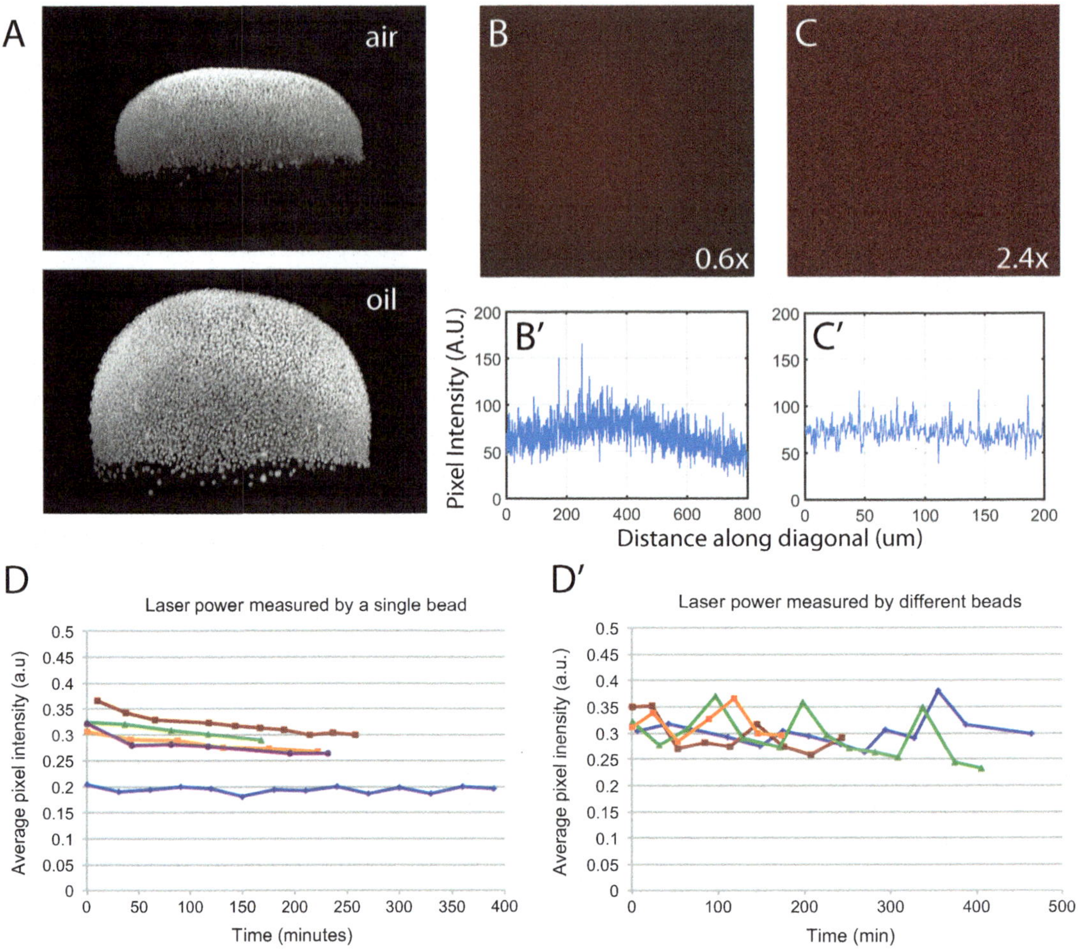

Fig. 3 Controlling for fluorescence artifacts. (**a**) Lateral max projection of nuclei in the same embryo stained with Sytox Green and imaged using a 20× air lens (top) or a LD LCI Plan-Achromat 25×/0.8 Imm Corr DIC M27 multi-immersion oil lens (bottom). (**b**, **c**) Alexa Fluor 647 dissolved in water reveals the nonuniformity across the *XY* field in a LD LCI Plan-Achromat 25×/0.8 Imm Corr DIC M27 multi-immersion lens at 0.6× (**b**) and 2.4× (**c**) digital zoom. (**b**′, **c**′) Intensity profiles plotted from the top-right to bottom-left corner of the above images revealing a 40% drop-off from the center to the edge at 0.6× zoom and 8% drop-off from the center to the edge at 2.4× zoom. (**d**, **e**) Laser power fluctuation vs. time when imaging a single bead (**d**) or of different beads (**e**) on a ThermoFisher cat. no. F369009, well A1 calibration slide

transforms targeting peaks with an intensity less than 10 units (8-bit) higher than background (*see* **Note 17**).

(b) Finding all maxima by using the "imregionalmax" function on the entire 1024 × 1024 × *Z* array (Fig. 4a‴). Maxima closer together than 6 pixels can be assumed to be in the same nucleus and therefore be combined into a single point.

3. Extract P-Smad5 intensities from the identified nuclei maxima locations (Fig. 4b). P-Smad5 distribution in each nucleus is

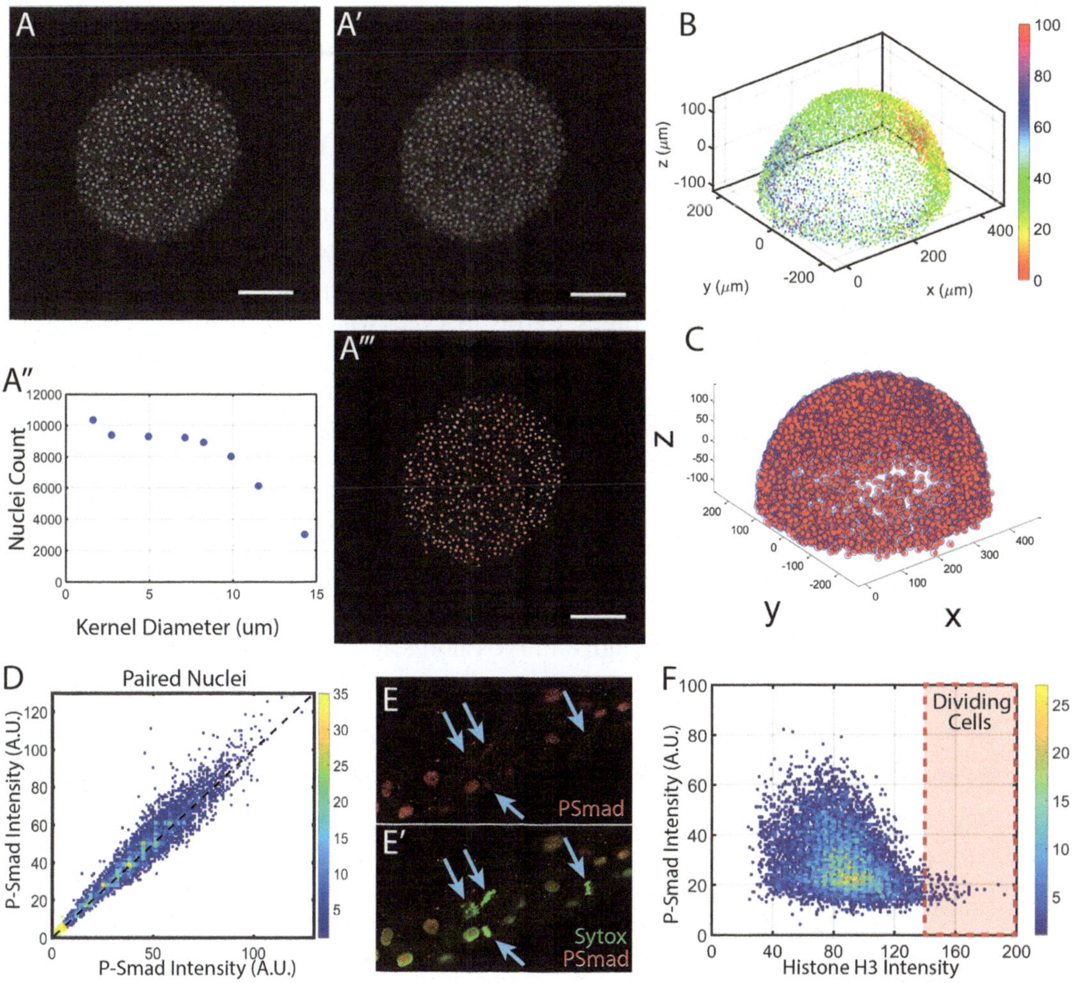

Fig. 4 Nuclear identification, P-Smad5 measurement, and embryo alignment. (**a**) Animal view of a single 2.2 μm Z-slice of fluorescently labeled histone H3. (**a**′) The same Z-slice after filtering with a Gaussian kernel 9 pixels (4.95 μm) by 9 pixels (4.95 μm) by 3 pixels (6.9 μm) with a standard deviation of 1.91. (**a**″) The number of nuclei identified in a single embryo plotted versus the *XY* kernel size. (**a**‴) Nuclei with centerpoints on the Z-slice are shown in red, while the Histone H3 stain after all smoothing shown in white. Note that the nuclei that appear to be missed have center points that fall on adjacent planes. Scale bar in a, a′, a‴ is 100 μm. (**b**) A 3-D display of all nuclei from a single embryo colored by a heatmap of P-Smad5 fluorescence. (**c**, **d**) An embryo was imaged, then repositioned, and imaged a second time to determine the overall image acquisition noise, by comparing P-Smad5 fluorescent intensity for each paired nucleus as shown. (**c**) Correspondence of nuclei for the embryo imaged twice (blue-original image, red-repositioned and reimaged). (**d**) The overall image acquisition noise includes positioning error and detector noise from the Hamamatsu R6357 photomultiplier tube on a Zeiss 710 confocal. (**e**) P-Smad5 (red) is absent in cells undergoing mitosis while Sytox Green (green) intensifies (blue arrowheads). (**f**) Histone H3 intensity vs. P-Smad5 intensity in all nuclei in a single gastrula. Cells with high histone H3 intensity and low P-Smad5 intensity are identified as dividing cells and removed from further analysis

approximately uniform, so a small sphere within each nucleus is averaged to attain the P-Smad5 intensity. A spherical 6 × 6 × 3 kernel is recommended (*see* **Note 18**).

3.7 Embryo Alignment and Processing

1. Remove cell types unresponsive to BMP signaling. P-Smad5 is not localized in the nucleus during cell division (Fig. 4e, e′), so dividing cells were removed from the analysis.
   (a) Dividing cells are identified by their bright and concentrated Histone H3 (or Sytox) fluorescence. Histone H3 becomes tightly packed when chromatin is condensed during cell division, making DNA stains such as fluorescent Histone H3 or Sytox green bright.
   (b) Cells with a bright DNA fluorescence staining (above 140% of the mean DNA fluorescence) were identified as dividing and eliminated from the analysis (Fig. 4f).
2. To calculate population-level statistics for similarities and differences in the mean P-Smad5 levels, embryos of similar stages need to be uniformly aligned along the AV and DV axes.
   (a) To align embryos along the AV axis using the nuclei center points, use the software to perform a planar regression on the nuclei centerpoints and then rotate the embryo until the plane aligns with the *XY* axis (Fig. 5a, b).
   (b) Use the software to align the embryos along the DV axis using the embryonic shield as a morphological marker of the dorsal side. We rotate the embryo so that the shield always faces the same direction. Before 6 hpf, when the shield is not present, the embryos can be aligned in the DV direction by fitting a polynomial regression to the P-Smad5 gradient around the margin and rotating until the peak maximum is ventral (Fig. 5c–e).
3. The shape of some embryos may be altered slightly in the fixation, the antibody staining, or the mounting procedures. To eliminate these imperfections, so that embryos can be compared to each other, we conform embryos to a template embryo using an affine coherent point drift (CPD) (Fig. 5f) [29]. The method aligns the nuclei point clouds as a probability density estimation problem that maximizes the likelihood of correspondence between the nuclei point clouds of the template embryo with the test embryo. The approach maintains the spatial arrangement of the points as a nonrigid point set allowing the point cloud being aligned to "stretch" a small degree to overlap the template [29]. The result of implementing this method is that all embryos are similarly oriented for comparative statistics from the resulting profiles. The template embryo should be an embryo from the dataset that is not deformed. Alignment of the embryos to a template or reference

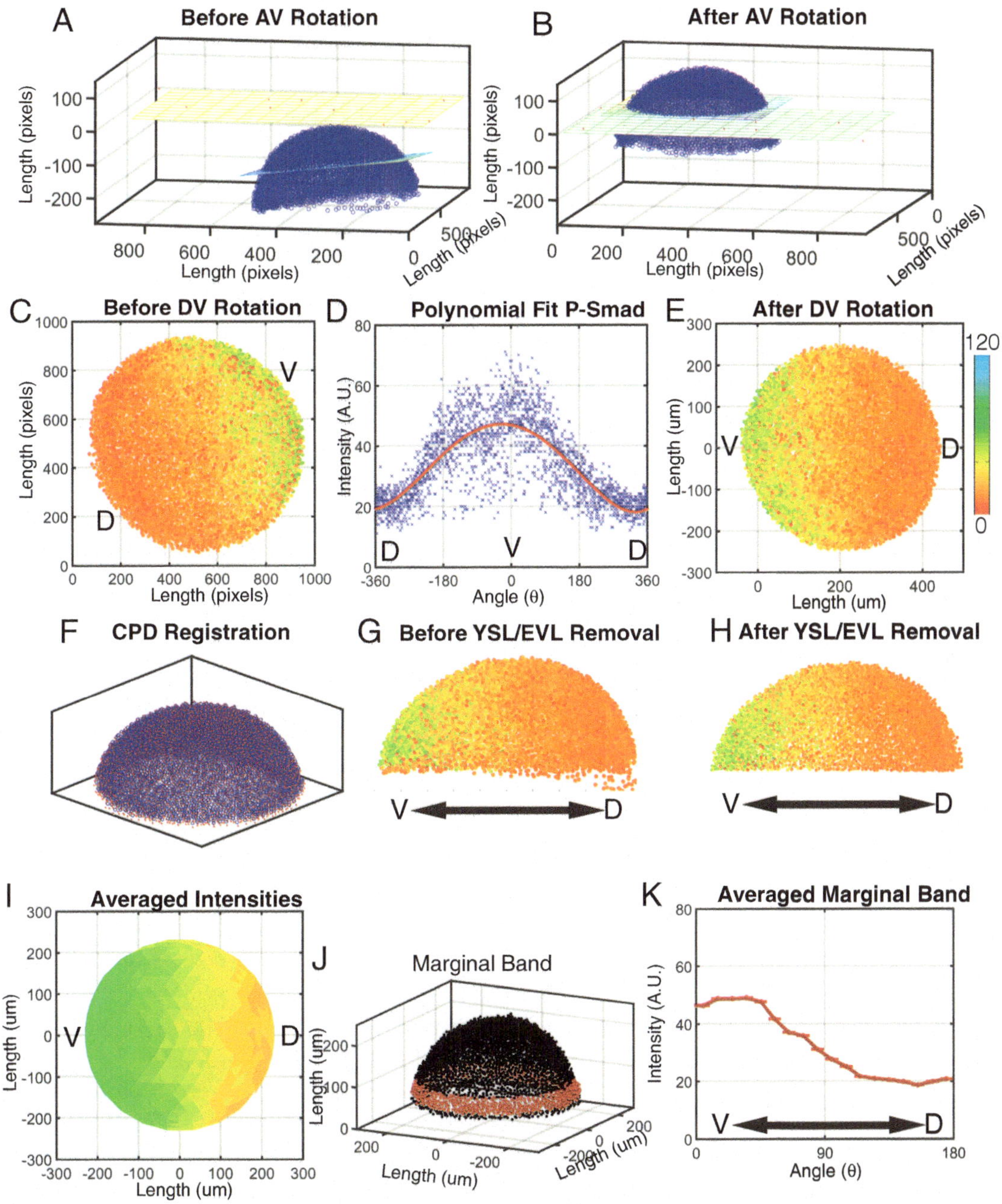

Fig. 5 Embryo alignment and display. (**a**, **b**) The nuclei of an embryo (blue dots) before and after alignment of the AV axis. (**c**) An animal view of an embryo before rotation to align the DV axis. Nuclei are colored based on their P-Smad5 intensities. (**d**) A polynomial curve (red line) fit to the nuclei around the margin of the embryo (blue crosses). (**e**) An animal view of an embryo after rotation to align the DV axis. (**f**) An embryo (blue dots) after being fit to a reference embryo (red dots) using coherent point drift. (**g**, **h**) Lateral view of an embryo before and after the removal of the YSL and EVL nuclei by a MATLAB program. (**i**) Animal view of the average intensities of 13 WT embryos at 5.7 hpf averaged onto a 1280 sided polyhedron. (**j**) A band of cells around the margin of the embryo. (**k**) The average intensity of P-Smad5 in 13 WT embryos around the margin of the embryo

embryo provides a consistent orientation for quantitative comparisons between embryos and for statistically testing hypotheses of the measured gradient properties. The goal of CPD is to fine tune the embryo alignment and also correct for any distortions in embryo shape that may have occurred during fixation and staining (*see* **Notes 19** and **20**).

4. The extraembryonic yolk syncytial layer (YSL) nuclei and the enveloping layer (EVL) cells do not respond to BMP signaling in the same manner as the deep cells. The sparse YSL nuclei located below the vegetal margin were eliminated by hand (Fig. 5g, h). Use the software to remove the remaining YSL nuclei and the EVL cells by removing the inner and outer layer of identified nuclei, approximately 15% of the total nuclei (Fig. 5g, h). The remaining cells were assumed to be nondividing deep cells.

3.8 Displaying the Data

1. The individual nuclei are then displayed in 3-D in MATLAB using the *XYZ* coordinates and associated P-Smad5 intensities (Fig. 5c, e, g, and h).
2. One useful method for data visualization is to project the average P-Smad5 values from individuals or populations to a 3D mesh representing the zebrafish embryo. The nuclei across multiple embryos are averaged together and projected onto a polyhedron with many sides (4800 in this case). The diameter of the polyhedron is estimated from the diameter of the embryos. To develop the 3D representation, first, all the nuclei are converted from X, Y, Z coordinates into spherical coordinates (r, θ, ϕ). The ϕ and θ measurements are then used to determine into which face of the polyhedron each nuclear point falls. This is done for all nuclei across multiple embryos. Averaging the nuclei in each face together creates an averaged virtual embryo (Fig. 5i).
3. Another useful representation for visualization, modeling, and testing hypotheses of fluorescent intensity data are one-dimensional graphs extracted in different regions of the embryo data set. To represent the P-Smad5 gradient as a one-dimensional graph around the margin or over the top of the embryo, the P-Smad5 intensities of the closest neighboring cells are averaged. For example, a band of cells approximately 40 μm thick around the vegetal margin can be chosen (Fig. 5j). Cells within that band can be grouped into 10° intervals and averaged together to form 36 individual points. The left and right side of the gradient can be averaged together into a single ventral to dorsal profile (Fig. 5k). Following extraction of the data from the individual embryos, hypothesis testing can be performed to measure the significance of differences between embryos.

4 Notes

1. Once embryos are dechorionated, only use a glass Pasteur pipette to transfer embryos since they stick to plastic.
2. It is efficacious to remove the yolk at early stages (4–7 hpf) because the embryos are spherical and difficult to consistently orient for imaging. Without the yolk, embryos naturally settle with their animal poles downward. However, do not remove the yolk after 7 hpf (65–70% epiboly) because doing so will damage embryo morphology.
3. The Histone H3 antibody is an optional, additional stain for nuclei. Nuclei can also be stained with Sytox Green (in Sect. 3.3 **step 5**) alone or in combination with the H3 antibody. We find that the Histone H3 antibody marks DNA most evenly and, when combined with its secondary antibody, remains more stable (has little to no background) than Sytox Green when imaging for extended periods of time (>4 h, see next section). Sytox Green can still be sufficient for identifying single nuclei, though its efficacy in our hands decreases after three or more freeze-thaw cycles. We recommend storing Sytox Green at −80 °C in small (2–5 μL) aliquots.
4. Nuclei can be stained with Sytox Green alone or in addition to Histone H3. If using the Histone H3 antibody, it may still be necessary to stain embryos with Sytox Green so they are visible after being cleared for imaging (see next section). Sytox Green is much brighter and may be more easily seen under the dissecting fluorescent scope, depending on its power and resolution.
5. Embryos tend to stick to the plastic of the tube after dehydration. Be wary of inverting the tubes after dehydration has begun.
6. BABB should be mixed fresh just prior to use, and the BABB solution should be clear. Benzyl benzoate (BB) and benzyl alcohol (BA) stocks must be kept tightly closed and away from light to preserve their tissue-clearing ability. Since BABB dissolves plastics, it should be mixed in a glass vial, and embryos should not be left in BABB in plastic tubes for a prolonged period of time (>2 h).
7. The embryos will no longer be visible to the naked eye. To see the embryos, use a fluorescent dissecting microscope with a green light filter cube to illuminate the Sytox Green.
8. Method 2 is best for embryos at 70% epiboly and later stages. The ring of grease enables positioning the embryo directly against the surface of the coverslip, as opposed to suspended above it when using the silicon wafer. This is important at later

stages because the volume occupied by the cells of the embryo is larger and will challenge the working distance of the objective during imaging. If the embryo is not close to the surface of the coverslip, a large proportion of the embryo will not be imaged.

9. Once the coverslip touches the BABB in the well, it will seal tightly against the silicon, and removing it will no longer be possible. It may be possible to slide it slightly into place depending on the amount of suction created by the well.

10. We used the following equipment and settings: an LSM 710 confocal microscope with a LD LCI Plan-Apochromat 25×/0.8 Immersion Corr DIC M27 multi-immersion lens, which has a free working distance of 0.57 mm. We used a 2.4× digital zoom with 4 × 4 tiling and 256 × 256 pixels in each tile, making the stitched image 1024 × 1024 pixels. The *X* and *Y* pixel length was 0.55 μm. A pinhole size of 29.5 μm gave a Z-slice thickness of 2.2 μm (for a 633 nm wavelength). We used 488, 561, and 633 nm excitation lasers. The 633 nm laser, which we used to measure P-Smad5 fluorescence, is a 5 mW laser made by Lasos, model LGK 7628-1F. For the channel used to measure P-Smad5, we used 8-bit images, a pixel dwell time of 1.58 μs, a gain between 800 and 950, and no digital offset. We used laser power of 7% for the 633 laser. Z-slices were spaced by 2.2 μm, and approximately 70–190 slices were needed per embryo. Total embryo scan times ranged between 12 min at 30% epiboly and 30 min at the end of gastrulation.

11. To minimize light scattering by the cells and yolk, we cleared embryos with BABB. If the refractive index (RI) of the immersion lens does not match the RI of the media of the sample, spherical aberrations can distort the *Z*-dimension of the embryo and cause nonlinear drop-off of intensity along the *Z*-axis. These effects can be easily visualized by comparing confocal stacks of an embryo emerged in BABB (R.I. ≈ 1.56) taken with a 25× oil-immersion (R.I. = 1.518) lens and a 20× air lens (R.I. = 1) (Fig. 3a). We used the oil-immersion setting of the 25× lens, since immersion oil, the coverslip, and BABB all have similar R.Is. Also note that most lenses have recommended coverslip thicknesses that must be used to avoid spherical aberration. Here we used Fisherbrand microscope cover glass 12-544-C 24 × 40 – 1.5.

12. Many lenses, especially high magnification lenses, do not transmit intensity equally across the field of view, and therefore objects near the periphery of the image appear dimmer than they actually are. To determine if this phenomenon is significant to a lens, image a fluorophore dissolved in water or a calibration slide (e.g., Thermo Fisher cat. no. F369009, well

A1) with a uniform intensity and measure the difference between pixels near the center and the edge of the image. To decrease this effect, increase the digital zoom on the microscope. For example, with the LD LCI Plan-Achromat 25×/0.8 Imm Corr DIC M27 multi-immersion lens at a 0.6× digital zoom, there is ≈40% drop-off from the center to the edge of the image (Fig. 3b, b′). When increased to a 2.4× digital zoom, there is only an 8% drop-off from the center to the edge of the image (Fig. 3c, c′).

13. Using a higher digital zoom will decrease the field of view and may require tiling to keep the entire embryo in view. We used 2.4× zoom and 4 × 4 tiling with 256 × 256 pixels in each tile, making the stitched image 1024 × 1024 pixels.

14. It is important to ensure that the chosen laser power and pixel dwell time does not bleach the sample. Image the same embryo twice and compare the mean intensity to ensure there is little to no drop-off due to bleaching. A drop-off of less than 5% is ideal.

15. Depending on the type and age of the laser being used (we used a 5 mW laser by Lasos, model LGK 7628-1F), laser power may fluctuate over time. Large fluctuations in laser power will result in systematic differences in fluorescence intensity and must be controlled for when imaging P-Smad5 fluorescence. In general, laser intensity will stabilize after 1–4 h of being turned on and the rate of power drop-off can be measured using an optical power meter (e.g., Thor Labs, PM100A). It is necessary to measure laser power between imaging samples to correct for any change in laser power. This can be done using a test slide containing fluorescent beads (e.g., ThermoFisher cat. no. F369009, well A1). Measuring the fluorescence intensity of an inert fluorescent bead enables the quantification of fluctuations in laser power. Importantly, ***the same bead must be imaged each time*** with identical settings (number of slices, gain) since fluorescent beads can differ in their intensity, which will introduce variability in intensity measurements (Fig. 3d, d′).

16. To recover embryos from mounting Method 1, carefully pry the coverslips apart under a dissecting microscope. This process is tedious due to the suction in the well and the invisibility of the embryo. The silicon wafers are reusable after being carefully cleaned with ethanol. To recover embryos from mounting Method 2, carefully lift the short coverslip off of the vacuum grease. Either method may be performed with or without a fluorescent dissecting microscope. If not using a fluorescent microscope, drip methanol into the opened imaging well to make the embryo visible again. Practicing this step beforehand is highly recommended to prevent loss of embryos.

17. Kernel diameter is important to properly identify nuclei. The kernel diameter should not exceed the diameter of the nucleus or nuclei will be merged. Conversely, too small of a diameter will oversample the same nucleus. Ideally, the kernel diameter should match the average diameter of the nuclei. When the kernel size exceeds the size of the nuclei by so much that it begins including neighboring nuclei, nuclei become merged, and the total number drops off rapidly (Fig. 4a″).
18. To calculate the overall image acquisition error for our system, we imaged the same embryo twice by repositioning the slide, reidentifying the embryo, and redetermining the z-stack limits. We paired individual nuclei from the first and second image by finding the nonrepeating correspondence of the closest nuclei from each image (Fig. 4c). Once nuclei had been paired, we measured the difference between them (Fig. 4d). After removing drop-off from photobleaching, an average overall image acquisition noise had a normal distribution with a $\sigma = 6.4\% \pm 0.3\%$, which includes positioning error and detector noise (Fig. 4d).
19. To decrease the amount of processing power needed and increase accuracy, embryos should first be aligned in the AV and DV direction BEFORE attempting CPD to fit experimental embryos to a template embryo.
20. Embryos stained on different days or imaged with different settings can be normalized by multiplying the entire set by a single scalar value. Embryos stained in the same tube and imaged using the same confocal settings on the same day do not need to be subjected to normalization, provided that laser power fluctuations do not occur (*see* Fig. 3d′). To determine the normalization scalar, control WT embryos should always be imaged in conjunction with each experimental condition. The scalar normalization value is determined by minimizing the sum of the error between the control WT embryos imaged on different days.

Acknowledgment

This work was supported by grant NIH R01GM056326 and NIH R01HD073156.

References

1. Gregor T, Wieschaus EF, McGregor AP, Bialek W, Tank DW (2007) Stability and nuclear dynamics of the bicoid morphogen gradient. Cell 130(1):141–152
2. Teleman AA, Cohen SM (2000) Dpp gradient formation in the Drosophila wing imaginal disc. Cell 103(6):971–980

3. Coppey M, Boettiger AN, Berezhkovskii AM, Shvartsman SY (2008) Nuclear trapping shapes the terminal gradient in the Drosophila embryo. Curr Biol 18:915–919
4. Gregor T, Tank DW, Wieschaus EF, Bialek W (2007) Probing the limits to positional information. Cell 130(1):153–164
5. Reeves GT, Trisnadi N, Truong TV, Nahmad M, Katz S, Stathopoulos A (2012) Dorsal-ventral gene expression in the Drosophila embryo reflects the dynamics and precision of the dorsal nuclear gradient. Dev Cell 22(3):544–557. https://doi.org/10.1016/j.devcel.2011.12.007
6. Jaeger J, Surkova S, Blagov M, Janssens H, Kosman D, Kozlov KN, Manu ME, Vanario-Alonso CE, Samsonova M, Sharp DH, Reinitz J (2004) Dynamic control of positional information in the early Drosophila embryo. Nature 430(6997):368–371. https://doi.org/10.1038/nature02678
7. Umulis DM, Shimmi O, O'Connor MB, Othmer HG (2010) Organism-scale modeling of early Drosophila patterning via bone morphogenetic proteins. Dev Cell 18(2):260–274. https://doi.org/10.1016/j.devcel.2010.01.006
8. Kanodia JS, Rikhy R, Kim Y, Lund VK, DeLotto R, Lippincott-Schwartz J, Shvartsman SY (2009) Dynamics of the Dorsal morphogen gradient. Proc Natl Acad Sci U S A 106(51):21707–21712. https://doi.org/10.1073/pnas.0912395106
9. Tuazon FB, Mullins MC (2015) Temporally coordinated signals progressively pattern the anteroposterior and dorsoventral body axes. Semin Cell Dev Biol 42:118–133. https://doi.org/10.1016/j.semcdb.2015.06.003
10. Dubrulle J, Jordan BM, Akhmetova L, Farrell JA, Kim SH, Solnica-Krezel L, Schier AF (2015) Response to Nodal morphogen gradient is determined by the kinetics of target gene induction. Elife 4:e05042. https://doi.org/10.7554/eLife.05042
11. Muller P, Rogers KW, Jordan BM, Lee JS, Robson D, Ramanathan S, Schier AF (2012) Differential diffusivity of Nodal and Lefty underlies a reaction-diffusion patterning system. Science 336(6082):721–724. https://doi.org/10.1126/science.1221920
12. Harvey SA, Smith JC (2009) Visualisation and quantification of morphogen gradient formation in the zebrafish. PLoS Biol 7(5):e1000101. https://doi.org/10.1371/journal.pbio.1000101
13. van Boxtel AL, Chesebro JE, Heliot C, Ramel MC, Stone RK, Hill CS (2015) A temporal window for signal activation dictates the dimensions of a nodal signaling domain. Dev Cell 35(2):175–185. https://doi.org/10.1016/j.devcel.2015.09.014
14. Shimozono S, Iimura T, Kitaguchi T, Higashijima S, Miyawaki A (2013) Visualization of an endogenous retinoic acid gradient across embryonic development. Nature 496(7445):363–366. https://doi.org/10.1038/nature12037
15. Sosnik J, Zheng L, Rackauckas CV, Digman M, Gratton E, Nie Q, Schilling TF (2016) Noise modulation in retinoic acid signaling sharpens segmental boundaries of gene expression in the embryonic zebrafish hindbrain. Elife 5:e14034. https://doi.org/10.7554/eLife.14034
16. Shimmi O, Umulis D, Othmer H, O'Connor MB (2005) Facilitated transport of a Dpp/Scw heterodimer by Sog/Tsg leads to robust patterning of the Drosophila blastoderm embryo. Cell 120(6):873–886. https://doi.org/10.1016/j.cell.2005.02.009
17. Bollenbach T, Pantazis P, Kicheva A, Bokel C, Gonzalez-Gaitan M, Julicher F (2008) Precision of the Dpp gradient. Development 135(6):1137–1146. https://doi.org/10.1242/dev.012062
18. Gavin-Smyth J, Wang YC, Butler I, Ferguson EL (2013) A genetic network conferring canalization to a bistable patterning system in Drosophila. Curr Biol 23(22):2296–2302. https://doi.org/10.1016/j.cub.2013.09.055
19. Tucker JA, Mintzer KA, Mullins MC (2008) The BMP signaling gradient patterns dorsoventral tissues in a temporally progressive manner along the anteroposterior axis. Dev Cell 14(1):108–119. https://doi.org/10.1016/j.devcel.2007.11.004
20. Hashiguchi M, Mullins MC (2013) Anteroposterior and dorsoventral patterning are coordinated by an identical patterning clock. Development 140(9):1970–1980. https://doi.org/10.1242/dev.088104
21. Diaspro A, Federici F, Robello M (2002) Influence of refractive-index mismatch in high-resolution three-dimensional confocal microscopy. Appl Opt 41(4):685–690
22. Azaripour A, Lagerweij T, Scharfbillig C, Jadczak AE, Willershausen B, Van Noorden CJ (2016) A survey of clearing techniques for 3D imaging of tissues with special reference to connective tissue. Prog Histochem Cytochem 51(2):9–23. https://doi.org/10.1016/j.proghi.2016.04.001
23. Keller PJ, Schmidt AD, Wittbrodt J, Stelzer EH (2008) Reconstruction of zebrafish early

embryonic development by scanned light sheet microscopy. Science 322(5904):1065–1069. https://doi.org/10.1126/science.1162493

24. Myronenko A, Song X (2010) Point set registration: coherent point drift. IEEE Trans Pattern Anal Mach Intell 99(RapidPosts)
25. Myronenko A, Song X, Carreira-Perpinan MA (2007) Non-rigid point set registration: coherent point drift. Adv Neural Inf Proces Syst 19:1009–1016
26. Truett GE, Heeger P, Mynatt RL, Truett AA, Walker JA, Warman ML (2000) Preparation of PCR-quality mouse genomic DNA with hot sodium hydroxide and tris (HotSHOT). Biotechniques 29(1):52, 54
27. Smith SM, Maughan PJ (2015) SNP genotyping using KASPar assays. Methods Mol Biol 1245:243–256. https://doi.org/10.1007/978-1-4939-1966-6_18
28. Neff MM, Neff JD, Chory J, Pepper AE (1998) dCAPS, a simple technique for the genetic analysis of single nucleotide polymorphisms: experimental applications in *Arabidopsis thaliana* genetics. Plant J 14(3):387–392
29. Myronenko A, Song X (2010) Point set registration: coherent point drift. IEEE Trans Pattern Anal Mach Intell 32(12):2262–2275

Chapter 11

An Adult Zebrafish Model of Fibrodysplasia Ossificans Progressiva

Melissa LaBonty and Pamela C. Yelick

Abstract

Fibrodysplasia ossificans progressiva (FOP) is a rare human skeletal disease caused by constitutively activating mutations in the gene *ACVR1*, which encodes a type I BMP/TGFβ family member receptor. FOP is characterized by progressive heterotopic ossification (HO) of fibrous tissues, including skeletal muscle, tendons, and ligaments, as well as malformation of the big toes, vertebral fusions, and osteochondromas. Surgical interventions in patients often result in enhanced HO, which can exacerbate rather than improve diagnostic outcomes. As a result of these difficulties, a variety of animal models are needed to study human FOP. Here we describe the methods for creating and characterizing zebrafish conditionally expressing $Acvr1l^{Q204D}$, the first adult zebrafish model for FOP.

Key words Fibrodysplasia ossificans progressive, Disease model, Zebrafish, Acvr1l

1 Introduction

The TGFβ/BMP signaling pathways are involved in many cellular and developmental processes, from early embryonic axis formation, to left-right asymmetry, to organogenesis (Reviewed in [1]). Endochondral bone formation is one such process that is uniquely regulated by the BMP and TGFβ signaling pathways [2]. The type I TGFβ/BMP receptor family member ACVR1 is a key player in the BMP signaling that promotes endochondral ossification. Constitutively activating mutations in ACVR1 cause the rare human skeletal disease, fibrodysplasia ossificans progressiva (FOP) [3, 4]. Patients with FOP develop heterotopic ossification (HO) throughout skeletal muscle, tendons, and ligaments, and also display a number of other bone-related defects, including digit malformations, vertebral fusions, and osteochondromas [4, 5]. Overactive BMP signaling caused by constitutive activation of ACVR1 leads to the formation of endochondral bone in HO lesions in FOP [6–8].

In the decade since *ACVR1* was named as the causative gene for human FOP, researchers have developed a number of animal

Melissa B. Rogers (ed.), *Bone Morphogenetic Proteins: Methods and Protocols*, Methods in Molecular Biology, vol. 1891, https://doi.org/10.1007/978-1-4939-8904-1_11,

models to learn more about FOP disease etiology and progression [6, 8–13]. Each of these models has provided critical insight in the FOP field, including altered ligand affinity and signaling activation resulting from FOP-associated human ACVR1 mutations [9, 10, 12], the importance of injury and inflammation in promoting HO formation [6, 11], the identification of an abnormal mechanism of activation of ACVR1^{R206H} by the ligand activin A [8], and the identification of progenitor cell populations that contribute to HO formation in FOP patients [13]. While each of these studies has made significant contributions to our understanding of FOP, many questions remain.

The zebrafish model also offers numerous advantages to study FOP (reviewed in [14]). Their small size, rapid and transparent embryonic development, and high fecundity make them ideal for numerous large-scale screening methods, including in vivo drug and genetic modifier screening. In addition, a large number of transgenic reporter lines are readily available to perform cell lineage tracing experiments. Finally, zebrafish exhibit strong conservation of key signaling pathways that drive the development of human tissues and organ systems, an important consideration for accurately modeling FOP etiology and progression.

The functional ortholog of human *ACVR1* in zebrafish is called *acvr1l* or *alk8* [15]. Embryonic overexpression of constitutively active Acvr1l results in ventralization of the embryo due to upregulated BMP signaling, resulting in early embryonic lethality [16–18]. Therefore, to study roles for activated Acvr1l in adult FOP zebrafish models, we created heat-shock inducible expression constructs for *acvr1^{Q204D}*, which we used to express Acvr1l^{Q204D} after early embryonic patterning was completed. Here we describe the establishment of long-term heat-shock experiments in adult *Tg (hsp70l:acvr1l_Q204D-mCherry)* zebrafish as a method for modeling FOP. We also describe imaging and staining techniques used to analyze the development of FOP-like phenotypes in Acvr1^{Q204D}-expressing zebrafish.

2 Materials

For all methods, use deionized water unless otherwise noted.

2.1 Creating a Zebrafish Model for Human FOP

1. Zebrafish carrying *Tg(Bre:GFP)*, a reporter construct containing five tandem BMP response elements driving GFP expression [19].
2. Heat-shock-inducible Acvr1l^{Q204D}-expressing zebrafish: zebrafish carrying *Tg(Bre:*GFP*)* and *Tg(hsp70l:acvr1l_Q204D-mCherry)* were described in LaBonty et al. 2017 [20]. Acvr1l^{Q204D} is expressed in response to heat shock for 1 h at 38 °C.

2.2 Heat-Shock System

1. Pro-Heat IC 50 W (Won Brothers, Fredericksburg, VA, USA): a submersible heater-controller system recalibrated to heat to 38 °C (*see* **Note 1**).
2. Four-channel temperature logger with digital display (UX120-006 M, Onset Computer, Bourne, MA, USA).
3. Four air/water/soil temperature sensors with a one-foot long cable (TMC1-HD, Onset Computer, Bourne, MA, USA) attached to the four-channel temperature logger.
4. Power strip with GFCI plug (*see* **Note 2**).
5. Outlet timer (TN111, Intermatic, Spring Grove, IL, USA).

2.3 Fixation of Whole Zebrafish

1. 250 mg/L tricaine methanesulfonate (Sigma Aldrich) in water.
2. Modified Davidson's fixative (MDF): 22 mL 37% formaldehyde, 11.5 mL glacial acetic acid, 33 mL 95% ethanol, 33.5 mL water.
3. 10% formalin in phosphate-buffered saline (PBS).
4. PBS.

2.4 MicroCT Analyses

1. Fixed zebrafish stored in PBS at 4 °C.
2. Skyscan 1176 high-resolution microCT scanner (Bruker, Allentown, PA, USA).
3. Plastic wrap (important as does not interfere with imaging).
4. Plastic spoon.

2.5 Paraffin Embedding and Sectioning of Fixed Zebrafish

1. CalEx decalcifying solution (Sigma).
2. Graded ethanol series.
3. Graded tert-butyl alcohol series.
4. Xylene.
5. Shaker.
6. Paraplast Plus paraffin.
7. Disposable plastic base molds (*see* **Note 3**).
8. Embedding cassettes.
9. Microtome.
10. SuperFrost Plus charged glass microscope slides (Fisher Scientific).
11. Water bath.
12. Slide warmer.
13. Blunt scalpel.
14. Forceps.

2.6 Histological Reagents

1. Hematoxylin.
2. Dilute hydrochloric acid: 40 drops of concentrated (~37%) hydrochloric acid in 250 mL of water.
3. Ammonium water: 4 mL of ammonium hydroxide in 250 mL of water.
4. Eosin.
5. 1% Safranin O in water.
6. 0.02% Fast Green in 0.2% acetic acid.
7. Celestine Blue stock solution: Mix 25 g ferric ammonium sulfate and 2.5 g Celestine Blue (C7143, Sigma-Aldrich, St. Louis, MO, USA) in 250 mL water, stirring over gentle heat, until powder is fully dissolved. Cover with aluminum foil, boil for 3 min, cool, and filter. Add water to up to 500 mL, then add 70 mL of glycerol.
8. Celestine Blue working solution: add 50 mL of Celestine Blue stock solution to 450 mL of water. Make fresh every month.
9. Mayer's hematoxylin.
10. Acid alcohol: 0.5% glacial acetic acid in 80% ethanol.
11. Alcian Blue solution: Mix equal parts of filtered 1% Alcian Blue (05500, Sigma-Aldrich, St. Louis, MO, USA) in water and 1% glacial acetic acid in water. Make fresh every month.
12. 1% phosphomolybdic acid in water, filtered. Make fresh for every use.
13. 0.5% Direct Red (CI 28160, 195,251, Sigma-Aldrich, St. Louis, MO, USA) in water, filtered. Make fresh every month.
14. Permount mounting media.
15. Glass cover slips.

3 Methods

3.1 Heat-Shock Induction of Acvr1^{Q204D} in Adult Zebrafish

1. Cross adult zebrafish pairs of *Tg(Bre:*GFP*)* and *Tg(Bre:*GFP*); Tg(hsp70l:acvr1l_Q204D-mCherry)* lines and raise progeny to 14 days post fertilization (dpf).
2. While raising embryonic zebrafish, set up long-term automated heat-shock systems [20, 21]. Submerge calibrated heaters in system tanks filled with system water and plug into power strip. Connect power strip to outlet timer and set timer to turn on for 1 h a day (*see* **Note 4**). Install four-channel temperature logger near tanks with heaters and extend temperature sensor into each individual tank to allow for daily temperature monitoring.

3. When zebrafish reach 14 dpf, carefully transfer 10 *Tg(Bre:*GFP*)* or *Tg(Bre:*GFP*) or Tg(hsp70l:acvr1l_Q204D-mCherry)* zebrafish to their respective heat-shock system tanks (*see* **Note 5**). Establish a low flow rate into each heat-shock tank of 10–20 mL/min, which is needed to allow the heaters to reach and maintain optimal heat-shock temperature. After 24 h, check the logger to confirm that the temperature in each tank was successfully raised to 38 °C.
4. At desired collection time points, lethally anesthetize zebrafish using 250 mg/L tricaine methanesulfonate in water and then fix. Fixation and processing methods are described in further detail in Subheading 3.3 below.

3.2 MicroCT to Visualize FOP-Like Phenotypes

1. Euthanized, fixed adult zebrafish can be analyzed by microCT prior to processing for paraffin embedding and sectioning. The easiest way to get clear images of fixed zebrafish by microCT is to mount them in the machine as follows: tightly wrap a piece of plastic wrap around the machine's specimen base to create a flat surface for the fish to rest upon. Transfer the fish in PBS to the plastic wrap surface using a plastic spoon (*see* **Note 6**).
2. Image each fish using the following parameters: 200 μm aluminum filter, 9 μm resolution, 360 views at 0.5° increments, 2 frame averages. We have used these parameters with success on a Skyscan 1176 high-resolution microCT scanner.
3. Reconstruct 3D data using NRecon (Bruker, Allentown, PA, USA) and generate 3D volume renderings using CTVox (Bruker, Allentown, PA, USA). Keep all reconstruction and rendering settings constant between samples to allow for comparison.

3.3 Paraffin Embedding and Sectioning

1. MDF fixation and processing method [20] (*see* **Note 7**):
   (a) Fixation: Immerse animals in MDF for 2 days at room temperature, then rinse with PBS 3–4 times and store in PBS at 4 °C.
   (b) Processing: Immerse samples for 1 h each in graded ethanol (25% and 50% in PBS, 75% in water, 100%), followed by xylene.
2. Fixed and processed samples can be embedded in fresh molten paraffin in disposable plastic base molds, with an embedding cassette placed on top and filled with molten paraffin to serve as a secure adaptor for future sectioning. Allow samples to harden at room temperature.
3. Serially section each paraffin block at 7 μm using a microtome and collect paraffin ribbons in order of cutting. Carefully label rows of paraffin ribbons in numerical order to keep track of position within sample.

4. To mount sections, transfer a series of 2–3 paraffin sections using forceps or a blunt scalpel to the surface of a 48 °C water bath. Wick sections off of the surface of the water using Super-Frost Plus charged glass microscope slides. Let sections dry on slide for at least 10 min, move to a slide warmer at 45 °C for 1 h, and then turn the slide warmer up to 54 °C overnight.

3.4 Histological Analysis of FOP-Like Phenotypes

1. For all staining methods, begin with a standard protocol to deparaffinize and rehydrate the sections: immerse slides in xylene for 5 min twice, then for 2 min each in 100% ethanol, 95% ethanol, 75% ethanol, and 50% ethanol. Immerse in water for at least 2 min.
2. H&E staining: Stain with hematoxylin for 1 min, then rinse gently with tap water for 2.5 min. Dip once in dilute hydrochloric acid and three times in ammonia water. Immerse in water for 2 min, then transfer to eosin for 20 s. Proceed to dehydration and cover slipping (**step 5**) (*see* **Note 8**).
3. Safranin O staining: Stain with 0.02% Fast Green in 0.2% acetic acid for 15 s, then stain in 1% Safranin O in water for 30 s (*see* **Note 9**).
4. Hall's and Brunt's Quadruple (HBQ) staining [22]: stain in the following solutions, in order: Celestine Blue working solution for 30 s; Mayer's hematoxylin for 1 min; acid alcohol for 2 min; Alcian Blue solution for 5 min; phosphomolybdic acid for 5 min; and Direct Red for 4 min (*see* **Note 10**).
5. All stained samples should be dehydrated after staining is complete: 6 dips each in 95% ethanol #1, 95% ethanol #2, 100% ethanol #1, and 100% ethanol #2. Immerse slides in xylene for 5 min twice. Use Permount mounting media to secure a cover slip over top of the sections and dry overnight.

4 Notes

1. To recalibrate the Pro-Heat IC 50 W submersible heaters, unscrew the controller box to open and turn the capacitor between 90 and 180° clockwise. Next, manually calibrate each individual heater to the desired temperature by mounting the heater in a fish tank filled with system water, turning on the heater, and using a thermometer to monitor the point at which the heater reaches 38 °C. Mark this point on the heater using a permanent marker. This procedure is described in further detail with useful images by Duszynski et al. 2011 [21].
2. It is imperative to use power strips with GFCI plugs in aquatic facilities to reduce the risk of electric shock when in contact with water.

3. We recommend 37 × 24 × 10 mm molds for adult zebrafish ranging in size from 15 to 25 mm in length (27147-6, Ted Pella, Inc., Redding, CA, USA).
4. We recommend setting the heat-shock outlet timer to turn on during the regular facility light cycle so that proper heating can be monitored daily. We heat shock from 11 am to 12 pm daily to avoid interfering with our fish feeding schedule. Be sure to make adjustments as needed for annual Daylight Savings time changes.
5. In our preliminary experiments to establish our long-term heat-shock systems, we found that zebrafish that were heat-shocked starting at 14 dpf showed good survival to adulthood. Starting heat shock at 7 dpf resulted in a loss of at least 50% of animals by 14 dpf.
6. When performing microCT, we found that placing the zebrafish in a small pool of PBS helped to avoid any movements that could occur if the animal began to dry out while imaging. PBS introduced only minimal background noise during microCT imaging that was easily removed during post-processing.
7. We have used two different methods for fixation and processing of adult zebrafish and have identified certain pros and cons with each of these methods. When using the MDF method, we found that zebrafish sections stained well with common histological stains including Safranin O and HBQ, but that some tissue integrity could be lost due to chatter encountered during paraffin sectioning. When using the Webb lab method, we maintained good tissue integrity while sectioning, as needed for performing immunohistochemistry (IHC) or immunofluorescence (IF), but noticed that some histological stains did not stain as well as compared to the MDF method. Given these findings, we recommend testing each method on your samples first to determine the ideal conditions for you based on your desired outcomes. We have provided the Webb lab method here.

   Webb lab method [23]:

   (a) Fixation: Immerse animals in 10% formalin in PBS for 2 days at room temperature, then rinse with PBS 3–4 times, and store in PBS at 4 °C.

   (b) Processing: Decalcify samples in CalEx and then rinse in PBS for 2 h. Immerse samples for 1 h in 50% ethanol, overnight in 75% ethanol, and then in a graded ethanol/water/t-butyl alcohol series for 1 h each: (1) 20:20:10, (2) 25:7.5:17.5, (3) 22.5:0:27.5, (4) 12.5:0:37.5, and (5) 100% t-butyl alcohol twice. Transfer samples to molten Paraplast Plus paraffin overnight. t-Butyl alcohol freezes at 20 °C but also has a flash point of 38 °C. We recommend setting a water bath to 25 °C and storing the t-butyl alcohol in it throughout the processing steps to keep the alcohol in liquid form.

8. Hematoxylin provides a dark blue-purple nuclear stain, while eosin provides a bright pink cytoplasmic stain.
9. Safranin O stains cartilaginous proteoglycans a dark red color, while Fast Green provides a light green-blue cytoplasmic counter stain.
10. Celestine Blue and hematoxylin both create a dark blue nuclear stain, Alcian Blue stains acidic mucopolysaccharides found in mucus and cartilage a bright blue, and Direct Red stains the collagen in the bone and basement membranes a bright pink-red.

All experimental procedures on zebrafish were approved by the Tufts University Institutional Animal Care and Use Committee (IACUC) and Ethics Committee.

Acknowledgments

This work was supported by NIH/NIDCR R01DE018043 (PCY), R21AR065761 (PCY), and NSF GRFP NS9344 (ML).

References

1. Wu MY, Hill CS (2009) Tgf-beta superfamily signaling in embryonic development and homeostasis. Dev Cell 16:329–343. https://doi.org/10.1016/j.devcel.2009.02.012
2. Wu M, Chen G, Li Y-P (2016) TGF-β and BMP signaling in osteoblast, skeletal development, and bone formation, homeostasis and disease. Bone Res 4:16009. https://doi.org/10.1038/boneres.2016.9
3. Shore EM, Xu M, Feldman GJ et al (2006) A recurrent mutation in the BMP type I receptor ACVR1 causes inherited and sporadic fibrodysplasia ossificans progressiva. Nat Genet 38:525–527. https://doi.org/10.1038/ng1783
4. Kaplan FS, Xu M, Seemann P et al (2009) Classic and atypical fibrodysplasia ossificans progressiva (FOP) phenotypes are caused by mutations in the bone morphogenetic protein (BMP) type I receptor ACVR1. Hum Mutat 30:379–390. https://doi.org/10.1002/humu.20868
5. Pignolo RJ, Shore EM, Kaplan FS (2011) Fibrodysplasia ossificans progressiva: clinical and genetic aspects. Orphanet J Rare Dis 6:80. https://doi.org/10.1186/1750-1172-6-80
6. Chakkalakal SA, Zhang D, Culbert AL et al (2012) An Acvr1 R206H knock-in mouse has fibrodysplasia ossificans progressiva. J Bone Miner Res 27:1746–1756. https://doi.org/10.1002/jbmr.1637
7. Bagarova J, Vonner AJ, Armstrong KA et al (2013) Constitutively active ALK2 receptor mutants require type II receptor cooperation. Mol Cell Biol 33:2413–2424. https://doi.org/10.1128/MCB.01595-12
8. Hatsell SJ, Idone V, Wolken DMA et al (2015) ACVR1R206H receptor mutation causes fibrodysplasia ossificans progressiva by imparting responsiveness to activin A. Sci Transl Med 7:303ra137. https://doi.org/10.1126/scitranslmed.aac4358
9. Twombly V, Bangi E, Le V et al (2009) Functional analysis of saxophone, the Drosophila gene encoding the BMP type I receptor ortholog of human ALK1/ACVRL1 and ACVR1/ALK2. Genetics 183:563–579. 1SI–8SI. https://doi.org/10.1534/genetics.109.105585

10. Le VQ, Wharton KA (2012) Hyperactive BMP signaling induced by ALK2(R206H) requires type II receptor function in a Drosophila model for classic fibrodysplasia ossificans progressiva. Dev Dyn 241:200–214. https://doi.org/10.1002/dvdy.22779
11. Yu PB, Deng DY, Lai CS et al (2008) BMP type I receptor inhibition reduces heterotopic [corrected] ossification. Nat Med 14:1363–1369. https://doi.org/10.1038/nm.1888
12. Haupt J, Deichsel A, Stange K et al (2014) ACVR1 p.Q207E causes classic fibrodysplasia ossificans progressiva and is functionally distinct from the engineered constitutively active ACVR1 p.Q207D variant. Hum Mol Genet 23 (20):5364–5377. https://doi.org/10.1093/hmg/ddu255
13. Lees-Shepard JB, Yamamoto M, Biswas AA et al (2018) Activin-dependent signaling in fibro/adipogenic progenitors causes fibrodysplasia ossificans progressiva. Nat Commun 9:471. https://doi.org/10.1038/s41467-018-02872-2
14. LaBonty M, Yelick PC (2018) Animal models of fibrodysplasia ossificans progressiva. Dev Dyn 247:279–288. https://doi.org/10.1002/dvdy.24606
15. Yelick PC, Abduljabbar TS, Stashenko P (1998) zALK-8, a novel type I serine/threonine kinase receptor, is expressed throughout early zebrafish development. Dev Dyn 211:352–361. https://doi.org/10.1002/(SICI)1097-0177(199804)211:4<352::AID-AJA6>3.0.CO;2-G
16. Payne TL, Postlethwait JH, Yelick PC (2001) Functional characterization and genetic mapping of alk8. Mech Dev 100:275–289
17. Mintzer KA, Lee MA, Runke G et al (2001) Lost-a-fin encodes a type I BMP receptor, Alk8, acting maternally and zygotically in dorsoventral pattern formation. Development 128:859–869
18. Bauer H, Lele Z, Rauch GJ et al (2001) The type I serine/threonine kinase receptor Alk8/Lost-a-fin is required for Bmp2b/7 signal transduction during dorsoventral patterning of the zebrafish embryo. Development 128:849–858
19. Alexander C, Zuniga E, Blitz IL et al (2011) Combinatorial roles for BMPs and Endothelin 1 in patterning the dorsal-ventral axis of the craniofacial skeleton. Development 138:5135–5146. https://doi.org/10.1242/dev.067801
20. LaBonty M, Pray N, Yelick PC (2017) A zebrafish model of human fibrodysplasia ossificans progressiva. Zebrafish 14:293–304. https://doi.org/10.1089/zeb.2016.1398
21. Duszynski RJ, Topczewski J, LeClair EE (2011) Simple, economical heat-shock devices for zebrafish housing racks. Zebrafish 8:211–219. https://doi.org/10.1089/zeb.2011.0693
22. Hall BK (1986) The role of movement and tissue interactions in the development and growth of bone and secondary cartilage in the clavicle of the embryonic chick. J Embryol Exp Morphol 93:133–152
23. Webb JF, Shirey JE (2003) Postembryonic development of the cranial lateral line canals and neuromasts in zebrafish. Dev Dyn 228:370–385. https://doi.org/10.1002/dvdy.10385

Chapter 12

Generation and Identification of Genetically Modified Mice for BMP Receptors

Jingwen Yang and Yuji Mishina

Abstract

BMP signaling is critical in embryogenesis and in the development of numerous tissues. Many genetically modified (knockout and transgenic) mice have been established to study BMP function in development and disease. Mice with altered BMP receptor genes (including global knockout, conditional knockout, and conditional constitutively active transgenic mouse lines) have been particularly informative. In this chapter, we describe how the genetically modified mice were generated and introduce genotyping methods. These methods include regular PCR and genomic real-time PCR using specific primers based on different constructs in different mice strains.

Key words Transgenic, BMP receptors, Regular PCR, Genomic real-time PCR, Primers

1 Introduction

Bone morphogenetic proteins (BMPs) are members of the transforming growth factor-β (TGF-β) superfamily. Similar to TGF-β, BMPs signal through type I and type II transmembrane serine/threonine kinase receptors. In response to the binding of BMP ligands, type I and type II BMP receptors form a membrane-bound heterotetrameric complex. Then, the constitutively active type II receptor transphosphorylates the type I receptor at a glycine-serine rich motif (GS domain). Consequently, the Smad signal transducers are phosphorylated, and the downstream signal is propagated [1–3]. There are three type I BMP receptors (BMPR-1A or ALK3, BMPR-1B or ALK6, ACVR-1 or ALK2) and three type II BMP receptors (BMPR-2, ACVR-2A, ACVR-2B). Both type I and type II receptors are required to form a heterotetrameric complex for BMP signal transduction [1–3]. However, the mechanism of the heterotetrameric signaling complex formation can vary. For example, BMP-2 and BMP4 interact with type I receptors and recruit type II receptors, whereas BMP6 and BMP7 preferentially bind to type II receptors and recruit type I receptors [4]. More

Melissa B. Rogers (ed.), *Bone Morphogenetic Proteins: Methods and Protocols*, Methods in Molecular Biology, vol. 1891,
https://doi.org/10.1007/978-1-4939-8904-1_12,

importantly, the binding of a BMP ligand to preformed receptor complexes activate signaling pathways that differ from those activated by a receptor complex whose assembly was stimulated by BMP binding [5].

Genetic studies into the function of these complex receptors are essential for clarifying the role of BMP signaling in development and disease. As of now, several genetically modified BMP receptor mice have been established, including global knockout (KO), conditional knockout (cKO), and conditional constitutively active (ca) mice [6–17]. BMP receptor gene modifications were observed to result in embryonic lethality [18, 19] or abnormalities in many tissues, including the skeleton [20–22], craniofacial [23, 29], heart [24], vascular [25], lung [26], eye [27], and tooth [28], thus establishing that BMP signaling is critical for normal embryogenesis.

In this chapter, we first describe how the genetically modified mice were generated and then introduce genotyping methods. These methods include regular PCR and genomic real-time PCR using specific primers based on different constructs in different mice strains.

2 Materials

Unless otherwise noted, all solutions are prepared in water purified by double distillation or other methods.

2.1 Regular PCR

1. Lysis buffer: 100 mM NaCl, 1 mM Tris pH 7–8, 0.1 mM EDTA, 0.1% Triton X-100 in distilled water. Store at room temperature. Add 1/50 volume of 40 mg/mL proteinase K (0.8 mg/mL final concentration) immediately before use.
2. 40 mg/mL proteinase K.
3. PCR oil.
4. Primers for genotyping of each receptor mutations are listed in Table 1. The positions of the primers are marked in the maps of targeted alleles shown in Figs. 1 and 2.
5. 0.5 unit/μL Taq DNA polymerase.
6. 1× Taq buffer: 10 mM Tris-HCl, 50 mM KCl, and 1.5 mM MgCl2, pH 8.3.
7. 5 mM dNTP mix: 5 mM each of dATP, dCTP, dGTP, and dTTP.
8. 25 mM $MgCl_2$.
9. Thermo cycler.
10. 10× Tris/Borate/EDTA (TBE) buffer: 1 M Tris, 0.9 M boric acid, and 0.01 M EDTA. Dilute 100 mL 10× TBE buffer in

Table 1
Primers used for genotyping of different BMP receptors genetically modified mice

Ref	Modified gene	Primers	Primer ratios	PCR products	Annealing
Global knockout mutation					
Fig. 1a [6]	*Alk2* null	A2-5 5′-ATGCTAGACCTGGGCAGCCATA-3′ PGK-Mot 5′-CGTGTGTAAGGTGTAGGTG GCC-3′	A2-5:PGK-Mot: A2-3 = 2:1:1	195 bp (*Alk2* null)	65 °C, 40 cycles
		A2-5 5′-ATGCTAGACCTGGGCAGCCATA-3′ A2-3 5′-CATGCTAGCAGCTCGGAGAAAC-3′		371 bp (wild type)	
Fig. 1b [18]	*Alk3* null	fx3 5′-AGACTGCCTTGGGAAAAGCGC-3′ fx5 3′-GGACTATGGACACACAATGGC-3′	fx3:fx5:fx0 = 1:2:1	190 bp (*Alk3* null)	65 °C, 40 cycles
		fx5 5′-GGACTATGGACACACAATGGC-3′ fx0 5′-CTCTGAATTTCTAGTCCACATCTGC-3′		280 bp (wild type)	
Fig. 1c [7]	*Alk6* null	Alk6-I 5′-TGGTGAGTGGTTACAACAAGATC AGCA-3′ Neo-Al2 5′-GAAAGAACCAGCTGGGGCTC GAG-3′	Neo-Al2: Alk6-I: Alk6-E = 1:2:1	300 bp (*Alk6* null)	65 °C, 40 cycles
		Alk6-I 5′-TGGTGAGTGGTTACAACAAGATC AGCA-3′ Alk6-E 5′-CTCGGCCCAAGATCCTACGTTG-3′		350 bp (wild type)	
Fig. 1d [14]	*Bmpr2* null	A 5′-GCTAAAGCGCATGCTCCAGACTGCC TT-3′ C 5′-AGGTTGGCCTGGAACCTGAGGAAATC-3′	Primer A:B:C = 2:1:1	260 bp (*Bmpr2* null)	
		A 5′-GCTAAAGCGCATGCTCCAGACTGCC TTG-3′ B 5′-TCACAGCATGAACATGATGGAGGCGG-3′		200 bp (wild type)	

(continued)

Table 1 (continued)

	Ref	Modified gene	Primers	Primer ratios	PCR products	Annealing
Fig. 1e	[16, 32]	*Acvr2a* null	A 5′-TGGGAAGACAATAGCAGGCATGC-3′ B 5′-GCAGAGTGTGACCCGTACCCAC-3′ C 5′-GTTGGTACCCGGAGGTATATGGC-3′ D 5′-CCCTTACCATCTGCAGCAGTGCA-3′	1:1:1:1	900 bp (*Acvr2A* null) 140 bp (wild type)	70 °C, 30 cycles
		Acvr2b null	A 5′-ATGAACTGCAGGACGAGGCAGCG-3′ B 5′-GGCGATAGAAGGCGATGCGCTG-3′ C 5′-CCGACAGCCCCCACCCTGCTCA-3′ D 5′-GGCCCACCAGAGGGGATGGGGG-3′	1:1:1:1	600 bp (*Acvr2B* null) 241 bp (wild type)	
Conditional knockout mutation						
Fig. 2a	[29]	*Alk2* flox	Alk217F 5′-CCCCCATTGAAGGTTTAGAG AGAC-3′ Alk217R2 3′-CTAAGAGCCATGACAGA GGTTG-3′	Alk217F: lk217R2 = 1:1	160 + 90 bp (flox) (*see* **Note 4**), 250 bp (wild type)	65 °C, 40 cycles
		Alk2-recombined	Alk2RecF 5′-GAATTGCTAGAAGCCCATA GGC-3′ Alk2WTR 5′-TGAGATTGTTCTAGCACTGC CC-3′ Alk217F 5′-CCCCCATTGAAGGTTTAGAGA GAC-3′ Alk2WTR 5′-TGAGATTGTTCTAGCACTGC CC-3′	Alk2RecF:Alk2WTR: Alk217F = 1:2:1	625 bp (recombined) 530 bp (wild type)	60 °C, 40 cycles
Fig. 2b	[8]	*Alk3* flox	fx2 5′-GCAGCTGCTGCTGCAGCCTCC-3′ fx4 3′-TGGCTACAATTTGTCTCATGC-3′	fx2:fx4 = 1:1	230 bp (flox) 150 bp (wild type)	65 °C, 40 cycles
Fig. 2b		*Alk3*-recombined (*see* **Note 6**)	fx1 5′-GGTTTGGATCTTAACCTTAGG-3′ fx4 5′-TGGCTACAATTTGTCTCATGC-3′ BMP2-A 5′-AGCATGAACCCTCATGTGTT GG-3′ BMP2-B 5′-GTGACATTAGGCTGCT GTAGCA-3′	fx1:fx4:BMP2-A: BMP2-B = 1:1:1:1	180 bp (recombined) 322 bp (wild type)	55 °C, 40 cycles

Fig. 2c	[15]	*Bmpr2* flox	5F 50-GGCAGACTCTGACTTTGACGCTAG-3′ C 5′-TTATTGTAAGTACACTGTTGCTGTC-3′ A 5′-CACACCAGCCTTATACTCTAG ATAC-3′ 6R 5′-CACATATCTGTTATGAAACTTGAG-3′ A 5′-CACACCAGCCTTATACTCTAG ATAC-3′ C 5′-TTATTGTAAGTACACTGTTGCTGTC-3′		315 bp (2loxP and 3loxP) 260 bp (wild type) 350 bp (2loxP) 270 bp (wild type) 500 bp (1loxP)	60 °C, 40 cycles
Conditional constitutively active transgenic line						
Before recombination						
Fig. 2d	[12]	*caAlk*	TF41 5′-GTGCTGGTTATTGTGCTGTCTC-3′ TF61 5′-GACGACAGTATCGGCCTCAGGAA-3′	TF41:TF61:LbnFR1:LbnRev3 = 1:1:1:1	580 bp (*caAlk* transgene)	65 °C, 40 cycles
			LbnFR1 5′-GAGGACGCAGTCCAGTACCT-3′ LbnRev3 5′-TAGCCTCTGCCTCACGCCCTGC-3′		334 bp (Internal control)	
After recombination						
Fig. 2d	[11]	*caAlk2*	TF41 (common) 5′-GTGCTGGTTATTGTGCTGTCTC-3′ TF9 (Alk2) 5′-CGAACACTACAGAGAGAATAATG-3′	TF41:TF9 = 1:1	750 bp (*caAlk2* transgene)	65 °C, 40 cycles
	[12]	*caAlk3*	TF41 (common) 5′-GTGCTGGTTATTGTGCTGTCTC-3′ TF17 (Alk3) 5′-CGGCGTAGCTGGGCTTTTGGA-3′	TF41:TF17 = 1:1	300 bp (*caAlk3* transgene)	
		caAlk6	TF41 (common) 5′-GTGCTGGTTATTGTGCTGTCTC-3′ TF25 (Alk6) 5′-GACATCCAGAGGTGACAACAG-3′	TF41:TF25 = 1:1	300 bp (*caAlk6* transgene)	

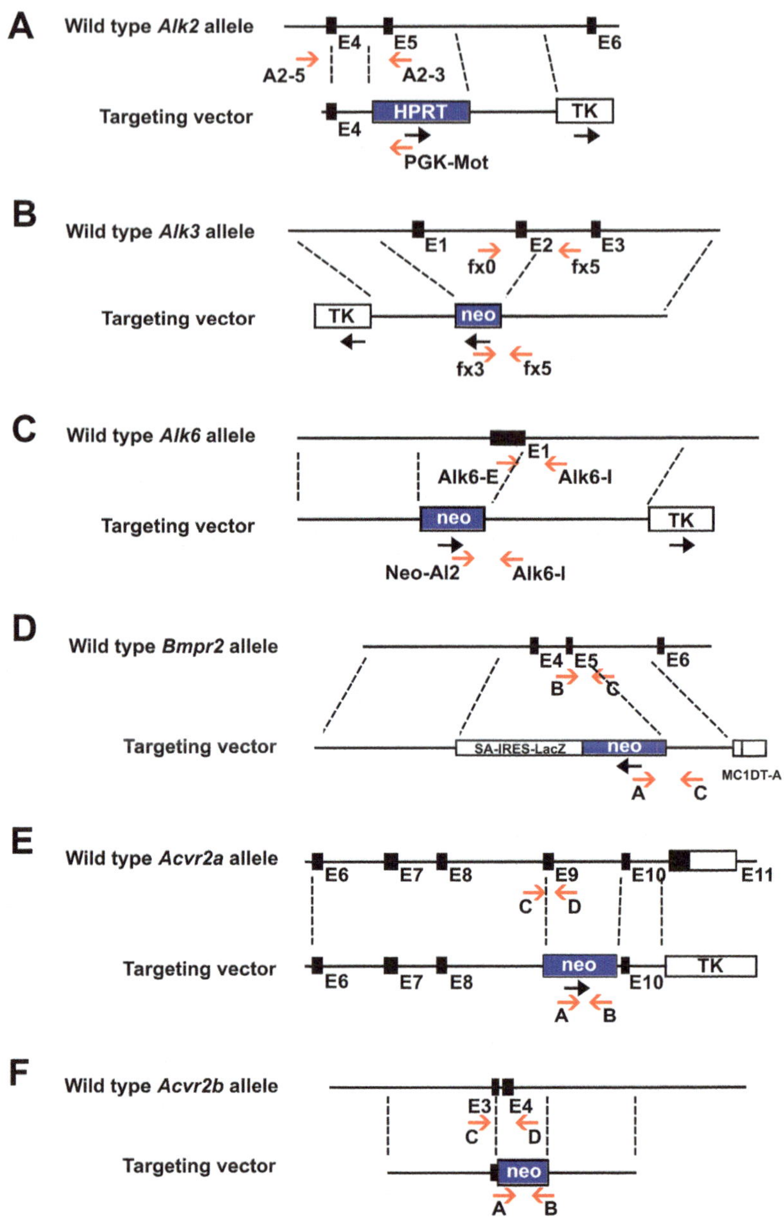

Fig. 1 Structure representation of each BMP receptor KO mutation and PCR genotyping strategy. Schematic diagrams showing the wild-type locus of the *Alk2* (**a**), *Alk3* (**b**), *Alk6* (**c**), *Bmpr2* (**d**), *Acvr2a* (**e**), and *Acvr2b* (**f**) gene, and the targeting vector for each gene. The positions (red arrows) of PCR primers for genotyping are indicated below the locus

900 mL water to make 1000 mL 1× TBE for agarose gel electrophoresis.

11. 3% agarose gel prepared in 1× TBE.
12. 10 mg/mL ethidium bromide stock.

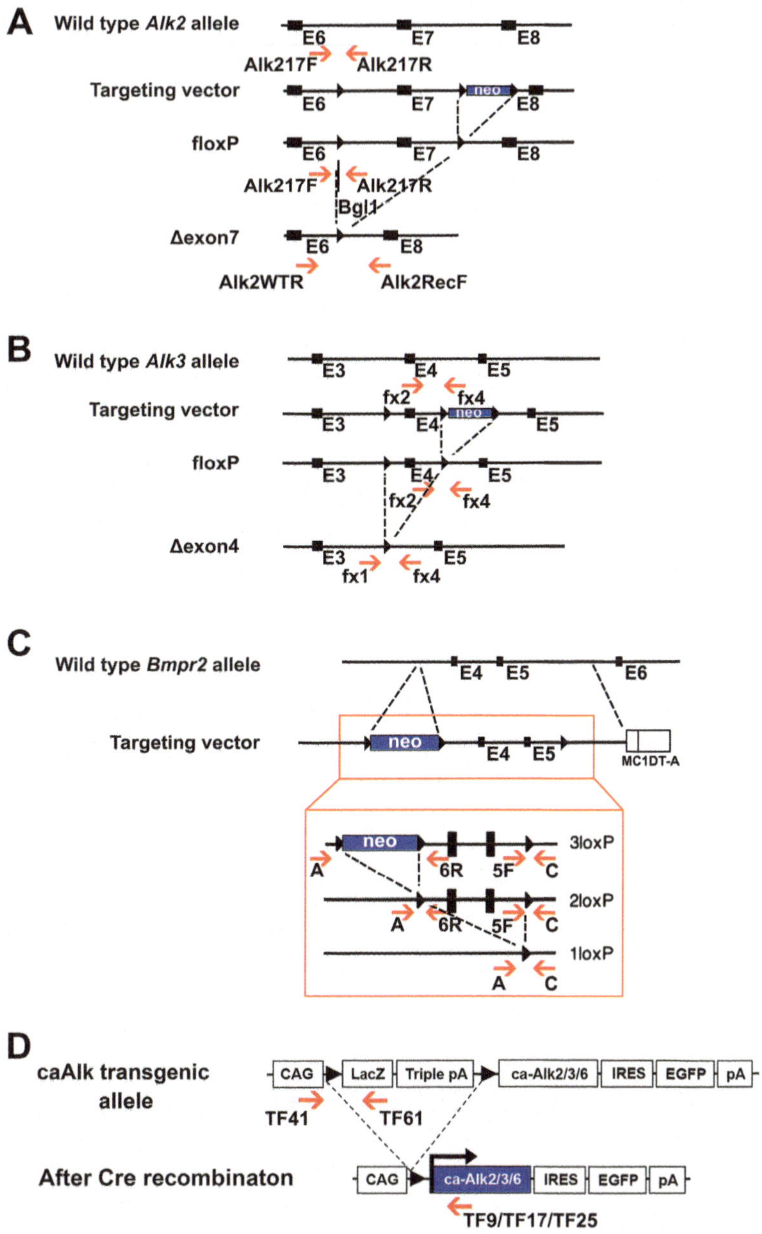

Fig. 2 Structure representation of each BMP receptor cKO mutation and ca transgene and PCR genotyping strategy. (**a**–**c**) Schematic diagrams of the wild-type *Alk2* (**a**), *Alk3* (**b**), and *Bmpr2* (**c**) locus, targeting vector, and mutant alleles after recombination. The positions of primers for genotyping by PCR are indicated by red arrows. (**d**) Schematic representation of the ca transgene of *Alk2*, *Alk3*, or *Alk6*. The primers used for genotyping are shown by red arrows below the locus

Table 2
TaqMan primers used for genomic real-time PCR of different BMP receptors genetically modified mice

Ref	Modified gene	Primers
Conditional knockout mutation		
[31]	*Alk2* (exon 7) (*see* **Note 5**)	AIKAL5S_F 5′-CTCACTACTCTGGATACGGTTAGCT –3′, AIKAL5S_R 5′-GGGTCCCAAATATCTCTATGTGCAA-3′, AIKAL5S_M FAM 5′-CTATGGACAGTACAATCCG-3′
[30, 31]	*Alk3* (exon 4) (*see* **Note 5**)	AI89LJ8_F 5′-GACCAGAAGAAGCCAGAAAATGGA-3′, AI89LJ8_R 5′-TGTCCTGAGCAATAGCACTTTAAGAA-3′, AI89LJ8_M FAM 5′-CCTCTGGTGCTA AAGTC-3′
Conditional constitutively active transgenic line		
	Egfp (*see* **Note 5**)	5′-GAGCGCACCATCTTCTTCAAG-3′, 5′-TGTCGCCCTCGAACTTCAC-3′, FAM 5′-ACGACGGCAACTACA-3′

2.2 Genomic Real-Time PCR

1. TaqMan® Universal PCR Master Mix (Thermo Fisher, Cat: 4334437).
2. TaqMan primer sets of each receptor mutations are listed in Table 2.
3. Optical 96-well reaction plates compatible with your PCR machine.
4. Optical adhesive film.
5. Real-time PCR system.

3 Methods

3.1 Regular PCR

1. Collect small piece (less than 1 mm^3) of tissues from the ear (ear notch), tail, or any organs. For embryos, the yolk sac or amniotic membrane may be used (do not use the placenta for genotyping). Place tissues into 96-format PCR tubes (do not cap).
2. Add 50 μL lysis buffer in each tube, overlaid with PCR oil.
3. Incubate at 55 °C for 6 h or more, then incubate at 85 °C for 30 min to inactivate proteinase K.
4. Take 4 μL DNA solution to mix with 76 μL of water to dilute samples, then use 4 μL of the diluted samples to set up 10 μL PCR reaction. Reaction mixture will be made as follows:

10× Taq buffer	1.0 μL
5 mM dNTP mix	1.0 μL
0.5 unit/μL Taq DNA polymerase	1.5 μL

(continued)

25 mM $MgCl_2$	1.5 μL
50 μM 5′ primer (Table 1)	0.1 μL
50 μM 3′ primer (Table 1)	0.1 μL
Water	0.8 μL
Template DNA (20× diluted)	4.0 μL

The conditions for thermal cycling are as follows (*see* **Note 1**):

Initial denaturation for 94 °C for 5 min followed by 30 to 40 cycles of denaturation at 94 °C for 30 s, annealing at 50–70 °C for 30 s, and extension at 72 °C for 1 min (30–-40 cycles), then ending with 72 °C for 5–10 min followed by a cool down.

The number of cycles and annealing temperatures for different primer sets is shown in Table 1.

5. Run 3% agarose gel at 250–300 V, stain the gel with ethidium bromide, and photograph.

3.2 Genomic Real-Time PCR for In Vivo Deletion Efficiency

The in vivo deletion efficiency of conditional knockout by Cre-recombinase and other recombinases can vary. This protocol describes the quantification of *Alk2* or *Alk3* deletion in conditional knockout mice by genomic real-time PCR using custom-designed primer set [30, 31]. 5′ primers and 3′ primers are designed to amply the flox regions (exon 7 for *Alk2*, exon 4 for *Alk3*). FAM-labeled probes are designed within the PCR amplicons for detection (*see* Table 2).

1. Extract genomic DNA as in of Subheading 3.1, **steps 1–3**.
2. Mix the Gene Expression Master Mix thoroughly by swirling the bottle. Thaw Alk2 or Alk3 frozen primer set (Table 2) and templates DNA on ice. When thawed, vortex and then centrifuge the tubes briefly (*see* **Note 2**).
3. Prepare the PCR reaction mix (20 μL reactions):

TaqMan PCR Master Mix (2×)	10 μL
Primer set (20×)	1 μL
Template DNA (20× diluted)	5 μL
Water	4 μL

Perform three replicates of each reaction. Then vortex the tubes briefly to mix the solutions, centrifuge the tubes briefly to spin down the contents, and eliminate any air bubbles from the solutions.

4. Transfer 20 μL of each reaction mixture to each well of an optical plate.
5. Cover the plate with an optical adhesive film. Centrifuge the plate briefly to spin down the contents and eliminate air bubbles from the solutions.
6. Run the plate on a real-time PCR instrument using the following thermal cycling conditions: initial incubation at 50 °C for 2 min then denaturation at 95 °C for 10 min followed by 40 cycles of denaturation at 95 °C for 15 s and annealing at 60 °C for 1 min, then ending with a cool down.
7. Determine the threshold cycles (CT) for the amplification curves. Use the comparative CT method to analyze target gene levels normalized to *Gapdh* level as described by the manufacturer of the instrument.

3.3 Genomic Real-Time PCR to Determine Homozygosity vs. Heterozygosity of Conditional Constitutively Active Transgenes

To generate conditional constitutively active transgenic mice, the EGFP cassette was inserted into the constructs (Fig. 2d) [10]. Therefore, the copy number of target genes can be quantified based on the copy number of *Egfp* via real-time PCR. This is a common strategy for the caALK2, caALK3, and caAlk6 mouse lines. The *Egfp* primer set is shown in Table 2.

1. Same as Subheading 3.2, **steps 1** and **2**.
2. Prepare the PCR reaction mix (20 μL reactions):

TaqMan PCR Master Mix (2×)	10 μL
Egfp primer set (20×)	1 μL
Template DNA (20× diluted)	5 μL
Water	4 μL

Perform three replicates of each reaction. Then vortex the tubes briefly to mix the solutions, centrifuge the tubes briefly to spin down the contents, and eliminate any air bubbles from the solutions. Wild-type, *caAlk* het, and *caAlk* homo samples, which genotypes are known, should be used as controls.

3. Same as Subheading 3.2, **steps 4–7** (*see* **Note 3**).

4 Notes

1. Samples for which genotypes are known should be used as controls. For these primers and most others, these conditions work adequately. If not, try the following: (1) change the dilution of template DNA. The reaction will not work when the DNA concentration is too high; (2) optimize the annealing

temperature; (3) change Taq DNA polymerase to Taq hot start DNA polymerase; or (4) purify DNA further.

2. Protect all reagents from light in the freezer until you are ready to use them. Excessive exposure to light may affect the fluorescent probes.
3. The calculated copy number of the *Egfp* in *caAlk* homozygous mice should be about twice as that of heterozygous mice. Homozygosity also can be genetically confirmed by crossing the mice to wild-type mice. All F1 pups should carry the transgene.
4. PCR amplification generates a 250 bp from both flox and wild-type alleles. Bgl1 digestion, which uniquely digests the flox gene product into 160 bp and 90 bp fragments, can be used to distinguish flox and wild-type PCR products.
5. All three primers are pre-mixed into one tube by the manufacturer.
6. The primers BMP2-A and BMP2-B are for an internal control at the *Bmp2* locus. Please refer reference [33] for the specific positions of those primers.

Acknowledgments

We thank Drs. Ce Shi and Taocong Jin for designing probe sets for quantitative PCR. We are grateful to Dr. Kaitrin Kramer for critical reading of this manuscript. This work was supported by the National Institutes of Health (R01DE020843 to Y.M.), International FOP Association (Y.M.), the grant-in-aid from the National Natural Science Foundation of China (31500788 to J.Y.), and the Fundamental Research Fund for the Central Universities of China (410500114 to J.Y.).

References

1. Grafe I, Alexander S, Peterson JR, et al (2017) TGF-β family signaling in mesenchymal differentiation. Cold Spring Harb Perspect Biol 10 (5) https://doi.org/10.1101/cshperspect.a022202
2. Yadin D, Knaus P, Mueller TD (2016) Structural insights into BMP receptors: specificity, activation and inhibition. Cytokine Growth Factor Rev 27:13–34. https://doi.org/10.1016/j.cytogfr.2015.11.005
3. Ehrlich M (2016) Endocytosis and trafficking of BMP receptors: regulatory mechanisms for fine-tuning the signaling response in different cellular contexts. Cytokine Growth Factor Rev 27:35–42. https://doi.org/10.1016/j.cytogfr.2015.12.008
4. De Caestecker M (2004) The transforming growth factor-beta superfamily of receptors. Cytokine Growth Factor Rev 15(1):1–11
5. Nohe A, Hassel S, Ehrlich M et al (2002) The mode of bone morphogenetic protein (BMP) receptor oligomerization determines different BMP-2 signaling pathways. J Biol Chem 277 (7):5330–5338
6. Mishina Y, Crombie R, Bradley A et al (1999) Multiple roles for activin-like kinase-2 signaling during mouse embryogenesis. Dev Biol 213 (2):314–326

7. Yi SE, Daluiski A, Pederson R et al (2000) The type I BMP receptor BMPRIB is required for chondrogenesis in the mouse limb. Development 127(3):621–630
8. Mishina Y, Hanks MC, Miura S et al (2002) Generation of Bmpr/Alk3 conditional knockout mice. Genesis 32(2):69–72
9. Peterson JR, Eboda O, Agarwal S et al (2014) Targeting of ALK2, a receptor for bone morphogenetic proteins, using the Cre/lox system to enhance osseous regeneration by adipose-derived stem cells. Stem Cells Transl Med 3 (11):1375–1380. https://doi.org/10.5966/sctm.2014-0082
10. Yoon BS, Ovchinnikov DA, Yoshii I et al (2005) Bmpr1a and Bmpr1b have overlapping functions and are essential for chondrogenesis in vivo. Proc Natl Acad Sci U S A 102 (14):5062–5067
11. Fukuda T, Scott G, Komatsu Y et al (2006) Generation of a mouse with conditionally activated signaling through the BMP receptor, ALK2. Genesis 44(4):159–167
12. Komatsu Y, Yu PB, Kamiya N et al (2013) Augmentation of Smad-dependent BMP signaling in neural crest cells causes craniosynostosis in mice. J Bone Miner Res 28 (6):1422–1433. https://doi.org/10.1002/jbmr.1857
13. Yu X, Espinoza-Lewis RA, Sun C et al (2010) Overexpression of constitutively active BMP-receptor-IB in mouse skin causes an ichthyosis-vulgaris-like disease. Cell Tissue Res 342 (3):401–410. https://doi.org/10.1007/s00441-010-1077-2
14. Beppu H, Kawabata M, Hamamoto T et al (2000) BMP type II receptor is required for gastrulation and early development of mouse embryos. Dev Biol 221(1):249–258
15. Beppu H, Lei H, Bloch KD et al (2005) Generation of a floxed allele of the mouse BMP type II receptor gene. Genesis 41(3):133–137
16. Song J, Oh SP, Schrewe H et al (1999) The type II activin receptors are essential for egg cylinder growth, gastrulation, and rostral head development in mice. Dev Biol 213 (1):157–169
17. Mayeur C, Leyton PA, Kolodziej SA et al (2014) BMP type II receptors have redundant roles in the regulation of hepatic hepcidin gene expression and iron metabolism. Blood 124 (13):2116–2123. https://doi.org/10.1182/blood-2014-04-572644
18. Mishina Y, Suzuki A, Ueno N et al (1995) Bmpr encodes a type I bone morphogenetic protein receptor that is essential for gastrulation during mouse embryogenesis. Genes Dev 9(24):3027–3037
19. Park C, Lavine K, Mishina Y et al (2006) Bone morphogenetic protein receptor 1A signaling is dispensable for hematopoietic development but essential for vessel and atrioventricular endocardial cushion formation. Development 133(17):3473–3484
20. Salazar VS, Gamer LW, Rosen V (2016) BMP signalling in skeletal development, disease and repair. Nat Rev Endocrinol 12(4):203–221. https://doi.org/10.1038/nrendo.2016.12
21. Wu M, Chen G, Li YP (2016) TGF-beta and BMP signaling in osteoblast, skeletal development, and bone formation, homeostasis and disease. Bone Res 4:16009. https://doi.org/10.1038/boneres.2016.9.eCollection
22. Jing J, Hinton RJ, Feng JQ (2015) Bmpr1a signaling in cartilage development and endochondral bone formation. Vitam Horm 99:273–291. https://doi.org/10.1016/bs.vh.2015.06.001
23. Graf D, Malik Z, Hayano S et al (2016) Common mechanisms in development and disease: BMP signaling in craniofacial development. Cytokine Growth Factor Rev 27:129–139. https://doi.org/10.1016/j.cytogfr.2015.11.004
24. Gavrilov S, Lacy E (2013) Genetic dissection of ventral folding morphogenesis in mouse: embryonic visceral endoderm-supplied BMP2 positions head and heart. Curr Opin Genet Dev 23(4):461–469. https://doi.org/10.1016/j.gde.2013.04.001
25. García de Vinuesa A, Abdelilah-Seyfried S, Knaus P et al (2016) BMP signaling in vascular biology and dysfunction. Cytokine Growth Factor Rev 27:65–79. https://doi.org/10.1016/j.cytogfr.2015.12.005
26. Eblaghie MC, Reedy M, Oliver T et al (2006) Evidence that autocrine signaling through Bmpr1a regulates the proliferation, survival and morphogenetic behavior of distal lung epithelial cells. Dev Biol 291(1):67–82
27. Zhao Q, Zhao JY, Wu D et al (2012) Mutually inductive interactions between the lens and retina require ALK3 functions during mouse embryonic development. Int J Ophthalmol 5 (2):119–124. https://doi.org/10.3980/j.issn.2222-3959.2012.02.01
28. Yang Z, Hai B, Qin L et al (2013) Cessation of epithelial Bmp signaling switches the differentiation of crown epithelia to the root lineage in a beta-catenin-dependent manner. Mol Cell Biol 33(23):4732–4744. https://doi.org/10.1128/MCB.00456-13

29. Dudas M, Sridurongrit S, Nagy A et al (2004) Craniofacial defects in mice lacking BMP type I receptor Alk2 in neural crest cells. Mech Dev 121(2):173–182
30. Shi C, Zhang H, Louie K et al (2017) BMP signaling mediated by BMPR1A in osteoclasts negatively regulates osteoblast mineralization through suppression of Cx43. J Cell Biochem 118(3):605–614. https://doi.org/10.1002/jcb.25746
31. Pan H, Zhang H, Abraham P et al (2017) BmpR1A is a major type 1 BMP receptor for BMP-Smad signaling during skull development. Dev Biol 429(1):260–270. https://doi.org/10.1016/j.ydbio.2017.06.020
32. Oh SP, Li E (1997) The signaling pathway mediated by the type IIB activin receptor controls axial patterning and lateral asymmetry in the mouse. Genes Dev 11(14):1812–1826
33. Singh AP, Castranio T, Scott G et al (2008) Influences of reduced expression of maternal bone morphogenetic protein 2 on mouse embryonic development. Sex Dev 2(3):134–141. https://doi.org/10.1159/000143431

Chapter 13

Phenotypic Analyses of Genetically Modified Mice for BMP Receptors

Jingwen Yang and Yuji Mishina

Abstract

BMPs play important roles in the development, disease, and regeneration of many tissues. Genetically modified mice with altered BMP receptor genes are particularly informative for clarifying the role of BMP signaling. In this chapter, we introduce several selected protocols for in vivo functional characterization of BMP receptors in genetically modified mice, including immunohistochemistry of BMP downstream signaling (P-Smad1/5/9 or others), histological analysis, whole-mount skeletal staining for cartilage and bone tissues, and whole-mount cartilage staining.

Key words Transgenic, BMP receptors, Immunohistochemistry, Histological analysis, Whole-mount skeletal staining, Whole-mount cartilage staining

1 Introduction

Bone morphogenetic proteins (BMPs) are multifunctional paracrine growth factors that belong to the transforming growth factor β (TGFβ) superfamily [1–3]. BMPs play important roles in the development, disease, and regeneration of many tissues, including craniofacial [4], heart [5], vascular [6], lung [7], eye [8], and tooth [9], especially cartilage [10–12] and bone [13–17]. BMPs signal through the interaction with BMP-specific type I (BMPR-1A or ALK3, BMPR-1B or ALK6, ACVR-1 or ALK2) and type II (BMPR-2, ACVR-2A, ACVR-2B) serine/threonine kinase receptors and then start the signaling cascade via both Smad-dependent (canonical) and Smad-independent (noncanonical) signaling pathways. In the Smad-dependent signaling pathway, BMPs activate receptor-specific Smads, Smad1, Smad5, and Smad9 (P-Smad1/5/9), at a C-terminal SSXS motif. Then P-Smad1/5/9 forms a complex with Smad4, and this complex translocates to the nucleus where it functions as a transcription factor with coactivators and corepressors to regulate gene expression [1–3].

Melissa B. Rogers (ed.), *Bone Morphogenetic Proteins: Methods and Protocols*, Methods in Molecular Biology, vol. 1891, https://doi.org/10.1007/978-1-4939-8904-1_13, © Springer Science+Business Media, LLC, part of Springer Nature 2019

Both type I and type II receptors are required to form a heterotetrameric complex for BMP signal transduction [1–3]. However, the mechanism of the heterotetrameric signaling complex formation can vary [18, 19]. Therefore, different receptors may have similar or totally different functions in transducing BMP signals. Genetically modified mice with altered BMP receptor genes are particularly informative for clarifying the role of BMP signaling in development, disease, and regeneration. During embryo development, mice lacking *Alk2*, *Alk3*, or *Bmpr2* show defects in embryogenesis at the gastrulation stage [20–22]. In cartilage tissues, mice lacking *Alk3* and *Alk6* show achondroplasia in vivo [10, 11], while nasal cartilage morphogenesis is inhibited in mice with conditional constitutively active *Alk3* (*caAlk3*) in the neural crest [12]. In bone, mice lacking *Alk6* show decreased bone mass [13], whereas loss-of-function of *Alk2* in osteoblasts increases bone mass [14]. A mice model with constitutively active *Alk2* (*caAlk2*) that requires signaling complexes with type II receptors leads to ectopic bone formation [15]. During tissue regeneration, *caAlk2* in hair follicle stem cells attenuates cutaneous wound healing [23], and *caAlk2*-expressing adipose-derived stem cells significantly enhance bone regeneration [16]. The phenotypic analyses of such complex responses in mice with genetically modified BMP receptors require numerous methods.

In this chapter, we introduce several selected protocols that we use for phenotypic analysis of BMP receptors in genetically modified mice, including immunohistochemistry of BMP downstream signaling (P-Smad1/5/9, or others), histological analysis, whole-mount skeletal staining for cartilage and bone tissues, and whole-mount cartilage staining. We previously provide the detailed methods for in situ *hybridization* (both whole mount and section); thus, for people who are interested in gene expression studies, please refer to the previous chapter [24]. In this chapter, we will focus on immunohistochemistry and other phenotypic analyses.

2 Materials

Unless otherwise noted, all solutions are prepared in water purified by double distillation or other methods.

2.1 Immunohistochemical Examination of P-Smad1/5/9

1. 4% paraformaldehyde fixative.
2. 30% sucrose: 30 g sucrose dissolved PBS (phosphate-buffered saline, 8.0 g NaCl, 0.2 g KCl, 1.42 g Na_2HPO_4, 1.8 g KH_2PO_4 dissolved in water to 1000 mL, pH 7.4.), to a final volume of 100 mL.
3. Cryopreservative solution (OCT compound).
4. Cryostat.

5. Small artist's paint brushes.
6. Positively charged microscope slides.
7. 0.1% PBST (0.1% Triton-X 100 in PBS): dissolve 1 mL Triton-X 100 in PBS to make 1000 mL solution.
8. Citrate buffer: 10 mM sodium citrate, adjusted to pH 6.
9. Blocking buffer: 500 μL normal donkey serum or 500 μg bovine serum albumin diluted in 0.1% PBST to a final volume of 10 mL.
10. Primary antibody: Rabbit anti-P-Smad1/5/9 antibody.
11. Alexa fluor 488-anti-Rabbit secondary antibody.
12. Coverslips.
13. Antifade Mountant with DAPI.
14. Fluorescence microscope.

2.2 Phenotypic Examination of Mice with Genetically Modified BMP Receptors

2.2.1 Hematoxylin and Eosin Staining

1. Ethanol (*see* **Note 1**).
2. 7 g/L hematoxylin (Sigma, HHS32-1L).
3. 1% acid alcohol solution: hydrochloric acid 1 mL, add 70% ethanol to 100 mL.
4. 0.2% ammonia water solution (bluing): ammonium hydroxide (concentrated) 2 mL, add water to 1000 mL.
5. 0.1% (w/v) eosin (Sigma, HT110316-500ML).
6. Xylene.
7. Permount mounting medium.
8. Light microscope.

2.2.2 Safranin O Staining

1. Weigert's iron hematoxylin solution: stock solution A (1 g hematoxylin, 100 mL 95% ethanol), stock solution B (4 mL 29% ferric chloride in water, 95 mL water, 1 mL hydrochloric acid). Mix equal parts of stock solution A and B to make Weigert's iron hematoxylin working solution (*see* **Note 2**).
2. 0.02% fast green solution: 0.2 g fast green dissolved in water to make a final volume of 1000 mL.
3. 1% acetic acid solution: 1 mL glacial acetic acid, add water to 100 mL.
4. 0.1% safranin O solution: 0.1 g safranin O dissolved in water to a final volume of 100 mL.

2.3 Whole-Mount Skeletal Staining

1. 95% ethanol.
2. 0.015% Alcian Blue solution: 0.015 mg/mL Alcian Blue 8GX, 20% acetic acid, 80% ethanol (*see* **Note 3**).
3. 2% KOH (use plastic container).

4. 0.015% alizarin red S solution: 0.015 mg/mL alizarin red sodium sulfate in 1% KOH (use plastic container).
5. 1% KOH, 20% glycerol: 10 g KOH, 200 mL glycerol, add water to 1000 mL (use plastic container).
6. 47.5% ethanol, 50% glycerol: 475 mL 100% ethanol (or 500 mL 95% ethanol), 500 mL glycerol, add water to 1000 mL.
7. Filter paper.
8. Stereo microscope.

2.4 Whole-Mount Cartilage Staining

1. Bouin's solution: 300 mL saturated picric acid, 100 mL formaldehyde, and 20 mL glacial acetic acid. Stable for 1 year.
2. 0.1% NH_4OH: 1 mL 30% NH_4OH (open in the fume hood) in 210 mL of 100% ethanol, add water to 300 mL.
3. 5% acetic acid: Take 10 mL of 100% acetic acid (open in the fume hood) and add water to 200 mL.
4. 0.05% Alcian Blue solution: 25 mg Alcian Blue 8 GX dissolved in 5% acetic acid.
5. Methanol.
6. BABB solution: 100 mL benzyl alcohol mixed with 200 mL benzyl benzoate (use glass container and handle in fume hood).

3 Methods

3.1 Immunohistochemical Examination of P-Smad1/5/9

3.1.1 Dissection

For the analysis of whole embryos, euthanize the pregnant mother quickly by CO_2 inhalation followed by induction of bilateral pneumothorax or other IACUC approved methods of euthanasia. Remove whole uterus with embryos, and wash the embryos with cold PBS. Separate individual embryos, remove all membranes covering them, wash briefly in cold PBS, and transfer to fixative. Remove the skin from embryos older than 17.5 days before fixation.

3.1.2 Fixation, Embedding, and Sectioning

1. Fix tissue in 4% paraformaldehyde at 4 °C with gentle shaking for 4 h (*see* **Notes 4** and **5**).
2. Rinse tissue in PBS at 4 °C with gentle shaking for 12 h.
3. Cryoprotection with sucrose: transfer the samples into 30% sucrose at 4 °C with gentle shaking until the samples sink to the bottom.
4. Embedding: transfer the samples into an embedding mold filled with OCT. Use forceps to adjust the location and direction of the samples. The trimmed side of the samples must be

facing to the bottom of the embedding mold. Place the mold on dry ice until the OCT solidifies.

5. Cut 10 μm thick sections and place on microscope slides.
6. Air-dry the sections. Proceed to staining or store at −80 °C (*see* **Note 6**).

3.1.3 Immunohistochemistry of P-Smad1/5/9

1. Soak slides in 0.1% PBST, change solution every 5 min for three times (3 × 5 min).
2. Antigen retrieval (optional): Preheat steamer or water bath with staining dish containing citrate buffer until temperature reaches 95–100 °C. Immerse slides in the staining dish. Place the lid loosely on the staining dish and incubate for 10 min. Turn off steamer or water bath and remove the staining dish to room temperature and allow the slides to cool for at least 20 min (*see* **Note** 7) [25].
3. Rinse sections in 0.1% PBST, 3 × 3 min.
4. Apply 200 μL blocking solution to each slide and incubate at room temperature for 30 min (no rinse) (*see* **Note 8**).
5. Apply 200 μL primary antibody (P-Smad1/5/9, 1:100 diluted in blocking buffer) to each slide, and incubate slides at 4 °C overnight in a moisturized sealed container.
6. Rinse sections in PBS, 3 × 3 min.
7. Apply 200 μL secondary antibody (1:200 diluted in PBS) to each slide, and incubate slides at room temperature for 1 h (protect from light from this step forward).
8. Rinse sections in PBS 3 × 3 min.
9. Apply 1–2 drops mountant with DAPI to each slide, and cover the slide with coverslip.
10. Keep slides in cold room before taking pictures under a fluorescent microscope (Fig. 1a, b).

3.2 Phenotypic Examination of Mice with Genetically Modified BMP Receptors

*3.2.1 Hematoxylin and Eosin Staining (See **Note 9**)*

1. Place slides in a slide rack and wash under running tap water for 2 min.
2. Stain in 7 g/L hematoxylin for 2 min. Wash under running tap water for 2 min.
3. Place in acid alcohol bath for 10 seconds (s) to remove nonspecific blue staining. Wash under running tap water for 2 min.
4. Place in 0.2% ammonia water solution for 20 s to develop the remaining specific blue color. Wash under running tap water for 2 min.
5. Place in 0.1% eosin for 3 s, then immediately wash under running tap water for 20 s.

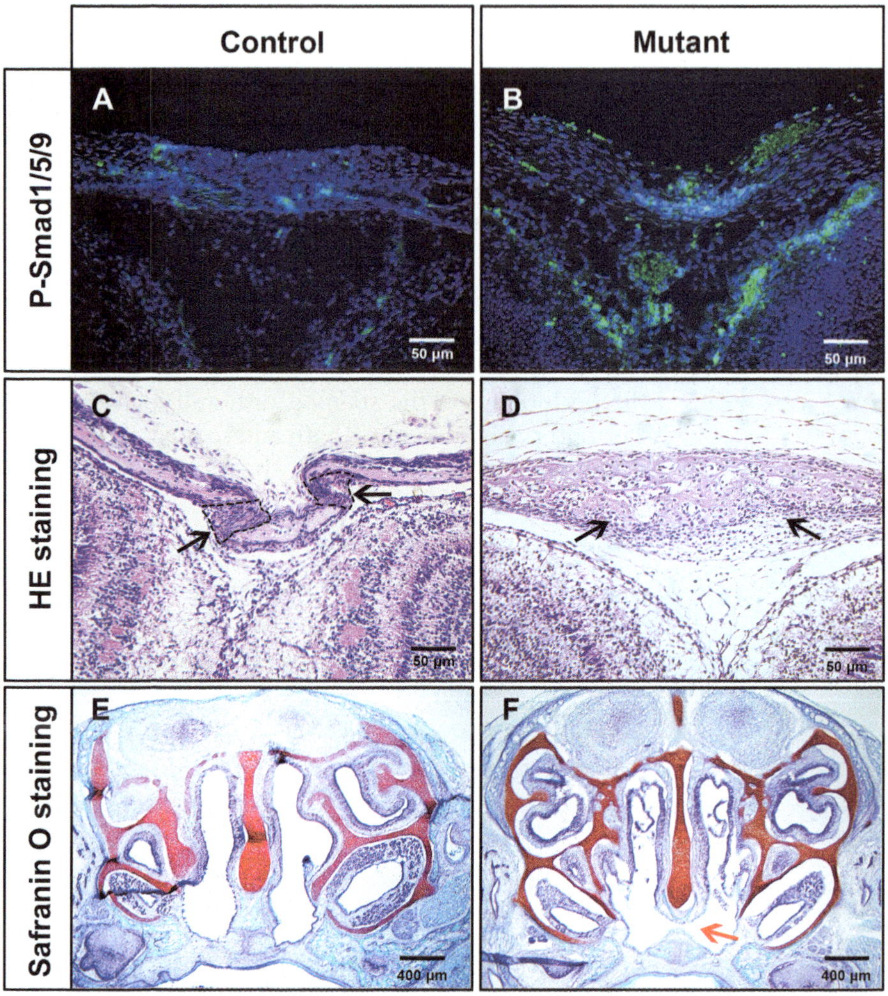

Fig. 1 Example of increased BMP signaling and embryonic phenotype in mice with a Cre-recombinase activated receptor gene. Conditional constitutively active *Alk3* (*caAlk3*) mice were crossed with *P0-Cre* mice (Mutant) to increase BMP signaling in neural crest cells (NCC) and NCC-derived cells. Control embryos had only the *P0-Cre* transgene. (**a**, **b**) P-Smad1/5/9 staining pattern in the skull of control (**a**), or mutant (**b**) embryos. (**c**, **d**) Hematoxylin and eosin staining (Subheading 3.2.1) of the sections from control (**c**) and mutant (**d**) heads, showing the disappearance of the suture (black arrows) in the mutant skull. (**e**, **f**) Safranin O staining (Subheading 3.2.2) of the sections from control (**e**) and mutant (**f**) nasal cavities, showing the nasal cartilage (red) and a failure of fusion between the nasal septum and the secondary palate in the mutant nasal cavity (**f**, red arrow)

6. Dehydrate in sequential 70%, 80%, 95%, 100%, and 100% ethanol solutions for 20 s each.
7. Place in xylene for 5 min. Repeat twice.
8. Mount each slide with Permount and a coverslip before taking pictures under a light microscope. The method will stain nuclei in blue, while the acidophilic cytoplasm in red, basophilic cytoplasm in purple, and erythrocytes in cherry red (Fig. 1c, d).

*3.2.2 Safranin O Staining (See **Note 9**)*

1. Hydrate frozen sections with water.
2. Stain in Weigert's iron hematoxylin for 5 min.
3. Rinse in water for 3 min.
4. Dip in 1% acid alcohol to destain hematoxylin.
5. Rinse in water for 3 min.
6. Stain in 0.02% fast green for 1 min (do not rinse).
7. Stain in 1% acetic acid for 10 s (do not rinse).
8. Stain in 0.1% safranin O for 20 min (do not rinse).
9. Rinse in 95% ethanol and then in 100% ethanol to dehydrate.
10. Rinse in xylene, then mount coverslip with Permount before taking pictures under a light microscope. This method will stain nuclei in black, cytoplasm in bluish green, and cartilage, mucin, mast cell granules in orange to red (Fig. 1e, f).

*3.2.3 Whole-Mount Skeleton Staining (See **Note 10**)*

This protocol covers the staining of the pre- and postnatal mouse skeleton using Alcian Blue and alizarin red to identify cartilage and bone, respectively.

1. Remove skin and internal organs. If possible, remove big muscles from shoulders, back, and limbs.
2. Fix remaining carcass overnight in 95% ethanol.
3. Stain in 0.015% Alcian Blue solution overnight for cartilage.
4. Put back in 95% ethanol for at least 3 h.
5. Treat with 2% KOH for 24 h or longer for adult mice (*see* **Note 11**).
6. Stain overnight in 0.015% alizarin red S solution for the bone.
7. Clear skeleton in 1% KOH, 20% glycerol solution for 2 days, or more (*see* **Note 12**).
8. Store the skeleton in a 1:1 mixture of glycerol and 95% ethanol before taking picture under a stereomicroscope. This method will stain bone tissue in red and cartilage tissue in blue (Fig. 2).

*3.2.4 Whole-Mount Cartilage Staining (See **Note 10**)*

1. Dissect embryos in cold PBS and remove extraembryonic membranes and skin (*see* **Note 13**).
2. Fix in Bouin's solution for 2 h or overnight.
3. Rinse with solution 1% NH_4OH diluted in 70% ethanol over a course of 24 h until embryos appear white. The solution may need to be changed six to eight times.
4. Equilibrate with 5% acetic acid for 1 h. Repeat twice.
5. Stain with 0.05% Alcian Blue solution for 2 h.
6. Rinse twice with 5% acetic acid, 1 h for each.

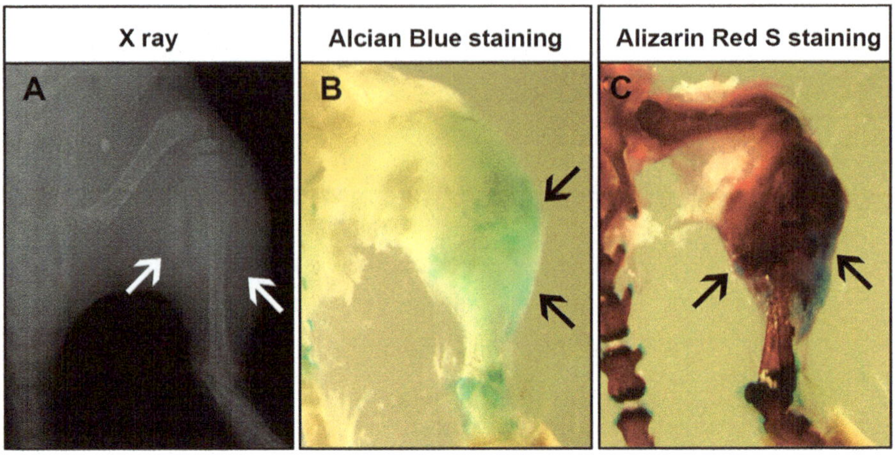

Fig. 2 Example of whole-mount skeletal staining of heterotopic ossification (HO). HO was induced by injection of Adenovirus Cre with cardiotoxin to the hind limb muscle of mice bearing a conditional constitutively active *Alk2* (ca*Alk2*) gene. (**a**) X-ray taken 10 days after induction, showing the increased density of tissue (white arrows) around the tibia. (**b**) After removing the skin, the limb was stained with Alcian Blue as described in Subheading 3.2.3. The black arrows show the ectopic cartilage tissue in the muscle. (**c**) The same tissue was further treated with KOH and stained with alizarin red as described in Subheading 3.2.3 to reveal heterotopic ossification in the muscle (black arrows)

7. Dehydrate the samples by incubating in 100% methanol for 1 h. Repeat twice.
8. Clear the samples in BABB solution before taking picture under a stereomicroscope (*see* **Note 14**). This method will stain cartilage in blue color (Fig. 3a, b).

4 Notes

1. Molecular sieve can be used to remove trace amount of water to maintain 100% purity of ethanol. This greatly improves penetrancy of xylene.
2. Weigert's iron hematoxylin working solution is stable for about 4 weeks.
3. Alcian Blue staining solution should be used within 3 weeks.
4. The duration for fixation should differ according to the size of the samples. Longer fixation is needed for larger samples. However, longer fixation may mask antigenic sites, which will increase false-negative results of immunohistochemistry signals. If this occurred, antigen retrieval before blocking might be required.
5. Here we describe cryosectioning followed by immunohistochemical detection of P-Smad1/5/9, which does not require antigen retrieval. Histological sections prepared after paraffin

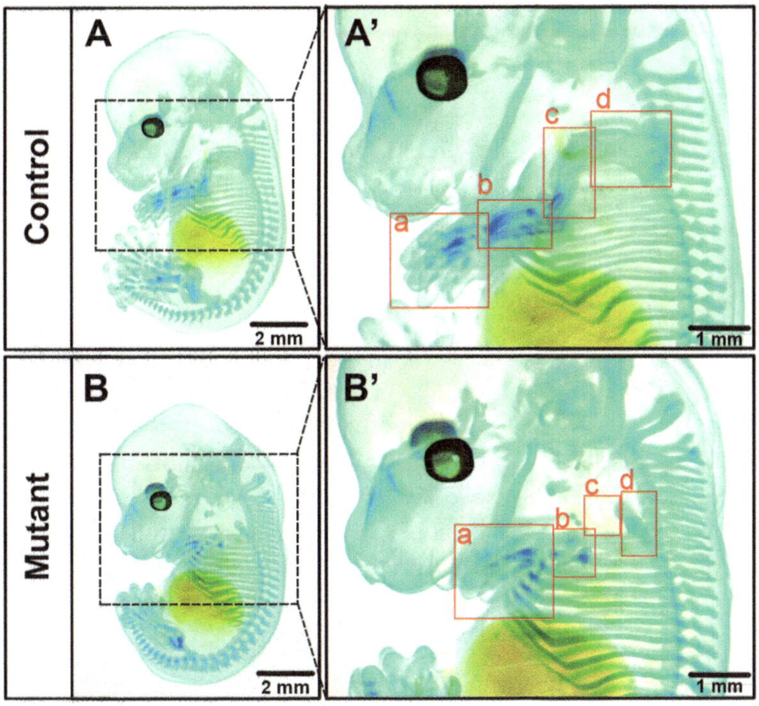

Fig. 3 Example of whole-mount cartilage staining of embryos with a Cre-recombinase activated receptor gene. Conditional constitutively active *Alk2* (*caAlk2*) mice were crossed with *Prx1-Cre* mice (Mutant) to increase BMP signaling in early limb bud mesenchyme cells (LBMCs) and LBMC-derived cells. Control embryos had only the *caAlk2* transgene. Embryos were stained with Alcian Blue and cleared in BABB solution as described in Subheading 3.2.4. Despite minimal differences in the autopod (red boxes (a) in A' and B'), mutant embryos showed major defects in the zeugopod (b), stylopod (c), and scapula (d) cartilage of the forelimb compared with control embryos

embedding also work for P-Smad1/5/9 detection; however, antigen retrieval is required.

6. A lower temperature maintains better morphology and antigenicity of sections.
7. Enzymes (e.g., trypsin, pepsin) and heat can be used for antigen retrieval. The concentration and treatment times of enzymatic retrieval must be tested and optimized to avoid damaging section. For the heat-induced antigen retrieval methods, the most commonly used buffers are 10 mM sodium citrate pH 6, Tris-EDTA pH 9, or EDTA pH 8. Heat-induced antigen retrieval is most often performed using a pressure cooker, microwave, steamer, or water bath. Unless the antigen retrieval method is stated on the antibody data sheet, the optimal method for each antibody/antigen combination must be determined.
8. Perform avidin/biotin blocking with 3% H_2O_2 if necessary.

9. Hematoxylin and eosin staining and Safranin O staining work well for both paraffin and cryosections.
10. Both fixatives for whole-mount skeletal (ethanol) and cartilage (Bouin's solution) work for Alcian Blue staining to reveal cartilage. The differences between those two methods are that (1) skeletal staining uses ethanol for fixation, allowing a sequential staining with alizarin red after Alcian Blue, while cartilage staining uses Bouin's solution for fixation, which is not compatible for alizarin red, (2) and samples for skeletal staining are cleared in KOH, while those for cartilage staining become transparent in BABB solution.
11. Do not move samples since the junction of bone parts becomes very fragile. Carefully remove the 2% KOH solution. If the tissues become very soft, removing the solution completely is not necessary.
12. If muscles are not cleared after 2 days, incubate in 1% KOH until they are cleared, or use forceps to remove muscles or other soft tissues.
13. In older embryos (E17 or older), removing the skin may be necessary.
14. Cartilage staining may last several years: however, photo documentation should be done as soon as possible. Use glass container when taking pictures, since plastic container cannot tolerate organic solvents.

Acknowledgments

We thank Dr. Ken-ichi Yamamura for *P0-Cre* mice. We also thank Drs. Haichun Pan and Satoru Hayano for sharing their results. We are grateful to Dr. Kaitrin Kramer for critical reading of this manuscript. This work was supported by the National Institutes of Health (R01DE020843 to Y.M.), International FOP Association (Y.M.), the grant-in-aid from the National Natural Science Foundation of China (31500788 to J.Y.), and the Fundamental Research Fund for the Central Universities of China (410500114 to J.Y.).

References

1. Grafe I, Alexander S, Peterson JR, et al (2017) TGF-β family signaling in mesenchymal differentiation. Cold Spring Harb Perspect Biol. https://doi.org/10.1101/cshperspect.a022202
2. Yadin D, Knaus P, Mueller TD (2016) Structural insights into BMP receptors: Specificity, activation and inhibition. Cytokine Growth Factor Rev 27:13–34. https://doi.org/10.1016/j.cytogfr.2015.11.005
3. Ehrlich M (2016) Endocytosis and trafficking of BMP receptors: Regulatory mechanisms for fine-tuning the signaling response in different cellular contexts. Cytokine Growth Factor Rev 27:35–42. https://doi.org/10.1016/j.cytogfr.2015.12.008

4. Graf D, Malik Z, Hayano S et al (2016) Common mechanisms in development and disease: BMP signaling in craniofacial development. Cytokine Growth Factor Rev 27:129–139. https://doi.org/10.1016/j.cytogfr.2015.11.004
5. Gavrilov S, Lacy E (2013) Genetic dissection of ventral folding morphogenesis in mouse: embryonic visceral endoderm-supplied BMP2 positions head and heart. Curr Opin Genet Dev 23(4):461–469. https://doi.org/10.1016/j.gde.2013.04.001
6. García de Vinuesa A, Abdelilah-Seyfried S, Knaus P et al (2016) BMP signaling in vascular biology and dysfunction. Cytokine Growth Factor Rev 27:65–79. https://doi.org/10.1016/j.cytogfr.2015.12.005
7. Eblaghie MC, Reedy M, Oliver T et al (2006) Evidence that autocrine signaling through Bmpr1a regulates the proliferation, survival and morphogenetic behavior of distal lung epithelial cells. Dev Biol 291(1):67–82
8. Zhao Q, Zhao JY, Wu D et al (2012) Mutually inductive interactions between the lens and retina require ALK3 functions during mouse embryonic development. Int J Ophthalmol 5 (2):119–124. https://doi.org/10.3980/j.issn.2222-3959.2012.02.01
9. Yang Z, Hai B, Qin L et al (2013) Cessation of epithelial Bmp signaling switches the differentiation of crown epithelia to the root lineage in a beta-catenin-dependent manner. Mol Cell Biol 33(23):4732–4744. https://doi.org/10.1128/MCB.00456-13
10. Yi SE, Daluiski A, Pederson R et al (2000) The type I BMP receptor BMPRIB is required for chondrogenesis in the mouse limb. Development 127(3):621–630
11. Yoon BS, Ovchinnikov DA, Yoshii I et al (2005) Bmpr1a and Bmpr1b have overlapping functions and are essential for chondrogenesis in vivo. Proc Natl Acad Sci U S A 102 (14):5062–5067
12. Hayano S, Komatsu Y, Pan H et al (2015) Augmented BMP signaling in the neural crest inhibits nasal cartilage morphogenesis by inducing p53-mediated apoptosis. Development 142(7):1357–1367. https://doi.org/10.1242/dev.118802
13. Shi C, Iura A, Terajima M et al (2016) Deletion of BMP receptor type IB decreased bone mass in association with compromised osteoblastic differentiation of bone marrow mesenchymal progenitors. Sci Rep 6:24256. https://doi.org/10.1038/srep24256
14. Kamiya N, Kaartinen VM, Mishina Y (2011) Loss-of-function of ACVR1 in osteoblasts increases bone mass and activates canonical Wnt signaling through suppression of Wnt inhibitors SOST and DKK1. Biochem Biophys Res Commun 414(2):326–330. https://doi.org/10.1016/j.bbrc.2011.09.060
15. Bagarova J, Vonner AJ, Armstrong KA et al (2013) Constitutively active ALK2 receptor mutants require type II receptor cooperation. Mol Cell Biol 33(12):2413–2424. https://doi.org/10.1128/MCB.01595-12
16. Peterson JR, Eboda O, Agarwal S et al (2014) Targeting of ALK2, a receptor for bone morphogenetic proteins, using the Cre/lox system to enhance osseous regeneration by adipose-derived stem cells. Stem Cells Transl Med 3 (11):1375–1380. https://doi.org/10.5966/sctm.2014-0082
17. Salazar VS, Gamer LW, Rosen V (2016) BMP signalling in skeletal development, disease and repair. Nat Rev Endocrinol 12(4):203–221. https://doi.org/10.1038/nrendo.2016.12
18. De Caestecker M (2004) The transforming growth factor-beta superfamily of receptors. Cytokine Growth Factor Rev 15(1):1–11
19. Nohe A, Hassel S, Ehrlich M et al (2002) The mode of bone morphogenetic protein (BMP) receptor oligomerization determines different BMP-2 signaling pathways. J Biol Chem 277 (7):5330–5338
20. Mishina Y, Suzuki A, Ueno N et al (1995) Bmpr encodes a type I bone morphogenetic protein receptor that is essential for gastrulation during mouse embryogenesis. Genes Dev 9(24):3027–3037
21. Mishina Y, Crombie R, Bradley A et al (1999) Multiple roles for activin-like kinase-2 signaling during mouse embryogenesis. Dev Biol 213 (2):314–326
22. Beppu H, Kawabata M, Hamamoto T et al (2000) BMP type II receptor is required for gastrulation and early development of mouse embryos. Dev Biol 221(1):249–258
23. Sorkin M, Agarwal S, Ranganathan K (2017) Hair follicle specific ACVR1/ALK2 critically affects skin morphogenesis and attenuates wound healing. Wound Repair Regen 25 (3):521–525. https://doi.org/10.1111/wrr.12549
24. Komatsu Y, Kishigami S, Mishina Y (2014) In situ hybridization methods for mouse whole mounts and tissue sections with and without additional β-galactosidase staining. Methods Mol Biol 1092:1–15. https://doi.org/10.1007/978-1-60327-292-6_1
25. Shi SR, Cote RJ, Taylor CR (2001) Antigen retrieval techniques: current perspectives. J Histochem Cytochem 49(8):931–937

Chapter 14

Immunofluorescent Visualization of BMP Signaling Activation on Paraffin-Embedded Tissue Sections

Maaike Alkema, Marie-José Goumans, and Boudewijn P. T. Kruithof

Abstract

Immunohistochemistry allows the detection of the presence, localization, and activation of proteins in biological tissues by using the ability of antibodies to bind to specific antigens. Cellular signaling can be visualized using antibodies raised against phosphorylated proteins. Phosphorylated Smad1, Smad5, and Smad9 are the activated signaling molecules of the BMP pathway that transfer BMP signals from the cell surface to the nucleus. Here we describe the detection of phospho-Smad1/5/9 on paraformaldehyde-fixed and paraffin-embedded tissue sections by immunofluorescence.

Key words BMP signaling, Immunohistochemistry, Phospho-Smad1, Phospho-Smad5, Phospho-Smad8/9, Tissue sections

1 Introduction

Smad1, Smad5, and Smad9 (also known as Smad8) are the intracellular signaling molecules of the BMP signaling pathway that, upon binding of BMP to its receptor, become phosphorylated and translocated to the nucleus to regulate gene expression [1]. Visualization of the cells expressing the phosphorylated form of these Smad proteins (pSmad) provides a means to determine the activation of the BMP signaling pathway at a cellular level within a tissue. Obtaining a highly specific and sensitive labelling of pSmads on tissue sections that allows discrimination of varying levels of activation can be challenging. Here, we demonstrate a method for optimal visualization of differential expression levels of the pSmad1, pSmad5, and pSmad9 proteins with the commonly used antibodies from Cell Signaling (#9511/#13820). Co-labeling with other signaling molecules (e.g., pSmad2) and specific cell markers allows comparisons with activation of other signaling pathways (e.g., the closely related TGF-β pathway) in specific cell types.

Melissa B. Rogers (ed.), *Bone Morphogenetic Proteins: Methods and Protocols*, Methods in Molecular Biology, vol. 1891, https://doi.org/10.1007/978-1-4939-8904-1_14,

2 Materials

Unless otherwise noted, all solutions are prepared in Milli-Q purified water.

2.1 Fixation

1. 4% paraformaldehyde (PFA) solution in phosphate-buffered saline (PBS). To prepare 1 L, dissolve 40 g of paraformaldehyde in water by magnetic stirring on heating plate (60 °C) (*see* **Note 1**). Add 15 drops of 1 M NaOH to clear solution. Cool solution to room temperature. Add 100 mL of 10× PBS and adjust the volume to 1 L with water. Filter the PFA solution through a 0.2 μm filter (*see* **Note 2**) and adjust to pH 7 if necessary. Aliquot the PFA solution and store at −20 °C.
2. 10× PBS solution. To prepare 1 L dissolve 80 g of NaCl, 2 g of KCl, 11.4 g of Na_2HPO_4, and 2 g of KH_2PO_4 in water.
3. 15 mL round bottom tube with cap.
4. 15 mL xylene-resistant tube with cap.
5. Roller apparatus.

2.2 Tissue Processing, Embedding, and Sectioning

1. Alcohol series for tissue processing: 50, 70, 90, 95, and 100% ethanol. For manual tissue processing, an amount of at least 10× the volume of the tissue is required. Automatic tissue processors can require up to 3 L.
2. Xylene. For manual tissue processing, an amount of at least 10× the volume of the tissue is required. Automatic tissue processors can require up to 3 L.
3. Paraffin.
4. Disposable plastic tissue embedding cassettes.
5. Stainless steel base molds.
6. Microtome blade (*see* **Note 3**).
7. Microtome.
8. Water bath.
9. Slide warmer.
10. Coated microscope slides (*see* **Note 4**).
11. Embedding station with cold plate.

2.3 Immunohistochemistry

1. Slide warmer.
2. Slide holder (polyethylene-terephthalate (PET)).
3. Slide staining set (12 dishes; PET).
4. Alcohol series: 100, 90, 70, and 50% ethanol in water for deparaffinization steps.

5. Xylene (*see* **Note 5**).
6. Liquid blocker, pen.
7. Humidified chamber, for example, a plastic slide box.
8. Antigen Unmasking Solution:10 mM Tris (pH 9), 1 mM EDTA, 0.05% Tween (*see* **Note 6**).
9. Pressure cooker.
10. Heating plate (1800 W).
11. Wash buffer: PBS/0.1% Tween-20.
12. Blocking solution: 1% bovine serum albumin (BSA) in wash buffer.
13. 3% H_2O_2 solution in water.
14. Rabbit anti Phospho-Smad1 (Ser463/465)/ Smad5 (Ser463/465)/Smad9 (Ser465/467). Antibody from Cell Signaling (#9511/#13820).
15. Horseradish peroxidase conjugated anti-rabbit antibody.
16. Tyramide Signal Amplification® biotin detection kit from Pelkin Elmer (NEL700A001KT).
17. Alexa fluor 488-conjugated streptavidin from ThermoFisher (#S32354).
18. ProLong Gold antifade reagent with DAPI.
19. Cover glasses.

    Optional:

20. Rabbit Anti-Phospho-Smad2 (Ser465/467) antibody from Cell Signaling (#3101/#3108).
21. Goat Anti-PECAM-1 antibody from Santa Cruz (SC-1506).
22. Donkey Anti-Rabbit Alexa Fluor 555 antibody from Thermo Fisher (#A-31572).
23. Donkey Anti-Goat Alexa Fluor 647 antibody from Thermo Fisher (#A-21447).

3 Methods

3.1 Fixation

1. Isolate your tissue of interest with maximum size of 5 mm.
2. Put tissue in 15 mL round bottom tube with 10 mL of 4% PFA solution and incubate overnight on roller apparatus at 4 °C (*see* **Note 7**).
3. Wash tissue twice in 10 mL of PBS for at least 10 min (*see* **Note 8**).

   The following three steps can also be done with the automatic tissue processor.

4. Dehydrate the tissue by sequential 1.5 h incubations in 10 mL of 50, 70, 90, 95, and 100% ethanol at 4 °C. Finish with one additional 1.5 h incubation in 100% ethanol (*see* **Note 9**).
5. Clear the tissue by incubating twice in 10 mL of xylene for 40 min in a xylene-resisted tube (*see* **Note 9**).
6. Infiltrate the tissue with paraffin by incubating three times for 30 min in melted paraffin at 60 °C (*see* **Note 10**).
7. Embed the tissue in paraffin by pouring melted paraffin in a pre-warmed mold, put the tissue in the mold with paraffin and position the tissue correctly, lay the embedding cassette on top of the mold, add some paraffin in the embedding cassette, and place the mold carefully on the cold plate of the embedding machine. Let the paraffin cool and solidify. Remove the mold from the paraffin block.
8. Place the paraffin block in the microtome and cut 6 μm sections (*see* **Note 11**).
9. Transfer the sections to 43 °C water bath to stretch the sections and collect on coated microscope slides (*see* **Note 4**).
10. Dry the sections overnight in a dry 37 °C incubator.

3.2 Immunohistochemistry

1. Melt the paraffin on the sections by briefly incubating on a 60 °C slide warmer (*see* **Note 12**).
2. Let the slide cool to room temperature, so the paraffin solidifies.

   In the following steps, use the slide staining dishes to hold 250 mL of the indicated solutions.
3. Transfer the slides to a slide holder and deparaffinize the slides by incubating them 7 min in xylene for three times (*see* **Notes 5** and **13**).
4. Dehydrate the sections by incubating the slides twice for 5 min in 100% ethanol, and 3 min in each of the 90, 70, and 50% ethanol solutions, and twice for 5 min in water.
5. Boil the slides in the Antigen Unmasking Solution for 35 min by using the pressure cooker as follows (*see* **Note 14**): Put the Antigen Unmasking Solution in a slide staining dish and add water to the pressure cooker till there is approximately 4 cm of water. Make sure both the Antigen Unmasking Solution and the water in pressure cooker are boiling hot using a microwave and heating plate, respectively (*see* **Note 15**). Put the slide holder with the slides in the slide staining dish with Antigen Unmasking Solution and put the slide staining dish in the pressure cooker. Close the lid of the pressure cooker and allow high-pressure build up. Start the timer when the steam starts to exit the pressure cooker. After 35 min, remove the

pressure cooker from the heating plate and carefully release the pressure from the pressure cooker. After all pressure is released (*see* **Note 16**), remove the lid from the pressure cooker. Beware of the hot steam.

6. Remove the dish from the pressure cooker and transfer the slides to another dish with water. Dip the slides ten times in the water. Immediately transfer the slides to another dish with water, dip the slides ten times, and incubate twice for 5 min in water (*see* **Note 17**).
7. Incubate the slides for 10 min in 3% H_2O_2 solution (*see* **Note 18**). Then wash twice for 5 min in water.
8. Incubate the slides for 5 min in PBS.
9. Remove the slides from the slide holder and incubate the sections in blocking solution for at least 30 min in a humidified chamber (*see* **Note 19**). For this, dry the slide carefully with tissue or by suctioning the liquid without touching the sections. Leave a film of fluid on the sections if possible. Draw a circle around the sections with the liquid blocker without touching the sections. Leave some space between the section and the liquid blocker (*see* **Note 20**). After the liquid blocker dries, carefully cover the section and a small surrounding area with blocking solution (30–100 μL).
10. Remove the blocking solution by careful swinging or suctioning. Cover the section with the primary antibody solution (30–100 μL), in which the pSmad1/5/9 antibody is diluted 1:200 in blocking solution (*see* **Note 21**). Incubate overnight at 4 °C in the humidified chamber.
11. Remove the primary antibody solution and wash twice for 5 min in wash buffer. Washing may be performed in a slide holder and slide staining dishes (*see* **Note 20**).
12. Remove excess wash buffer and cover the section with the secondary antibody solution as described in **step 9**. Dilute the HRP-conjugated anti-rabbit antibody 1:200 in blocking solution. Incubate 25 min in the humidified chamber at room temperature.
13. Remove the secondary antibody solution and wash twice for 5 min in wash buffer.
14. Remove excess wash buffer and cover the section with the biotin amplification reagent working solution, which is prepared by diluting the biotin amplification stock solution of the Tyramide Signal Amplification® biotin detection kit from Pelkin Elmer 1:50 in the amplification diluent. Incubate for 25 min in the humidified chamber at room temperature (*see* **Note 22**).

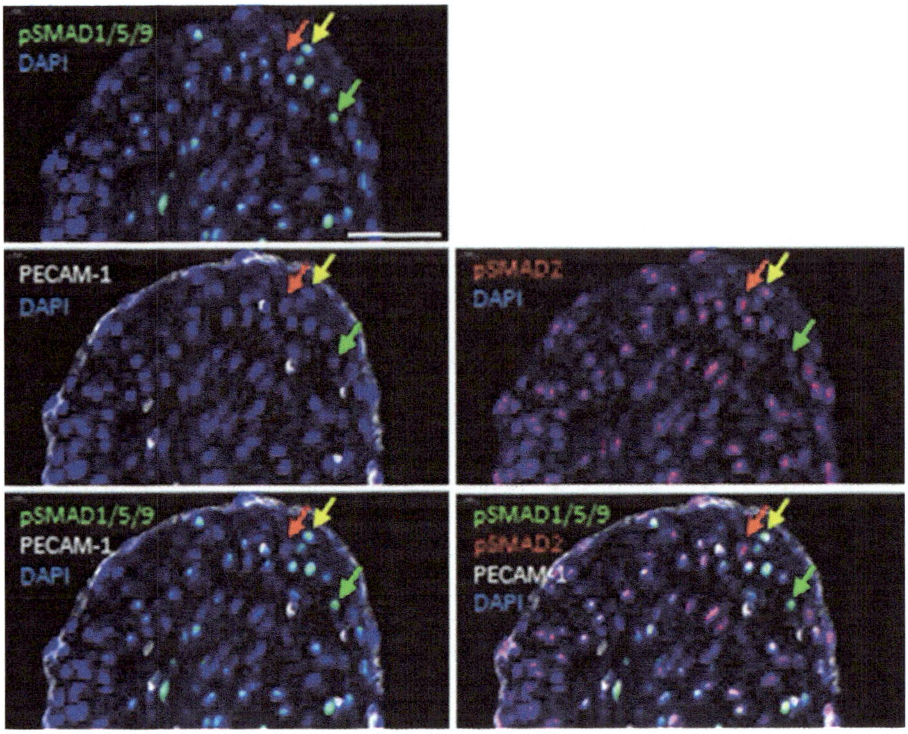

Fig. 1 Immunofluorescent detection of pSmad1/5/9 on PFA-fixed and paraffin-embedded tissue. The murine aortic valve cultured ex vivo as described in [2] shows cells expressing pSmad1/5/9 (green arrow), pSmad2 (red arrow), or both (yellow arrow). PECAM-1 expression indicates endothelial cells. Various levels of pSmad1/5/9 and pSmad2 expression can be observed. Scalebar: 50 μm

15. Remove the amplification solution and wash twice for 5 min in wash buffer.

16. Remove excess wash buffer and cover the section with Alexa-conjugated streptavidin solution, in which the Alexa-conjugated streptavidin is diluted 1:250 in blocking solution (*see* **Note 23**). Incubate for 90 min in the humidified chamber at room temperature.

    To proceed with the multi-labeling staining procedure (here with pSmad2 and PECAM-1; *see* Fig. 1), continue with **step 17**, otherwise go to **step 22**.

17. Remove the Alexa-conjugated streptavidin solution and wash twice for 5 min in wash buffer.

18. Remove excess wash buffer, cover the section with blocking solution, and incubate for 30 min in the humidified chamber at room temperature.

19. Remove the blocking solution and cover the section with a solution of primary antibodies that do not require amplification diluted in blocking solution. Here we dilute the pSmad2 and the PECAM-1 antibodies 1:200 and 1:100, respectively, in

the same blocking solution (*see* **Note 24**). Incubate overnight at 4 °C in the humidified chamber.

20. Remove the primary antibody solution and wash twice for 5 min in wash buffer.
21. Remove excess wash buffer and cover the section with Alexa-conjugated secondary antibody solution, in which the antibodies are diluted 1:250 in blocking solution (*see* **Note 23**). Here we use Donkey Anti-Rabbit Alexa Fluor 555 and Donkey Anti-Goat Alexa Fluor 647 antibodies. Incubate for 90 min in the humidified chamber at room temperature.
22. Remove the solution and wash 5 min in wash buffer and 5 min in PBS (*see* **Note 25**).
23. Dry the slide carefully with a tissue or by suctioning the liquid without touching the sections. Add a drop of ProLong Gold antifade reagent with DAPI on each section and lay a cover glass of the appropriate size on the slide. Gently press any potential air bubbles away from the sections. Let the slide dry overnight in a dark place at room temperature.

4 Notes

1. Prepare the solution in the fume hood as paraformaldehyde is carcinogenic.
2. Filtering out the small undissolved particles will reduce the non-specific staining.
3. Undamaged high-quality blades are essential for maintenance of tissue integrity.
4. High-quality microscope slides with a strong coating (like Super Frost Plus) are important. Sections on low-quality microscope slides will fall off during the 35 min pressure cooker incubation.
5. Xylene substitutes can be used for the deparafinization steps; however, not all xylene substitutes are as efficient in removing the paraffin. This might result in higher background and lower-specific staining.
6. A citrate-based solution at pH 6.0 can be used, when required for other primary antibodies using the multi-labeling procedure; however, this may result in overall lower pSmad1/5/9 signal.
7. For each tissue, use a fixation volume of at least ten times greater than your tissue. Continuous movement of the tissue in the fixation solution is crucial for homogenous fixation.

8. Remnants of PFA in the tissue can increase background staining.
9. Solutions must be clean. Automatic tissue processors recycle solutions which can result in "dirty" solutions and as a consequence in an inefficient exchange of solutions in the tissue. During manual processing the xylene-resistant tube containing the tissue should be put on a roller apparatus to allow continuous movement and proper penetration of the solution into the tissue. Incubation at 4 °C is not essential but is preferred.
10. Use clean paraffin. It is easy to obtain paraffin by using the paraffin dispenser of an embedding machine. Small glass containers can be used for this step. Paraffin can be kept warm by putting the glass containers in the paraffin bath of the embedding machine. Before and after changing the solutions, gently shake the glass container with the tissue in the paraffin solution.
11. To decrease background staining or to obtain sharper images, 3 μm sections can be made.
12. Melting the paraffin for a short time allows a better attachment of the section to the slide.
13. Incomplete removal of the paraffin from the sections can lead to higher background staining. Dipping the slides in the deparaffinization solution ten times before and after the incubation in the solution improves the paraffin removal.
14. Antigen retrieval can also be achieved using the microwave or a steamer; however, this may result in less optimal staining.
15. Preparing boiling hot solutions before adding the slides and closing the lid of the pressure cooker allows quicker pressure buildup and minimizes the loss of the Antigen Unmasking Solution from the slide staining dish. Sections must remain submerged.
16. Make sure that all pressure is released. Most pressure cookers have a feature that indicates when it is safe to remove the lid.
17. Dipping the slides several times in the water and transferring to freshwater solutions washes salt from the slides, which otherwise might form crystals on the sections. Alternatively, the slides may remain in the Antigen Unmasking Solution till the temperature has decreased to room temperature before transferring them to the water solution.
18. Exposure to H_2O_2 reduces the background staining caused by endogenous peroxidase activity. On PFA-fixed and paraffin-embedded tissue, this effect, however, is minimal. This step may be omitted if desired.

19. The humidified chamber prevents the drying out of the sections, which can greatly increase background staining.
20. The liquid blocker must not touch the section as this interferes with the staining. This method is used to minimize the required volume of expensive reagents such as antibodies and biotin amplification reagent. Steps such as washing may be performed using slide racks and holders.
21. Depending on the tissue, the dilution of the first antibody can vary between 1:100 and 1:2000.
22. Each batch of the Tyramide Signal Amplification® biotin detection kit from Pelkin Elmer differs slightly in the extent of amplification. Therefore, the primary antibody dilution and incubation time of the biotin amplification reagent working solution might have to be optimized with each new batch.
23. Centrifuge the Alexa-conjugated streptavidin or secondary antibody stock solution for 5 min at 20238 RCF. Take the required amount from the top part of the solution. Removal of small (undissolved) particles in the solution by centrifugation will reduce non-specific staining.
24. Although the pSmad1/5/9 and pSmad2 antibodies are both raised in rabbit, co-labelling can be obtained by using the amplification method for the pSmad1/5/9 antibody (Fig. 1). Because visualization of the pSmad1/5/9 antibody used at the recommended dilution requires amplification, secondary antibodies raised against the same species can be used to visualize primary antibodies that do not need amplification (here pSmad2 antibody). The labeling procedure for primary antibodies that do not need amplification should start after completing the labeling of the primary antibody that does require amplification.
25. Removing the Tween-20 improves the mounting in the next step.

Acknowledgment

We thank Karien Wiesmeijer for expert technical assistance. This work is supported by the Netherlands Institute for Regenerative Medicine (grant No. FES0908), SMARTCARE (grant No. P1.04) of the BioMedical Materials Institute, co-funded by the Dutch Ministry of Economic Affairs, Agriculture and Innovation, and the Netherlands Heart Foundation.

References

1. Goumans MJ, Liu Z, ten Dijke P (2009) TGF-beta signaling in vascular biology and dysfunction. Cell Res 19(1):116–127. https://doi.org/10.1038/cr.2008.326

2. Kruithof BP, Lieber SC, Kruithof-de Julio M, Gaussin V, Goumans MJ (2015) Culturing mouse cardiac valves in the miniature tissue culture system. J Vis Exp (105):e52750. https://doi.org/10.3791/52750

Chapter 15

Spatial and Quantitative Detection of BMP Activity in Mouse Embryonic Limb Buds

Marcelo Rocha Marques and Jean-Denis Bénazet

Abstract

Modulation of bone morphogenetic protein (BMP) activity is essential to the progression of limb development in the mouse embryo. Genetic disruption of BMP signaling at various stages of limb development causes defects ranging from complete limb agenesis to oligodactyly, polydactyly, webbing, and chondrodysplasia. To probe the state of BMP signaling in early limb buds, we designed two sets of primers to measure both spatially and quantitatively the transcription of nine key genes indicative of canonical BMP activity. One set is used to generate digoxigenin (DIG)-labeled antisense RNA probes for whole-mount mRNA in situ hybridization, while the second set is used for SYBR® Green-based quantitative PCR on limb bud cDNA. Here we describe step-by-step protocols for both methods around this specific set of genes.

Key words Limb bud, BMP signaling, Grem1, SHH, FGF, Whole-mount mRNA in situ hybridization, Quantitative PCR, Mouse

1 Introduction

The early steps of limb skeletal development consist in the patterning of the expanding chondrogenic population prefiguring the adult appendage skeleton. Between embryonic days E9.0 and E12.5 (18–60 somites) in the mouse, cells in the fore- and hindlimb mesenchymes integrate signals from multiple pathways specifying their positional information along the dorsoventral, proximodistal, and anteroposterior axis, as well as their chondrogenic fate and proliferative potential [1, 2]. Genetic analysis coupled to mathematical modeling revealed that this process is highly dynamic. It depends on a gene network centered on the transcriptional regulation of *Grem1*, an extracellular BMP antagonist, by Sonic Hedgehog (SHH), fibroblast growth factor (FGF), and BMP signaling feedback loops. This network architecture based on the modulation of BMP signaling allows the progression of limb patterning and outgrowth in a self-regulatory manner, from initiation to propagation and termination phases. This signaling system is characterized

Melissa B. Rogers (ed.), *Bone Morphogenetic Proteins: Methods and Protocols*, Methods in Molecular Biology, vol. 1891, https://doi.org/10.1007/978-1-4939-8904-1_15, © Springer Science+Business Media, LLC, part of Springer Nature 2019

by a burst of SHH and FGF morphogenetic signaling driven by an oscillation of opposing canonical BMP signaling. This oscillation of BMP activity delineates the three distinct phases of early limb development [3].

Limb outgrowth and patterning rely on a positive feedback loop between two signaling centers, namely, the zone of polarizing activity (ZPA) located in the posterior limb bud mesenchyme and expressing *Shh*, and the apical ectodermal ridge (AER), a cell population within the epithelium covering the limb bud mesenchyme and expressing notably *Fgf8* [4, 5]. During the initiation phase (E9.0–E9.75 in forelimbs, E9.5–E10.25 in hindlimbs), BMP4 from the early mesenchyme signals to the epithelium to promote the formation of the *Fgf8*-expressing AER [3, 6–9], which in turn positively regulates the onset of *Shh* transcription in the ZPA [10, 11]. In the mesenchyme, BMP4 activates the expression of its own extracellular antagonist *Grem1* through a BMP4/Grem1 negative feedback loop, triggering the auto-downregulation of BMP activity in this compartment [3, 12, 13].

Subsequently, the positive input of SHH signaling on *Grem1* transcription engages the limb bud in the propagation phase (E9.75–E11.5 in forelimbs, E10.25–E12.0 in hindlimbs) [3, 14–16]. During this phase, the aforementioned SHH/Grem1/FGF positive feedback loop between the ZPA and AER progressively reaches its maximum activity as revealed by the increasing expression of SHH and FGF signaling readouts such as *Gli1* and *Dusp6*, respectively [3, 17, 18]. In contrast, the opposing BMP activity produced by the *Bmp2*, *Bmp4*, *and Bmp7* genes is inhibited by increasing *Grem1* expression and activity [3]. *Msx2*, a direct target of canonical BMP signaling in the limb bud, is reduced to low expression levels. Its expression is repressed in the distal mesenchyme and restricted to the limb anterior and posterior margins [3, 12, 13, 15, 16]. Although BMP inhibition is driving the propagation of the SHH and FGF morphogenetic signaling, minimal levels of BMP activity are required to constrain AER length and overall proliferation in the limb bud [3, 12, 13, 19].

Finally, increasing FGF signaling triggers an FGF/Grem1 negative feedback loop that establishes a *Grem1* negative domain in the distal and posterior mesenchyme. As a result of limb bud growth, this cell population proggressively expands and physically disrupts the SHH/Grem1/FGF feedback loop. The separation of the ZPA and AER from the *Grem1* expression domain in the distal mesenchyme allows the return of the opposing BMP activity, engaging limb bud patterning in its termination phase (E11.5–E12.0 in forelimbs, E12.0–E12.5 in hindlimbs) [3, 20, 21]. Accordingly, *Msx2* expression is restored in the distal mesenchyme. SHH and FGF morphogenetic signaling is shut down as revealed by the downregulation of *Fgf8*, *Dusp6*, and *Gli1* expressions and the termination of *Shh* expression [3, 6, 12, 13, 21–24]. The comeback of

BMP activity in the mesenchyme is essential for digit ray induction and growth [12, 13, 25–27] and for the condensation of all limb chondrogenic anlagen, laying the template for endochondral ossification and growth of the future limb skeleton [12, 13, 25, 28, 29]. In addition, BMP activity in the AER is then required for the induction of cell death in the interdigit territories [6, 9, 30, 31].

In summary, the wide range of phenotypes associated with genetic disruptions of BMP signaling directly illustrates the multiple requirements of this pathway over time and space as well as the dynamic of its modulation. It also points to the importance of precise staging when comparing wild-type and mutant embryos, both along the limb development sequence and across samples.

In order to assess BMP activity in a precise and robust manner, we present a method coupling two types of messenger RNA detection (mRNA), namely, whole-mount mRNA in situ hybridization (WMISH) and quantitative real-time polymerase chain reaction (qPCR). WMISH is a suitable technique for the detection of transcripts in small and relatively undifferentiated tissues such as the embryonic limb bud up to E12.5. Here we describe a protocol using digoxigenin (DIG)-labeled riboprobes that hybridize to nine specific mRNAs after permeabilization of the plasma membrane of the limb bud cells. The hybridized riboprobes are detected using Fab fragments from polyclonal anti-digoxigenin antibodies coupled to the alkaline phosphatase. Sample incubation with BM Purple, a chromogenic substrate for the alkaline phosphatase, leads to the coloration of limb bud cells expressing the targeted transcripts.

While WMISH is a technique of choice to detect spatial changes in gene expressions between wild-type and mutant embryos, it is only semiquantitative. In order to precisely measure expression levels of these nine specific transcripts, we describe a SYBR® Green-based protocol for qPCR. After RNA purification from limb buds, mRNA are retrotranscribed to complementary DNA (cDNA) using oligo dT. Relative transcript abundance is measured by real-time PCR and normalized to two reference genes.

To perform these assays, we provide two sets of PCR primers to use on limb bud cDNA. The first set is to generate double-stranded DNA templates containing the T3 RNA polymerase promoter for RNA probe synthesis in the WMISH protocol. The second set is a list of validated primers for the qPCR assay [3, 12, 22, 32]. These two sets of primers target the three BMP ligands expressed in the limb bud, *Bmp2*, *Bmp4*, and *Bmp7*, and two transcriptional targets of canonical BMP signaling: the BMP antagonist *Grem1* and the transcription factor *Msx2*. Primers for *Shh* and transcriptional target *Gli1* as well as *Fgf8* and transcriptional target *Dusp6* are also provided. The opposing functions of the BMP pathway on the induction and subsequent activities of the limb signaling centers make SHH and FGF signaling essential readouts of BMP activity during early limb development.

2 Materials

2.1 General

Care must be taken to avoid contamination with ribonucleases.

1. Calibrated pipettes with barrier tips unless otherwise specified.
2. Ultrapure water prepared by purifying deionized water to attain a resistivity of 18.2 MΩ-cm at 25 °C.
3. Sterile, nuclease-free 1.5 mL microcentrifuge tubes.

2.2 Dissection

1. 10× phosphate-buffered saline (PBS) with Ca^{2+} and Mg^{2+} or PBS tablets diluted in ultrapure water before use.
2. Stereomicroscope.
3. RNA stabilization reagent.
4. 20 mL glass vials with tightly sealed caps.
5. 4% (w/v) paraformaldehyde in PBS (dissolve by heating to 65 °C, filter, aliquot, and store at −20 °C). For optimal embryo fixation, use 4% paraformaldehyde within 48 h after thawing, store at 4 °C.

2.3 Primer Mixes

1. 10 mM Tris–HCl pH 8: 1 M Tris–HCl pH 8 diluted in nuclease-free water.
2. Desalted primers from set 1 (Table 1) and set 2 (Table 2) (*see* **Note 1**).

2.4 cDNA Preparations

1. RNeasy® extraction kits or an equivalent system (*see* **Note 2**).
2. β-Mercaptoethanol.
3. 25-gauge needles and 1 mL syringes
4. Molecular biology grade ethanol.
5. Nuclease-free PCR tubes.
6. Oligo $(dT)_{20}$ (50 μM).
7. 10 mM dNTP Mix (dATP, dGTP, dCTP, and dTTP at 10 mM each).
8. Superscript™ III RT (200 units/μL) with 5× first-strand buffer and 0.1 M DTT or an equivalent system (*see* **Note 3**).
9. RNase inhibitor (40 units/μL).
10. Nuclease-free water.
11. Thermocycler.
12. Apparatus for nucleic acid quantification.

2.5 RNA Probe Synthesis

1. Q5® high-fidelity DNA polymerase or an equivalent system (*see* **Note 4**).
2. Nuclease-free water.

Table 1
Primer set 1 for RNA probe template amplification. Expected amplicon sizes are indicated in the first column

Template	Forward primer	*Tail*-T3 RNA polymerase promoter-reverse primer
Bmp2 691 bp	GTTTGGCCTGAAGCAGAGAC	*GCGCGA*AATTAACCCTCACTAAAGG**TGACGC TTTTCTCGTTTGTG**
Bmp4 636 bp	AGGAGGAGGAGGAAGAGCAG	*GCGCGA*AATTAACCCTCACTAAAGG**ATGC TTGGGACTACGTTTGG**
Bmp7 833 bp	TCCAGGGAAAGCATAATTCG	*GCGCGA*AATTAACCCTCACTAAAGG**AGGTC TCGGAAGCTGACGTA**
Grem1 629 bp	CCACGGAAGTGACAGAATGA	*GCGCGA*AATTAACCCTCACTAAAGG**TGGCTA TGACAGGCTGAATG**
Msx2 687 bp	GACAAGAAGCCGCCCAAG	*GCGCGA*AATTAACCCTCACTAAAGG**TCTGG TCCATCTGGTCTTCC**
Shh 839 bp	TACTCGCAGCTGCTCTACCA	*GCGCGA*AATTAACCCTCACTAAAGG**G TGGCGGTTACAAAGCAAAT**
Gli1 648 bp	CTGGAGAACCTTAGGCTGGA	*GCGCGA*AATTAACCCTCACTAAAGG**TCCTCC TGTAGCCCCCTAGT**
Fgf8 671 bp	TGTTGCACTTGCTGGTTCTC	*GCGCGA*AATTAACCCTCACTAAAGG**GCGGG TAGTTGAGGAACTC**
Dusp6 678 bp	TTGAATGTCACCCCCAATTT	*GCGCGA*AATTAACCCTCACTAAAGG**ACAGCA TGTCCTGTCCTTCC**

Italic is a primer tail to allow the polymerase to bind properly, Underlined is the promoter, Bold is the primer part bindind to DNA

Table 2
Primer set 2 for qPCR

cDNA	Forward primer	Reverse primer
Bmp2	ATGTGGAGACTCTCTCAATG	ACGCTAGAAGACAGCGGGTC
Bmp4	AGCCGAGCCAACACTGTGA	GTTCTCCAGATGTTCTTCGTGATG
Bmp7	TGTGGCAGAAAACAGCAGCA	TCAGGTGCAATGATCCAGTCC
Grem1	CCCACGGAAGTGACAGAATGA	AAGCAACGCTCCCACAGTGTA
Msx2	ATACAGGAGCCCGGCAGATACT	TCCGGTTGGTCTTGTGTTTCC
Shh	GATGACTCAGAGGTGCAAAGACAA	TGGTTCATCACAGAGATGGCC
Gli1	CAAGTGCACGTTTGAAG	CAACCTTCTTGCTCACACATGTAAG
Fgf8	TAATTGCCAAGAGCAACGGC	GCACGATCTCTGTGAATACGCA
Dusp6	GGCTGCTGCTCAAGAAACTCAA	AAGCCACCTTCCAGGTAGAACG
Rpl19	ACCCTGGCCCGACGG	TACCCTTTCCTCTTCCCTATGCC
Hmbs	TCTAGATGGCTCAGATAGCATG	TGGACCATCTTCTTGCTGAACA

Italic is a primer tail to allow the polymerase to bind properly, Underlined is the promoter, Bold is the primer part bindind to DNA

3. Nuclease-free PCR tubes.
4. Thermocycler.
5. Electrophoresis grade agarose.
6. 0.5× TBE, dilute 10× TBE: 1 M Tris-base, 1 M boric acid, 0.02 M EDTA in ultrapure water.
7. DNA ladder.
8. 6× loading buffer, 30% (v/v) glycerol, 0.25% (w/v) bromophenol blue, 0.25% (w/v) xylene cyanol FF.
9. Clean electrophoresis apparatus.
10. Clean gel documentation system.
11. Nuclease-free 2 mL tubes.
12. Gel extraction kit and necessary reagents.
13. 10× DIG RNA labeling mix, 10 mM ATP, CTP, GTP (each) 6.5 mM UTP, 3.5 mM DIG-11-UTP
14. T3 RNA polymerase (20 units/μL) with 10× transcription buffer.
15. 0.1 M DTT.
16. Commercial RNase inhibitor (40 units/μL) used as instructed.
17. TE: 10 mM Tris–HCl pH 8, 1 mM EDTA.
18. 5 M LiCl.

2.6 Whole-Mount In Situ Hybridization

1. Disposable 3 mL plastic transfer pipettes.
2. PBT: 1× PBS, 0.1% (v/v) Tween-20.
3. Technical grade methanol.
4. Binocular stereomicroscope.
5. Sharp tungsten needle (handmade or commercial).
6. 30% hydrogen peroxide.
7. 10 mg/mL proteinase K: resuspend in PBS, store aliquots at −20 °C.
8. Glycine.
9. 50% glutaraldehyde.
10. 20× SSC pH 4.5: 3 M NaCl, 0.3 M Na_3 citrate: $2H_2O$ in ultrapure water. Adjust pH to 4.5 with citric acid. Autoclave for long-term storage.
11. Hybridization buffer: 50% deionized extrapure formamide, 5× SSC, pH 4.5, 2% Roche blocking reagent (or similar), 0.1% (v/v) Tween-20, 0.5% (w/v) CHAPS, 50 μg/mL yeast tRNA, 50 μg/mL heparin, 5 mM EDTA completed with ultrapure water.
    (a) Heat to 70 °C to dissolve the blocking reagent.
    (b) Store at −20 °C.

12. Round-bottom, boiling-proof 2 mL tubes.
13. 2× SSC: dilute 20× SSC pH 4.5 in ultrapure water.
14. RNase A (100 mg/mL): resuspend lyophilized RNase A in ultrapure water. Aliquot and store at −20 °C.
15. Rocking platform.
16. Maleic acid buffer: 100 mM maleic acid, 150 mM NaCl, pH 7.5 in ultrapure water. Autoclave for long-term storage.
17. 20× TBS: 0.5 M Tris–HCl pH 7.5, 2.8 M NaCl, 0.06 M KCl in ultrapure water. Autoclave for long-term storage.
18. TBST: dilute 20× TBS to 1× in ultrapure water, 1% (v/v) Tween-20. Make fresh.
19. Heat inactivated sheep serum. Aliquot and store at −20 °C.
20. Anti-digoxigenin-alkaline phosphatase (AP)-conjugated, Fab fragments (150 U/200 μL).
21. NTMT: 100 mM Tris–HCl pH 9.5, 100 mM NaCl, 50 mM $MgCl_2$, and 1% (v/v) Tween-20. Make fresh.
22. BM Purple.
23. Stereomicroscope with color camera and gooseneck lights.

2.7 Quantitative PCR

1. 384-well quantitative PCR plates.
2. Optical leads.
3. 2× SYBR® Green mix.
4. Centrifuge with a plate rotor.
5. Quantitative PCR instrument.

3 Methods

3.1 Generation of Primer Mixes

1. Centrifuge tubes with desalted primers briefly at maximum speed to pellet lyophilized primers.
2. Resuspend the content of each tube to a 100 μM concentration in 10 mM Tris–HCL pH 8 to generate primer stocks. For complete resuspension incubate 30 min at 50 °C in a heat block.
3. For each primer pair of set 1, generate a primer mix from the primer stocks in a 1.7 mL microfuge tubes:

Forward primer	5 μL
Reverse primer	5 μL
Nuclease-free water	90 μL

4. For each primer pair of set 2, generate a primer mix from the primer stocks in a 1.7 mL microfuge tube

Forward primer	7.5 μL
Reverse primer	7.5 μL
Nuclease-free water	180 μL

5. Store primer stocks and primer mixes at −20 °C. Aliquot to avoid freeze and thaw cycles.

3.2 cDNA Preparation from Mouse Embryonic Limb Buds

3.2.1 Sample Collection and Storage

1. Sacrifice timed-pregnant females according to your workplace guidelines.
2. Dissect embryos in ice-cold PBS, save the embryonic membrane in 1.5 mL tubes for genotyping if needed.
3. For each embryo isolate the limb buds using clean and sharp forceps in a separate dish (*see* **Note 5**).
4. Transfer the limb buds into a 1.5 mL tube with a glass pipettes.
5. Remove most of the PBS using a P200 pipette. Use a P10 pipette for the last drops.
6. Cover the limb buds with 200 μL of RNA stabilization reagent. The samples have the tendency to float. Briefly spin down and store at 4 °C.
7. The next day, check all tubes and if samples still float, flick the tubes and briefly spin until samples sink in the RNA stabilization reagent.
8. Store samples at −20 °C.
9. For more details consult the RNA stabilization reagent user manual.

3.2.2 RNA Purification

The following instructions refer to the RNeasy® kit (*see* **Note 2**).

1. Remove RNA stabilization reagent using pipettes (use a P10 pipette for the last drops).
2. To disrupt and homogenize the limb buds, pass the sample in RLT buffer with β-mercaptoethanol 15 times through a 25G needle using a 1 mL syringe. Change syringe and needle for each sample.
3. Load the samples on the columns and proceed as instructed.
4. Perform the on-column DNA digestion with the RNase free DNase set.
5. Elute in a minimal volume (*see* **Note 2**)..
6. Measure RNA concentration.
7. Store the purified total RNA at −80 °C.

3.2.3 cDNA Synthesis

1. Add the following components to a nuclease-free PCR tube (*see* **Notes 3** and **6**):

Oligo(dT)$_{20}$ (50 μM)	1 μL
Total RNA (5 μg total RNA max)	11 μL
10 mM dNTP Mix	1 μL

2. Mix by pipetting gently up and down.
3. Heat mixture to 65 °C for 5 min in a thermocycler and incubate on ice for at least 1 min.
4. Collect the contents of the tube by brief centrifugation and add.

5× First-strand buffer	4 μL
0.1 M DTT	1 μL
RNase inhibitor (40 units/μL)	1 μL
Superscript™ III RT (200 units/μL)	1 μL

5. Mix by pipetting gently up and down.
6. Incubate at 50 °C for 60 min in a thermocycler.
7. Inactivate the reaction by heating at 70 °C for 15 min in a thermocycler.
8. The produced cDNA can now be used as a template for amplification by PCR. Store cDNA at – 20 °C.
9. For further details see Superscript™ III RT user manual.

3.3 Whole-Mount mRNA In Situ Hybridization

3.3.1 Probe Synthesis

1. Generate a cDNA preparation from four wild-type E11.5 hindlimbs according to Subheading 3.2 (*see* **Notes 2**, **3** and **5**).
2. Set up the following PCR reactions for the nine primer pairs (set 1) using the high-fidelity Q5® enzyme (*see* **Note 4**):

5× Q5® reaction buffer	10 μL
10 mM dNTPs	1 μL
Primer mix (set 1)	5 μL
cDNA template	2 μL
Q5® high-fidelity DNA polymerase	0.5 μL
Nuclease-free water	31.5 μL

3. Use the following PCR program.

Step:	Temp:	Time:
Initial denaturation	98 °C	30 s
40 cycles:		
• Denaturation	98 °C	10 s
• Annealing	59 °C	30 s
• Elongation	72 °C	1 min 20 s
Final elongation	72 °C	2 min

4. Prepare a 1.5% agarose gel in 0.5× TBE for electrophoresis.
5. Add loading buffer and load the nine PCR reactions and DNA ladders on the gel.
6. Run the gel until PCR products are clearly separated form primers or unwanted PCR products, *see* Table 1 for expected PCR product sizes (*see* **Note 7**).
7. Cut and purify PCR products according to gel extraction kit instructions (*see* **Note 8**).
8. Measure DNA concentration. A minimum concentration of 250 ng of DNA probe template in 12 μL must be achieved (*see* **Note 9**).
9. Set up the in vitro transcription reaction in nuclease-free PCR tubes (*see* **Note 10**).

   For each purified probe template:

10× transcription buffer	2 μL
0.1 M DTT	2 μL
10× digoxigenin RNA labeling mix	2 μL
RNase inhibitor (40 U/μL)	1 μL
T3 RNA polymerase (20 U/μL)	1 μL
DNA template	250 ng
Nuclease-free water	to 20 μL

10. Incubate 2 h at 37 °C in a thermocycler (*see* **Notes 6** and **11**).
11. For each in vitro transcription reaction add:

TE	100 μL
5 M LiCl:	10 μL
Ethanol	300 μL

Mix well and incubate at −20 °C from 1 h to overnight to precipitate the probes.

12. Centrifuge at maximum speed at 4 °C for 15 min.
13. Remove the supernatant without dislodging the white RNA pellet.
14. Slowly add 1 mL of −20 °C 70% ethanol in nuclease-free water.
15. Centrifuge at maximum speed at 4 °C for 15 min.
16. Remove 70% ethanol and air-dry.
17. Resuspend the RNA probes in 100 μL of TE and store at −20 °C.

3.3.2 Embryo Collection and Storage

1. Sacrifice timed-pregnant females according to your workplace guidelines.
2. Dissect embryos in ice-cold PBS, save the embryonic membrane for genotyping if needed in 1.5 mL tubes.
3. Fix in 4% paraformaldehyde in PBS at 4 °C, overnight in 20 mL glass vials (*see* **Notes 12** and **13**).
4. Wash twice in PBT for 5 min at room temperature (*see* **Note 14**).
5. Dehydrate embryos in a methanol/PBT series (25%, 50%, 75% methanol, 5 min each at room temperature) (*see* **Note 13**).
6. Rinse twice in 100% methanol at room temperature and store in 100% methanol at −20 °C (*see* **Note 15**).

3.3.3 Day 1: Permeabilisation and Hybridization

1. Match control and mutant embryos by somite number and limb shape in 100% methanol under a binocular stereomicroscope (*see* **Note 16**).
2. Rehydrate by taking the embryos through the methanol series in reverse (*see* Subheading 3.3.2, **step 5**), rinse twice in PBT at room temperature.
3. Under a binocular stereomicroscope, split matched control and mutant embryos from head to tail with tungsten needles to use two probes per embryo (right and left halves).
4. Pool control half and mutant half in the same glass vial. Note morphological differences to recognize each sample after completion of the WMISH (*see* **Note 17**).
5. **Steps 6–12** are performed with gentle agitation on a rocking platform at room temperature.
6. Bleach samples with 6% hydrogen peroxide in PBT, 15 min.
7. Wash three times 5 min in PBT.
8. Treat with 5 μg/mL proteinase K in PBT for 4 min for ectodermal probes (*Bmp2*, *Bmp4*, *Bmp7*, *Msx2*, and *Fgf8*) or with

10 μg/mL proteinase K in PBT for 15 min for mesenchymal probes (*Bmp2*, *Bmp4*, *Bmp7*, *Grem1*, *Msx2*, *Shh*, *Gli1*, and *Dusp6*) (*see* **Note 18**).

9. Wash with freshly prepared 2 mg/mL glycine in PBT, 5 min.
10. Wash twice 5 min in PBT.
11. Fix with freshly prepared 0.2% glutaraldehyde/4% paraformaldehyde in PBT, 20 min (*see* **Note 13**).
12. Wash twice 5 min in PBT.
13. Transfer the control and mutant halves to a round bottom, boiling proof 2 mL tube containing 2 mL of hybridization buffer previously brought to room temperature.
14. Prehybridize by incubating the samples at 70 °C for 1 h.
15. Warm hybridization buffer containing the RNA probe at 70 °C (10–20 μL of synthetic probe per 1 mL of hybridization buffer).
16. Replace the buffer used for pre-hybridization on the samples with the 2 mL of warm probe in hybridization buffer (*see* **Note 19**).
17. Incubate overnight at 70 °C.

3.3.4 Day 2: Probes Washes and Immunodetection

1. Warm hybridization buffer (reuse hybridization buffer from Subheading 3.3.3.16), 2× SSC, 2× SSC, 0.1% (w/v) CHAPS and maleic acid buffer to 70 °C and 2× SSC, 0.1% (w/v) CHAPS to 37 °C. Steps at 70 °C and 37 °C are carried out in water baths. Use plastic Pasteur pipettes to change solutions in 2 mL tubes.
2. Remove the probes in hybridization buffer and store them at −20 °C for reuse (*see* **Note 20**).
3. Add 800 μL of warm hybridization buffer and incubate 5 min at 70 °C (*see* **Note 21**).
4. Add 400 μL of warm 2× SSC, incubate 5 min at 70 °C. Repeat two more times to reach a final volume of 2 mL (*see* **Note 21**).
5. Wash with warm 2× SSC, 0.1% (w/v) CHAPS twice for 30 min at 70 °C.
6. Treat with 20 μg/mL of RNaseA in 2× SSC, 0.1% (w/v) CHAPS for 45 min at 37 °C.
7. Wash with maleic acid buffer twice for 10 min at room temperature with gentle rocking (*see* **Note 22**).
8. Wash with maleic acid buffer twice for 30 min at 70 °C.
9. Wash three times with TBST for 5 min at room temperature with gentle rocking.
10. Block samples with 10% (v/v) heat inactivated sheep serum in TBST for at least 1 h at room temperature with gentle rocking.

11. Dilute the anti-digoxigenin-AP, Fab fragments in 1% sheep serum in TBST.
12. Replace blocking solution with 1.7 mL of antibody solution. Incubate overnight at 4 °C with gentle rocking.

3.3.5 Day 3: Antibody Washes

1. Wash three times in TBST for 5 min at room temperature with gentle rocking.
2. Wash five times in TBST for 1.5 h at room temperature with gentle rocking.
3. Wash in TBST overnight at 4 °C with gentle rocking.

3.3.6 Day 4: Detection

1. Wash three times in NTMT for 10 min at room temperature with gentle rocking.
2. During the washes, bring the necessary amount (1.7 mL/tube) of BM-Purple to room temperature after shaking carefully the stock bottle for homogenization.
3. Replace NTMT with BM-Purple; protect tube from light with aluminum foil, and incubate at room temperature with gentle rocking.
4. Every 30 min, monitor the progression of the chromogenic reaction. The different probes should reach optimum signal within 1–4 h.
5. Stop reaction by washing three times in NTMT at room temperature with gentle rocking (*see* **Note 23**).
6. Wash three times in PBT at room temperature with gentle rocking.
7. Wash three times in PBS at room temperature with gentle rocking and place at 4 °C for short-term storage (*see* **Note 24**).

3.3.7 Photography

1. Print a page of an appropriate background color.
2. Protect it from splashes with a clear plastic pocket.
3. Cut to resize to approximately 10 × 10 cm.
4. Place the background square on the stage of a stereomicroscope equipped with a color camera.
5. Place the lid of a 6 cm plastic dish on the background square and place the 6 cm dish filled with PBS on the lid. This way the background square will remain out of focus for a smooth and uniform background color.
6. Transfer the specimen into the imaging dish.
7. Dissect out flank tissue 4 somites above and below the limb bud. Remove ventral tissue to clear the limb bud.
8. Use the dissected pieces of embryonic tissue to orient the limb bud attached to the flank in a way that the apex is pointing either right or left and the dorsoventral axis is in the horizontal plane.

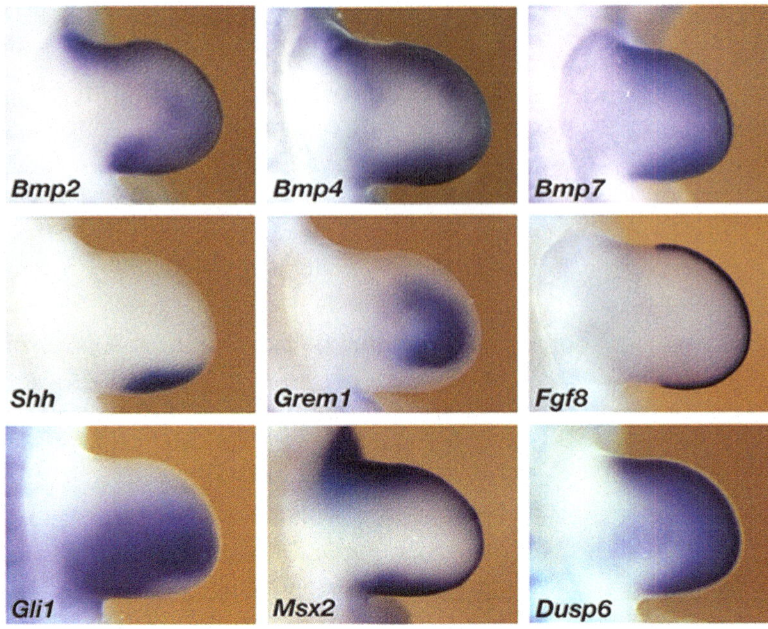

Fig. 1 expression pattern of *Bmp2*, *Bmp4*, *Bmp7*, *Grem1*, *Msx2*, *Shh*, *Gli1*, *Fgf8*, and *Dusp6* at E10.75 in forelimbs using riboprobes generated with primers in Table 1 and limb cDNA

9. Adjust the gooseneck lights to create raking light from both right and left.
10. Slightly adjust the gooseneck lights to remove shadows.
11. Take the picture (*see* **Note 25**).
12. Adjust brightness, contrast, and color balance in a similar manner on all picture files during figure processing (Fig. 1).

3.4 Quantitative PCR with SYBR® Green

To correct for small pipetting errors, we recommend running replicate qPCR reactions. In addition, extensively vortexing each reagent prior to pipetting is essential to minimize variation between reactions.

1. Prepare Primer/SYBR® Green mixes according to the number of reactions.

   For one reaction:

qPCR primer mix (set 2)	0.8 μL
2× SYBR® Green	5 μL

2. Prepare cDNA mixes according to the number of reactions.

   For one reaction:

cDNA	0.05 μL (1 limb bud pair E10.75)
Water	to 4.2 μL

(*see* **Notes 4** and **26**).

3. Distribute 5.8 μL of SYBR® Green mix and 4.2 μL of cDNA mix per well (total volume = 10 μL) in the appropriate 384-well plate. Dispense the SYBR® Green mixes before the cDNA mixes.
4. Cover the plate with appropriate optical adhesive lid.
5. Centrifuge the PCR plate at 900 × *g* on a swinging rotor for 5 min to remove of bubbles in the PCR reactions.
6. Use the following program (*see* **Note 27**):

Step:	Temp	Time
Initial denaturation	95 °C	2 min
40 cycles		
• Denaturation	95 °C	15 s
• Annealing, extension, and read fluorescence	60 °C	1 min
Melting curve	60–95 °C	0.5°C increment

Determine *Ct* (manually or using your qPCR software) and relative expression using the $2^{-\Delta\Delta Ct}$ method (*see* **Note 28**).

7. Use the mean of *Rpl19* and *Hmbs* Cts for normalization.
8. Measure gene expression in nine control and nine mutant age-matched limb buds (or limb bud pairs) for powerful statistical analysis.
9. Use one sample as reference to measure all other samples.
10. Normalize results that the mean expression value for each transcript is set to 100% in the control group.
11. Assess the statistical significance of mean expression differences using the two-tailed unpaired t-test when expression values in control and mutant groups have a normal distribution. If not, use the Mann-Whitney U test.

4 Notes

1. Good results are obtained with standard desalting; however, HPLC purification might be needed.
2. Both RNeasy® kits are suitable. The RNeasy® Micro kit can process up to 5 mg of starting material for a maximum yield of 45 μg of total RNA, a typical yield from a pair of E12.0 forelimbs or E12.5 hindlimbs. In addition, the RNeasy® Micro kit allows elution in smaller volumes (12 μL) which is convenient for subsequent cDNA synthesis. This kit also contains an RNA carrier that we recommend for very small samples such as single

E9.5 limb buds. The RNeasy® Mini kit can process up to 30 mg of starting material for a maximum yield of 100 μg of total RNA which can be useful for larger cDNA preparation during the generation of probe templates. The methods presented here have been tested only using RNeasy® kits. Other RNA purification systems of quality are available that may be used with adaptation.

3. The methods presented here have been performed only with Superscript RT III. Other reverse transcriptase systems of quality are available that may be used with adaptation.
4. The primers in Table 1 have been tested only with Q5® high-fidelity DNA polymerase and buffer. Other high-fidelity DNA polymerase systems of quality are available that may be used with adaptation.
5. For qPCR, isolate single limb buds or pairs. For probe synthesis, several pairs of limb buds (2 pairs of E11.5 hindlimbs) can be pooled to obtain concentrated cDNA.
6. Use a thermocycler rather than a water bath for better temperature control and to avoid contaminations.
7. Some primer pairs might generate unwanted PCR products. It is essential to eliminate them as well as the excess of primers for successful probe synthesis.
8. At this stage PCR fragments can be cloned for long-term storage and/or sequencing. Perform an A-tailing reaction with a regular Taq DNA polymerase and ATP and ligate into pGEM®-T or use TOPO cloning techniques.
9. If not enough PCR product is produced, or if the concentration is too low, repeat the PCR after increasing the amount of cDNA template. Alternatively, pool several PCR reactions and/or concentrate using a SpeedVac Concentrator.
10. These primers contain the minimal T3 RNA polymerase promoter sequence for efficient transcription. Alternatively, SP6 or T7 RNA polymerase promoter sequence can be used, but PCR optimization might be required.
11. Typical RNA probe synthesis involves a DNase I incubation step to remove the DNA probe template. In our experience this step is not necessary.
12. To transfer embryos use disposable plastic transfer pipettes with the tip cut off.
13. Dispose of toxic reagents according to your institution guidelines.
14. To avoid losing samples, always pour the content of the vial in a beaker on a black background to improve visibility. Retrieve

samples with a disposable plastic transfer pipette. Use this safety measure for all steps in glass vials.

Always leave a small amount of solution to keep the samples covered at all times. Drying will compromise the quality of the signal.

15. Embryos can be kept in storage for several years, but signal intensity will decrease. Use freshly collected samples if possible. If not, compare samples with similar storage times.
16. Precise developmental stage matching is essential for proper comparison of wild-type and mutant samples. In the mouse embryo, somitogenesis progresses at the average rate of 1 somite every 2 h. Somite counting is the most accurate way to age match samples. In addition, choosing limb buds with similar shapes improves matching accuracy.
17. This is the best way to ensure that samples are treated identically for direct comparison. More samples can be added to one tube.
18. *Bmp2*, *Bmp4*, *Bmp7*, and *Msx2* are expressed in both the limb bud mesenchyme and AER. Gentle handling throughout the procedure will preserve AER morphology and probe signal after a treatment of 15 min with 10 μg/μL of proteinase K. However, for best results we recommend running separate assays to assess mesenchyme and AER expressions.
19. The hybridization buffer can be reused in Subheading 3.3.4, **step 1**.
20. Reusing an RNA probe will reduce its background/unspecific signal levels, but detection time will be longer. Refresh the probe in hybridization buffer by adding a few microliter of synthetic probe after a few WMISH.
21. Use regular tips on a P1000. Change the tip for each tube, as high temperature will make pipetting inaccurate.
22. For all the following steps, fill the tube to about 1.7 mL; the air bubble should not displace the samples roughly during rocking. Adjust the rocking speed to preserve sample integrity.
23. Each probe requires individual monitoring as detection speed depends on transcript levels and probe concentrations. Chromogenic reactions should be stopped when strong signal is achieved or when unspecific background coloration just start to appear.
24. Tween-20 leftover might precipitate at 4 °C. Wash several times with PBS at room temperature until crystals disappear.
25. For the best comparisons, do not modify the lighting and photograph specimens sequentially. If comparing right and left limb buds, photograph one of the two samples with the

posterior margin up, keeping the limb apex on the same side of the picture to preserve lighting.

26. For accurate quantification, using cDNA concentrations corresponding to the primers' optimal efficiency (90–110%) is important. Threshold cycles (*Ct*) should be read between cycle 20 and cycle 32. Before running all targets for all samples, determine the required amount of cDNA per reaction to read Cts for *Rpl19* and *Hmbs* at around cycle 20 and cycle 25, respectively.
27. Use the melting curve to verify that only one PCR product is generated. Alternatively, load reactions on an agarose gel.
28. The provided primers have been generated using Primer3. They have been validated in dilution series and on wild-type and mutant limb cDNA. They have similar and optimal efficiencies [3, 12, 22, 32].

Acknowledgments

The authors are grateful Dr. Rolf Zeller for forwarding the invitation to write this chapter and to Dr. Licia Selleri for providing the material and reagents necessary for generating and testing the mRNA in situ hybridization probes.

References

1. Zeller R, López-Ríos J, Zuniga A (2009) Vertebrate limb bud development: moving towards integrative analysis of organogenesis. Nat Rev Genet 10:845–858
2. Bénazet JD, Zeller R (2009) Vertebrate limb development: moving from classical morphogen gradients to an integrated 4-dimensional patterning system. Cold Spring Harb Perspect Biol 1:a001339
3. Bénazet JD, Bischofberger M, Tiecke E et al (2009) A self-regulatory system of interlinked signaling feedback loops controls mouse limb patterning. Science 323:1050–1053
4. Niswander L, Jeffrey S, Martin GR et al (1994) A positive feedback loop coordinates growth and patterning in the vertebrate limb. Nature 371:609–612
5. Laufer E, Nelson CE, Johnson RL et al (1994) Sonic hedgehog and Fgf-4 act through a signaling cascade and feedback loop to integrate growth and patterning of the developing limb bud. Cell 79:993–1003
6. Benazet JD, Zeller R (2013) Dual requirement of ectodermal Smad4 during AER formation and termination of feedback signaling in mouse limb buds. Genesis 51:660–666
7. Ahn K, Mishina Y, Hanks MC et al (2001) BMPR-IA signaling is required for the formation of the apical ectodermal ridge and dorsal-ventral patterning of the limb. Development 128:4449–4461
8. Soshnikova N, Zechner D, Huelsken J et al (2003) Genetic interaction between Wnt/beta-catenin and BMP receptor signaling during formation of the AER and the dorsal-ventral axis in the limb. Genes Dev 17:1963–1968
9. Pajni-Underwood S, Wilson CP, Elder C et al (2007) BMP signals control limb bud interdigital programmed cell death by regulating FGF signaling. Development 134:2359–2368
10. Lewandoski M, Sun X, Martin GR (2000) Fgf8 signalling from the AER is essential for normal limb development. Nat Genet 26:460–463
11. Moon AM, Capecchi MR (2000) Fgf8 is required for outgrowth and patterning of the limbs. Nat Genet 26:455–459

12. Bénazet JD, Pignatti E, Nugent A et al (2012) Smad4 is required to induce digit ray primordia and to initiate the aggregation and differentiation of chondrogenic progenitors in mouse limb buds. Development 139:4250–4260
13. Ovchinnikov DA, Selever J, Wang Y et al (2006) BMP receptor type IA in limb bud mesenchyme regulates distal outgrowth and patterning. Dev Biol 295:103–115
14. Zúñiga A, Haramis AP, McMahon AP et al (1999) Signal relay by BMP antagonism controls the SHH/FGF4 feedback loop in vertebrate limb buds. Nature 401:598–602
15. Khokha MK, Hsu D, Brunet LJ et al (2003) Gremlin is the BMP antagonist required for maintenance of Shh and Fgf signals during limb patterning. Nat Genet 34:303–307
16. Michos O, Panman L, Vintersten K et al (2004) Gremlin-mediated BMP antagonism induces the epithelial-mesenchymal feedback signaling controlling metanephric kidney and limb organogenesis. Development 131:3401–3410
17. te Welscher P, Zuniga A, Kuijper S et al (2002) Progression of vertebrate limb development through SHH-mediated counteraction of GLI3. Science 298:827–830
18. Mariani FV, Ahn CP, Martin GR (2008) Genetic evidence that FGFs have an instructive role in limb proximal-distal patterning. Nature 453:401–405
19. Selever J, Liu W, Lu MF et al (2004) Bmp4 in limb bud mesoderm regulates digit pattern by controlling AER development. Dev Biol 276:268–279
20. Verheyden JM, Sun X (2008) An Fgf/Gremlin inhibitory feedback loop triggers termination of limb bud outgrowth. Nature 454:638–641
21. Scherz PJ, Harfe BD, McMahon AP et al (2004) The limb bud Shh-Fgf feedback loop is terminated by expansion of former ZPA cells. Science 305:396–399
22. Lopez-Rios J, Speziale D, Robay D et al (2012) GLI3 constrains digit number by controlling both progenitor proliferation and BMP-dependent exit to chondrogenesis. Dev Cell 22:837–848
23. Harfe BD, Scherz PJ, Nissim S et al (2004) Evidence for an expansion-based temporal Shh gradient in specifying vertebrate digit identities. Cell 118:517–528
24. Ahn S, Joyner AL (2004) Dynamic changes in the response of cells to positive hedgehog signaling during mouse limb patterning. Cell 118:505–516
25. Bandyopadhyay A, Tsuji K, Cox K et al (2006) Genetic analysis of the roles of BMP2, BMP4, and BMP7 in limb patterning and skeletogenesis. PLoS Genet 2:e216
26. Raspopovic J, Marcon L, Russo L et al (2014) Modeling digits. Digit patterning is controlled by a Bmp-Sox9-Wnt Turing network modulated by morphogen gradients. Science 345:566–570
27. Suzuki T, Hasso SM, Fallon JF (2008) Unique SMAD1/5/8 activity at the phalanx-forming region determines digit identity. Proc Natl Acad Sci U S A 105:4185–4190
28. Retting KN, Song B, Yoon BS et al (2009) BMP canonical Smad signaling through Smad1 and Smad5 is required for endochondral bone formation. Development 136:1093–1104
29. Yoon BS, Ovchinnikov DA, Yoshii I et al (2005) Bmpr1a and Bmpr1b have overlapping functions and are essential for chondrogenesis in vivo. Proc Natl Acad Sci U S A 102:5062–5067
30. Wong YL, Behringer RR, Kwan KM (2012) Smad1/Smad5 signaling in limb ectoderm functions redundantly and is required for interdigital programmed cell death. Dev Biol 363:247–257
31. Choi KS, Lee C, Maatouk DM et al (2012) Bmp2, Bmp4 and Bmp7 are co-required in the mouse AER for normal digit patterning but not limb outgrowth. PLoS One 7:e37826
32. Galli A, Robay D, Osterwalder M et al (2010) Distinct roles of Hand2 in initiating polarity and posterior Shh expression during the onset of mouse limb bud development. PLoS Genet 6:e1000901

Chapter 16

Pharmacologic Strategies for Assaying BMP Signaling Function

Teresa Dinter, Geoffrey A. Bocobo, and Paul B. Yu

Abstract

The bone morphogenetic protein (BMP) signaling pathway, a subset of the transforming growth factor β (TGF-β) signaling family, consists of structurally diverse receptors and ligands whose combinatorial specificity encodes autocrine, paracrine, and endocrine signals essential for regulating tissue growth, differentiation, and survival during embryonic patterning and postnatal tissue remodeling. Aberrant signaling of these receptors and ligands is implicated in a variety of inborn and acquired diseases. The roles of various receptors and their ligands can be explored using small molecule inhibitors of the BMP receptor kinases. Several BMP type I receptor kinase inhibitor tool compounds have been described that exhibit sufficient selectivity to discriminate BMP receptor signaling in vitro or in vivo, with various trade-offs in selectivity, potency, cell permeability, and pharmacokinetics. Several methods for assaying BMP function via pharmacologic inhibition are presented. Two in vitro methods, an In-Cell Western assay of BMP-mediated SMAD1/5/8 phosphorylation and an alkaline phosphatase osteogenic differentiation assay, represent efficient high-throughput methodologies for assaying pharmacologic inhibitors. Two in vivo methods are described for assaying the effects of BMP signaling inhibition in embryonic zebrafish and mouse development. Small molecule inhibitors of BMP receptor kinases represent an important complementary strategy to genetic gain- and loss-of-function and ligand-trap approaches for targeting this signaling system in biology and disease.

Key words Bone morphogenetic protein (BMP) signaling, Kinase inhibitor, Dorsomorphin, LDN-193189, LDN-214117, LDN-212854, SB-431452, DMH1, K02288a

1 Introduction

The bone morphogenetic protein (BMP) signal transduction pathway is a subset of the larger TGF-β signaling family, which includes 7 type I receptors, 5 type II receptors, and over 30 ligands of the BMP, TGF-β, activin, and growth and differentiation factor (GDF) ligand families [1]. Much progress has been made in understanding the multifaceted roles of BMPs and their signaling mechanisms as a result of genetic, molecular, biochemical, and structural biology tools. In addition to specifying developmental patterning of dorsoventral axis and key aspects of organogenesis, BMPs are now known

Melissa B. Rogers (ed.), *Bone Morphogenetic Proteins: Methods and Protocols*, Methods in Molecular Biology, vol. 1891, https://doi.org/10.1007/978-1-4939-8904-1_16,

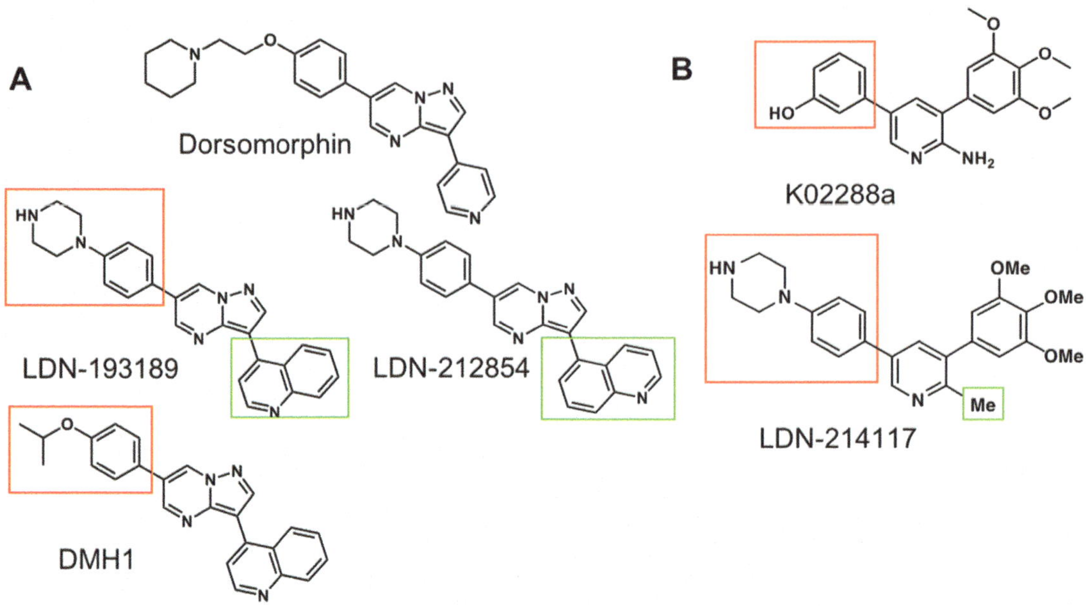

Fig. 1 Structures of BMP inhibitor tool compounds. (**a**) Dorsomorphin and pyrazolo[1,5-a]pyrimidine derivatives. Solvent-exposed domains of LDN-193189 and DMH1 are shown in red boxes. Deep hydrophobic pocket binding domains of LDN-193189 and LDN-212854 are shown in green. (**b**) K02288 and derivative LDN-214117 shown. Solvent-exposed domains are shown in red boxes and putative hinge binding groups (2-amino versus 2-methyl) shown in green boxes

to play a critical role in the maintenance of normal physiology beyond development, such as homeostasis of the vasculature, iron metabolism, and bone mineralization. A wide variety of inborn and acquired imbalances in the regulation of this signaling pathway can lead to disease, including hereditary hemorrhagic telangiectasia (HHT) syndromes, heritable pulmonary arterial hypertension (HPAH), juvenile familial polyposis (JP), and fibrodysplasia ossificans progressiva (FOP). Pharmacologic modulators of the BMP pathway are a useful tool in elucidating the role of these receptors and ligands in normal physiology and disease.

The first described small molecule inhibitor of BMP signaling was dorsomorphin (Fig. 1), a.k.a. Compound C, a pyrazolo[1,5-a] pyrimidine compound previously known to inhibit AMP-activated kinase at micromolar concentrations, identified for its potent dorsalization activity in a chemical biological screen examining embryonic zebrafish development [2]. Dorsalization is a phenotype that is known to reflect inhibition of BMP signaling during dorsoventral axis specification in multiple vertebrate species [3–5], leading to the inference that dorsomorphin might act as a BMP inhibitor. Biochemical analyses and co-crystallization with several type I receptor kinase domains subsequently identified the mechanism of this activity as resulting from the inhibition of BMP type I receptors kinases ALK2, ALK3, and ALK6 as an ATP-competitive antagonist [2, 6].

Cell-based assays confirmed sub-micromolar potency in inhibiting the BMP type I receptors, as well as a variety of other receptor tyrosine kinases, including VEGFR2 and PDGFRβ.

Multiple pharmaceutical modulators have since been designed to inhibit receptors of the BMP pathway with greater potency and/or selectivity, some of which are pyrazolo-pyrimidine derivatives, while others are based on novel scaffolds, each with distinct physiochemical properties and selectivity for individual BMP receptors. LDN-193189 is pyrazolo-pyrimidine derivative with modifications to the solvent-exposed and hydrophobic pocket domains (Fig. 1) that result in greater potency while retaining similar BMP selectivity as compared to dorsomorphin. LDN-193189 is a low nanomolar inhibitor of BMP signaling in cells with improved metabolic stability (>30 min in microsomal degradation assays), cell permeability, and in vivo bioavailability ($T_{1/2}$~3.7 h in mice by IP injection) [7]. LDN-193189 has been shown to inhibit BMP type I receptors ALK2 (IC_{50}~10 nM in cells), ALK3 (IC_{50}~11 nM), ALK1 (310 nM), and ALK6 in preference to TGF-β/GDF/activin type I receptors ALK5 (IC_{50}~1 μM in cells), ALK4, and ALK7. It exhibits improved selectivity for this family of receptors versus receptor tyrosine kinases and AMPK as compared to dorsomorphin yet still exhibits potent activity against RIPK2 and ABL family kinases. Despite its limitations in kinome selectivity, LDN-193189 has been a useful tool in probing BMP-mediated effects due to its high degree of potency, provided appropriate complementary methods are utilized for confirmation. The 2-aminopyridine compound K02288 (Fig. 1b), identified by a melting point shift screen with ALK2 kinase versus TGFβ type I kinase ALK5, shows comparable high selectivity for BMP versus TGF-β/activin type I receptors, as seen in LDN-193189 but with a substantially less off-target effects in the kinome [8]. K02288 exhibits greater potency in biochemical kinase assays than cell-based assays of signaling, suggesting that cell permeability may limit its activity in cells. Several derivative compounds of both scaffolds have been described as having a greater degree of selectivity for particular receptors or having altered physiochemical or drug-like properties. LDN-212854 is a structural isomer of LDN-193189 with similar physiochemical properties but by virtue of its 5-quinoline versus 4-quinoline pendant group has altered complementarity with residues in the deep hydrophobic binding pocket of ALK2 [8] and thus exhibits severalfold greater selectivity for ALK2 (IC_{50}~16 nM in cells) than for ALK3 (IC_{50}~166 nM in cells) and ALK1 (IC_{50}~100 nM in cells) as compared to LDN-193189. LDN-212854 has relatively similar kinome selectivity to LDN-193189 and similar off-target effects. DMH1 is an analog of LDN-193189 with its solvent-exposed domain modified to an isopropyl ether, resulting in improved selectivity for ALK2 (IC_{50}~107 nM) and very low activity against ALK5, VEGR2, and other targets in biochemical kinase assays [7–9]. The potency of

DMH1 in cell-based applications appears to be diminished in proportion to its biochemical activity, suggesting diminished permeability as a result of the solvent-exposed domain changes, and requires higher concentrations in the micromolar range to achieve its effects, potentially limiting the gains in selectivity. DMH2, DHM3, and DMH4 are other derivatives of LDN-193189 with altered solvent-exposed domains which result in further trade-offs in selectivity for potency [9].LDN-214117 is a methyl-pyridine derivative of K02288 (Fig. 1) which exhibits increased selectivity for ALK2, improved cell permeability, and in vivo bioavailability as compared to K02288 [10]. With careful dose titration experiments, compounds with greater selectivity within the BMP signaling family such as LDN-212854 and LDN-214117 may be used in vitro to distinguish between the effects of ALK2 ligand-mediated versus ALK3 and ALK1 ligand-mediated signaling. For example, LDN-214117 can be used to resolve the activity of classic BMP ligands BMP-2 and BMP4 transduced by ALK3 (IC_{50} ~1 μM in cell-based BRE-luciferase reporter assays) as distinct from BMP6 transduced by ALK2 (IC_{50} ~100 nM in the same assay) [10]. Due to their different off-target effects, compounds from the pyrazolo-pyrimidine versus 2-aminopyridine scaffold families may be used together to diminish the likelihood that their effects are due to inhibition of off-target kinases.

Two of the procedures provided represent potentially high-throughput in vitro approaches for demonstrating pathway-specific effects of these inhibitors and thus confirm effects of signaling inhibition in parallel with recombinant or overexpressed endogenous inhibitors (e.g., noggin) [11], recombinant BMP extracellular domain proteins acting as ligand traps such as ALK1-Fc or ALK3-Fc [12, 13], or with siRNA-mediated suppression of receptors and/or ligands [14, 15]. The other two procedures describe the application of inhibitors to disrupt or attenuate embryonic and early postnatal BMP signaling via delivery to embryonic zebrafish following fertilization, or to pregnant dams at a safe interval following fertilization, and offer the potential to confirm by pharmacologic epistasis the impact of BMP signaling upon a given developmental phenotype or process.

Procedure 1 demonstrates how inhibitors can be applied to various mesenchymal cell lineages which demonstrate the capacity for osteogenic-like differentiation in response to BMP signaling, a response that is measured by an increase in alkaline phosphatase activity based on a colorimetric assay. In some cases, the capacity of BMPs to induce alkaline phosphatase expression in these cells is a surrogate for the osteogenic effects of these BMPs to induce endochondral bone in vivo, whereas the inhibition of this activity by various BMP type I receptor inhibitors is a pathway-specific anti-osteogenic effect. This assay can be adjusted to different genetically modified cell lines (RNAi vs. global or conditional BMP receptor knockout cell lines) or in the presence of specific BMP ligands to yield insight into the osteogenic stimuli, responsible ligands, and receptors that mediate osteogenic differentiation in such cells [11, 16]. This procedure will work with C2C12, NIH3T3 cells, and other multipotent, mesenchyme-derived cell lines or primary cultures.

Procedure 2 describes a versatile In-Cell Western technique, in which levels of phosphorylated SMAD proteins, the immediate downstream effectors of canonical BMP signaling, can be detected as a measure of BMP signaling activation. This technique can be applied for the identification and characterization of novel inhibitors as a high-throughput screening methodology (Fig. 2a) and can be used to determine concentration-dependent effects of various BMP inhibitors (Fig. 2b) [7, 8].

Procedure 3 illustrates how various BMP modulators can be tested for their impact on dorsoventral axis development in zebrafish embryos. This assay is also easily modified for studying knockout zebrafish strains or for determining the contribution of BMP receptors in a given early embryonic or neonatal process, provided that the compounds have been shown to penetrate the embryo based on dorsalization, as has been shown for dorsomorphin, LDN-193189, LDN-212854, K02288a, and DMH1 [2, 7–9].

Procedure 4 describes an application of these compounds to inhibit BMP signaling during fetal and early neonatal murine development. This technique has been successfully applied to rescue skull malformations due to excessive BMP signaling in osteoblast lineages. The ability to modulate phenotypes in other tissues would need to be determined empirically and ideally would include demonstrating the engagement of target by detecting decreased abundance of phosphorylated SMAD1/5/8 or decreased expression of the BMP transcriptional target *Id1* in target tissues, as was shown previously for LDN-193189 in such an application [17].

These methods are meant to serve as examples of potential applications in which the effects of pharmacologic inhibition of BMP signaling can be assayed. It should be emphasized that given the various off-target effects of each of the published BMP inhibitor compounds [18] and concentration-dependent effects, confirmation with two distinct chemotypes (pyrazolo-pyrimidine versus pyridine scaffold-based inhibitors), recombinant protein inhibitors and ligand traps, or genetic manipulation of receptor expression should be used to raise confidence in the conclusion. Due to the limitations of recombinant and genetic approaches to antagonizing the redundant activities of this pathway with over 30 identified structurally diverse ligands, the pharmacologic approach to inhibition is most effective as an initial screen for ascertaining a potential role of BMP signaling in a given biological process.

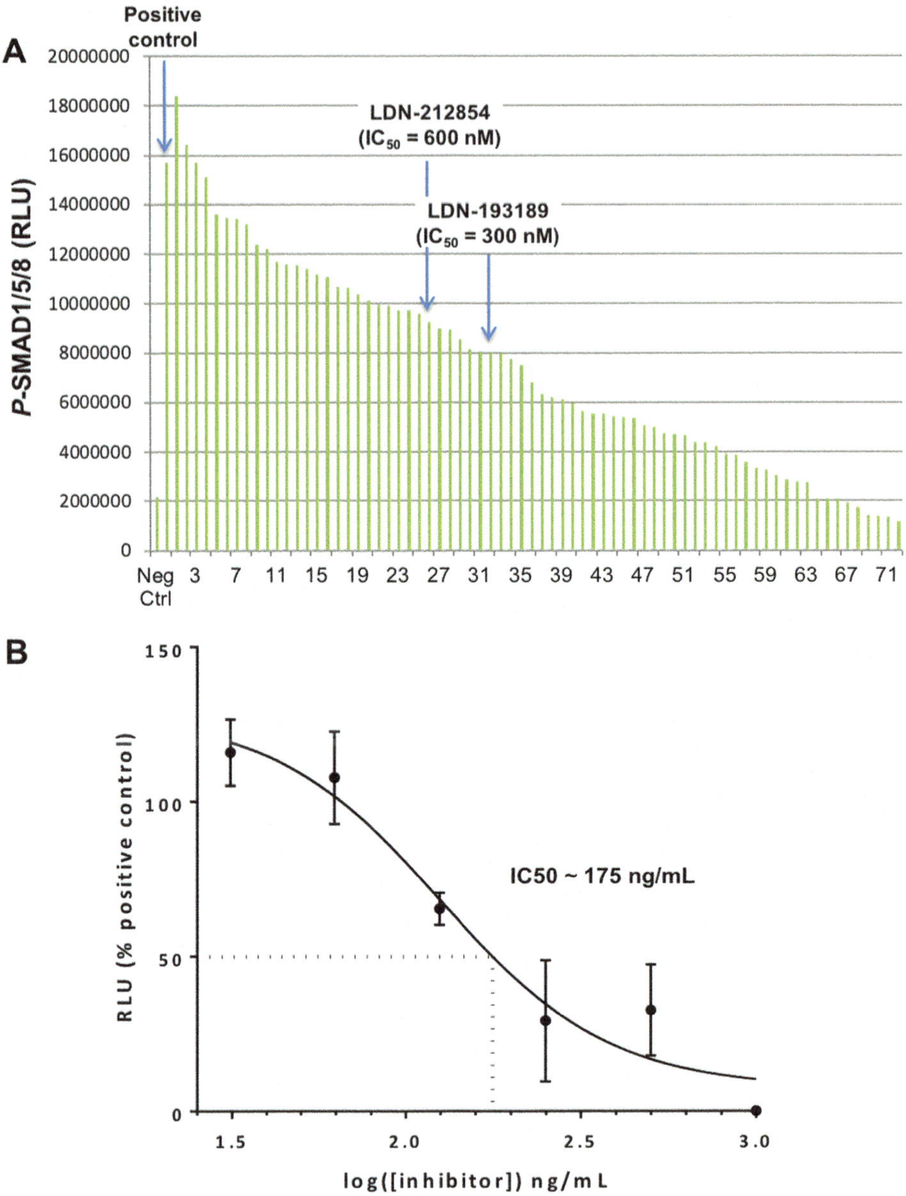

Fig. 2 Typical In-Cell Western assay data. (**a**) In-Cell Western data showing a high-throughput screen of dorsomorphin derivatives arranged by potency of inhibition of BMP4-mediated phosphorylation of SMAD1/5/8 in C2C12 cells measured in the presence of 100 nM of compounds and 10 ng/mL of BMP4, over 30′ at 37 °C. The residual phospho-SMAD1/5/8 activity detected after treatment with LDN-212854 and LDN-193189 under these conditions are indicated with their respective IC_{50} values measured in a similar cellular assay. (**b**) In-Cell Western is used to evaluate the inhibition of BMP9-mediated phosphorylation of SMAD1/5/8 in bovine aortic endothelial cells (BAEC) by varying concentrations of a BMP9 ligand trap, ALK1-Fc, measured in the presence of 1 ng/mL of BMP9 over 30′ at 37 °C. The IC_{50} for the activity of BMP9 indicated by dotted lines is approximately 175 ng/mL

2 Materials

2.1 Pharmacologic Modulators

1. LDN-214117 (Tocris Bioscience)
2. LDN-212854 (Tocris Bioscience)
3. LDN-193189 (Tocris Bioscience)
4. SB-431452 (Sigma)
5. DMH1 (Sigma)
6. K02288a (Sigma)
7. Dorsomorphin, a.k.a., Compound C (Tocris Bioscience)

2.2 Antibodies

1. Rabbit monoclonal anti-phospho-SMAD1/5/8 (Cell Signaling Technology #41D10)
2. Polyclonal goat anti-rabbit HRP conjugate (Cell Signaling Technology #7074)

2.3 Determining Kinetics of SMAD Signaling and Function In Vitro

1. Cells of interest with appropriate media (*see* **Note 1**)
2. 10% FBS-containing media for growth phase (*see* **Note 2**)
3. 1–2% FBS-containing media for starvation phase (*see* **Note 2**)
4. 1% Triton X-100/TBS: 1% Triton X-100, 1× Tris buffered saline (TBS)
5. Phosphate-buffered saline (PBS)
6. PnPP substrate Sigma P7998
7. Plate reader

2.4 In-Cell Western for Phosphorylated SMAD 1/5/8

1. 96-well clear-bottom plate with white walls, tissue culture treated. For cells that are poorly adherent, use appropriate coatings specific to cell type or consider using plate with improved binding such as lysine-coated or CellBIND (Corning) plates.
2. Cells of interest with appropriate media (*see* **Note 1**).
3. 10% FBS-containing media for growth phase (*see* **Note 2**).
4. 1–2% FBS-containing media for starvation phase (*see* **Note 2**).
5. Primary antibody anti-*p*-SMAD 1/5/8 Rabbit mAb, cell signaling #9516.
6. HRP anti-rabbit IgG, cell signaling.
7. ECL reagent (Bio-Fx Ultra Sensitive HRP Chemiluminescent Substrate, Surmodics #CHMI-0060).
8. 1× Phosphate-buffered saline (PBS).
9. 0.5% Glutaraldehyde (diluted from stock solution in PBS).
10. 2% BSA (bovine serum albumin) in PBS.
11. BMP and TGFβ ligands from R&D Systems diluted in 0.1% BSA PBS to desired concentrations (*see* **Note 3**).

2.5 Determining Kinetics of SMAD Signaling and Function In Vivo in Zebrafish

1. Zebrafish embryos isolated following fertilization
2. E3 buffer: 5 mM NaCl, 0.17 mM KCl, 0.33 mM $CaCl_2$, and 0.33 mM $MgSO_4$ [2]

2.6 Protocol for Murine In Utero Suppression of SMAD Signaling

1. Mice breeding pair with timed matings judged by plug
2. 31G insulin needle syringe combination for IP injection
3. CO_2/method for euthanasia

3 Methods

Inhibitor(s) to be used should be chosen based on the potency, selectivity, and cell permeability that are most suitable for the specific application or have been previously established empirically (Table 1, *see* **Notes 4** and **5**). Preliminary experiments to determine optimal dose ranges of inhibitors are recommended (*see* **Note 6**).

3.1 Determining Kinetics of BMP-Induced Osteogenic Differentiation Via Alkaline Phosphatase Expression In Vitro

Goal: To determine the temporal requirements of BMP-SMAD signaling in recruiting osteogenic differentiation

1. Seed 2000 cells/well in 10% FBS culture media. Incubate at 37 °C and 5% CO_2 overnight.
2. Remove media and replace with 2% FBS media in total volume of 50 μL in a 96-well plate and incubate overnight.
3. Stimulate with inhibitor at different time points (*see* **Note 6**).
   (a) T0, T1 h, T2 h, T4 h, T8 h, T16 h, T24 h, T32 h, T48 h, T72 h, T96 h
4. At T120 h, wash cells by replacing media with 200 μL PBS three times, discarding liquid after each wash, and then permeabilize/solubilize by the addition of 50 μL of 1% Triton X-100/TBS. Seal plate, and agitate or vortex for 30 s gently enough to avoid cross-contamination or loss of samples.
5. Transfer 20 μL of each sample to a clear 96-well microplate, and then add 100 μL of PnPP substrate to each well.
6. Incubate for 1 h at room temperature.
7. Read absorbance at 405 nm in a microplate spectrophotometer.

3.2 In-Cell Western for Phosphorylated SMAD 1/5/8

1. Grow adherent cells to confluence in a tissue-culture-treated clear bottom, white-walled 96-well plate. Seed the suspension of cells, in 100 μL 10% FBS media per well, and allow cells to achieve confluence at 37 °C and 5% CO_2 with humidification (*see* **Note 1**).

Table 1
BMP signaling modulators and basic information on commonly used concentrations of published data. Concentrations should be adapted for individual assays (*see* Notes 3 and 4). Please note that concentrations listed are not inclusive of all concentrations published but are representative of the common ranges used

Compound	Receptor for which compound has high specificity	Concentration in vitro	Concentration in vivo (mice)	Concentration in vivo (zebrafish)
LDN-214117	ALK2	21 nM–10 μM [10]		
LDN-212854	BMP Type 1 receptors	1 nM–10 μM [8]	To prevent HO 6 mg/kg, i.p. twice daily; 3 mg/kg, s.c. twice daily [23]	
LDN-193189	BMP Type 1 receptors: ALK1, ALK2, ALK3, ALK 6	1 nM–10 μM [8] 0.05–5 μM [21] (*see* **Note 8**)	To prevent HO: 6 mg/kg, i.p. twice daily; in utero: 2.5 mg/kg i.p. daily	
SB-431452	TGF-β Type 1 receptors: ALK4, ALK5, ALK7	0.05–5 μM [21]	10 mg/kg i.p. daily [24]	
DMH1	ALK2	5–10 μM [25] 3–100 μM [26] (*see* **Note 11**)		0.2 μM at 3 h.p.f. [9] (*see* **Note 11**)
K02288a	BMP Type 1 receptors: ALK1, ALK2, ALK3, ALK6	0.1 nM–10 μM [8]		8–10 μM at 12 h.p.f [20] 10–20 μM [27]
Dorsomorphin (Compound C)	ALK2, ALK3, ALK6	5 μM [21] 2 μM [11]		100 μM at 2 d.p.f. [27] 10 μM at 48 h.p.f. [20]

2. Wash confluent cells with 1× PBS in a sterile manner and serum-starve cells overnight in 45 μL 1% FBS media per well.
3. Add 5 μL of varying dilutions of inhibitor compounds in PBS or vehicle control, and incubate for 10–30 min at 37 °C and 5% CO_2 (*see* **Notes 7** and **8**), being careful not to allow plate to cool.
4. Add BMP/TGF-β ligands in a volume of 5.5 μL to wells and incubate for 30 min at 37 °C and 5% CO_2, being careful not to allow plate to cool.
5. Remove supernatant and add 50 μL ice-cold methanol. Incubate at room temperature for 15 min.

6. Remove methanol and add 50 μL 0.5% glutaraldehyde in PBS. Incubate at room temperature for 15 min.
7. Wash in plate washer four times with 200 μL PBS or equivalent.
8. Block with 200 μL per well 2% BSA in PBS for 1 h at room temperature on gentle rotator.
9. Remove blocking solution and add 50 μL per well primary antibody (rabbit mAb anti-*p*-SMAD 1/5/8), diluted 1:400 in 2% BSA in PBS. Incubate overnight at 4 °C with gentle rotation or rocking.
10. Wash in plate washer three times with 200 μL PBS or equivalent.
11. Add 50 μL per well secondary antibody (HRP-anti-rabbit IgG, cell signaling), diluted 1:400 in 2% BSA in PBS. Incubate 1 h at room temperature on rotator.
12. Wash plate in plate washer four times with 200 μL PBS or equivalent.
13. Add 30 μL ECL reagent and read plate in a sensitive microplate luminometer immediately. Do not let the plate develop more than 2–3 min prior to reading.

3.3 Determining Kinetics of SMAD Signaling and Function In Vivo in Zebrafish [2, 19]

Goal: To inhibit SMAD signaling in zebrafish embryos to determine BMP signaling role in development

1. Transfer newly fertilized eggs to 96-well microplate in 200 μL E3 buffer.
2. Add inhibitor at 5–10 μM at 4–8 h postfertilization (h.p.f) and incubate at 28.5 °C (*see* **Note 9**).
3. At 12, 24, and 48 h.p.f., examine embryos for dorsalization or ventralization of embryonic axis.

3.4 Protocol for Murine In Utero Suppression of SMAD Signaling [17]

Goal: To inhibit SMAD signaling during embryonic development

1. Day E-5 to E-1: Check for morning plugs in female mice in breeding cages.
2. Day E0 to E13: Remove pregnant dams to single cage housing and monitor pregnancy.
3. Day E14 to P0: Begin daily treatment with LDN-193189 2.5 mg/kg i.p. to pregnant dams (*see* **Note 10**).
4. P0 to P15: Continue treating lactating females with LDN-193189 at a dose of 2.5 mg/kg i.p.
5. P16: May euthanize pups for analysis or continue monitoring pups for phenotype.

4 Notes

1. For In-Cell Western, C2C12 cells (ATCC) are commonly for BMP 2,4,6, and 7 or TGF-β1/2/3 signaling experiments. Bovine aortic endothelial cells (BAECs; Lonza or Genlantis) are used for BMP9 or BMP10 experiments.
2. The recommended media for C2C12 cells is RPMI with 10% FBS. The recommended media for BAECs is DMEM supplemented with high glucose and glutamine with 10% FBS. For other mesenchymal-derived cell lines, more custom types of media can also be used, such as MCDB131.
3. For the In-Cell Western, concentrations close to the EC_{50} of the BMP ligands are often used. Common ranges tested are 10–25 ng/mL for BMP 2,4; 50 ng/mL for BMP6,7; and 0.5–1 ng/mL for BMP9/10 (specifically for endothelial cells).
4. To identify the optimal inhibitor for a given application, consider PK, IC_{50}, and selectivity data that have been previously published for LDN-214117, LDN-212854 [8], LDN-193189 [8, 20, 21], SB-431452 [22], DMH1 [8], K02288a [8, 20], and dorsomorphin (Compound C) [8, 21]).
5. At moderate to high micromolar concentrations, several of these BMP inhibitor compounds lose their selectivity and act as pan-BMP/TGF-β/activin type I receptor inhibitors [8, 10] with significant off-target effects against other kinases. This quality can be advantageous if blocking several or all of the BMP/TGFβ type I receptors is desirable but can generate nonspecific results.
6. Prior to the experiment, it may be helpful to run a pilot study testing varying concentrations of inhibitor that are relevant for the cell type, the complement of BMP/TGF-β/activin receptors being targeted, and the assay in question. The examination of signaling kinetics at different time points may also be helpful for optimizing the sensitivity of the assay and tailoring to biological endpoints.
7. In vitro, dorsomorphin and other compounds will attenuate steady-state levels of phosphorylated SMAD1/5/8 within minutes of addition to culture media. Given their solubility and high permeability, pre-treatment of cells for approximately 10–30 min. prior to the addition of ligand agonists is sufficient to abrogate effects of ligands in activating signaling [21].
8. Inhibitor compounds can be diluted in PBS. An 11-point ½-log scale of dilutions, i.e., 100 μM, 31.3 μM, 10 μM, 3.13 μM, 1 μM, 313 nM, 100 nM, 31 nM, 10 nM, 3.1 nM, and 1 nM, is sufficient to cover most biologically relevant concentrations of inhibitor within a standard row of a 96-well

plate, allowing one control well with no inhibitor. Triplicate measurements are recommended.

9. The effects of various inhibitors upon dorsalization in zebrafish embryos appear to be related to the time at which inhibitors are added to embryos, with most severe effects occurring at 3–4 h. p.f. and less severe effects occurring beyond 7–8 h.p.f. The desired degree of dorsalization can be tailored via the concentration to which embryos are exposed as well as the timing of exposure.
10. LDN-193189 plasma levels decrease to 50% of peak levels within 4 h after administration.
11. The LDN-193189 derivative DMH1 has excellent selectivity when assayed in biochemical kinase assays but loses some of this selectivity in cell assays apparently as a result of its reduced permeability in cells [8].

References

1. Wang RN, Green J, Wang Z et al (2014) Bone morphogenetic protein (BMP) signaling in development and human diseases. Genes Dis 1:87–105
2. Yu PB, Hong CC, Sachidanandan C et al (2008) Dorsomorphin inhibits BMP signals required for embryogenesis and iron metabolism. Nat Chem Biol 4:33
3. Steinbeisser H, Fainsod A, Niehrs C et al (1995) The role of gsc and BMP-4 in dorsal-ventral patterning of the marginal zone in Xenopus: a loss-of-function study using antisense RNA. EMBO J 14:5230–5243
4. Biehs B, François V, Bier E (1996) The Drosophila short gastrulation gene prevents Dpp from autoactivating and suppressing neurogenesis in the neuroectoderm. Genes Dev 10:2922–2934
5. Nagaso H, Suzuki A, Tada M et al (1999) Dual specificity of activin type II receptor ActRIIb in dorso-ventral patterning during zebrafish embryogenesis. Develop Growth Differ 41:119–133
6. Chaikuad A, Alfano I, Kerr G et al (2012) Structure of the bone morphogenetic protein receptor ALK2 and implications for fibrodysplasia ossificans progressiva. J Biol Chem 287:36990–36998
7. Yu PB, Deng DY, Lai CS et al (2008) BMP type I receptor inhibition reduces heterotopic [corrected] ossification. Nat Med 14:1363–1369
8. Mohedas AH, Xing X, Armstrong KA et al (2013) Development of an ALK2-biased BMP type I receptor kinase inhibitor. ACS Chem Biol 8:1291
9. Hao J, Ho JN, Lewis JA et al (2010) In vivo structure activity relationship study of Dorsomorphin analogs identifies selective VEGF and BMP inhibitors. ACS Chem Biol 5:245
10. Mohedas AH, Wang Y, Sanvitale CE et al (2014) Structure–activity relationship of 3,5-Diaryl-2-aminopyridine ALK2 inhibitors reveals unaltered binding affinity for fibrodysplasia ossificans progressiva causing mutants. J Med Chem 57:7900
11. Yu PB, Deng DY, Beppu H et al (2008) Bone morphogenetic protein (BMP) type II receptor is required for BMP-mediated growth arrest and differentiation in pulmonary artery smooth muscle cells. J Biol Chem 283:3877–3888
12. Cunha SI, Pardali E, Thorikay M et al (2010) Genetic and pharmacological targeting of activin receptor-like kinase 1 impairs tumor growth and angiogenesis. J Exp Med 207:85–100
13. Derwall M, Malhotra R, Lai CS et al (2012) Inhibition of bone morphogenetic protein signaling reduces vascular calcification and atherosclerosis. Arterioscler Thromb Vasc Biol 32:613–622
14. Yu PB, Beppu H, Kawai N et al (2005) Bone morphogenetic protein (BMP) type II receptor deletion reveals BMP ligand-specific gain of signaling in pulmonary artery smooth muscle cells. J Biol Chem 280:24443–24450

15. Bagarova J, Vonner AJ, Armstrong KA et al (2013) Constitutively active ALK2 receptor mutants require type II receptor cooperation. Mol Cell Biol 33:2413–2424
16. Yung L-M, Sánchez-Duffhues G, Ten Dijke P et al (2015) Bone morphogenetic protein 6 and oxidized low-density lipoprotein synergistically recruit osteogenic differentiation in endothelial cells. Cardiovasc Res 108:278–287
17. Komatsu Y, Yu PB, Kamiya N et al (2013) Augmentation of Smad-dependent BMP signaling in neural crest cells causes craniosynostosis in mice. J Bone Miner Res Off J Am Soc Bone Miner Res 28:1422
18. Vogt J, Traynor R, Sapkota GP (2011) The specificities of small molecule inhibitors of the TGFß and BMP pathways. Cell Signal 23:1831–1842
19. Williams CH, Hong CC (2015) High content screening for modulators of cardiovascular or global developmental pathways in zebrafish. Methods Mol Biol 1263:167–174
20. Sanvitale CE, Kerr G, Chaikuad A et al (2013) A new class of small molecule inhibitor of BMP signaling. PLoS One 8:e62721
21. Horbelt D, Boergermann JH, Chaikuad A et al (2015) Small molecules Dorsomorphin and LDN-193189 inhibit myostatin/GDF8 signaling and promote functional myoblast differentiation. J Biol Chem 290:3390
22. Inman GJ, Nicolás FJ, Callahan JF et al (2002) SB-431542 is a potent and specific inhibitor of transforming growth factor-β superfamily type I activin receptor-like kinase (ALK) receptors ALK4, ALK5, and ALK7. Mol Pharmacol 62:65–74
23. Dey D, Bagarova J, Hatsell SJ et al (2016) Two tissue-resident progenitor lineages drive distinct phenotypes of heterotopic ossification. Sci Transl Med 8:366ra163
24. Waghabi MC, de Souza EM, de Oliveira GM et al (2009) Pharmacological inhibition of transforming growth factor β signaling decreases infection and prevents heart damage in acute Chagas' disease. Antimicrob Agents Chemother 53:4694
25. Sheng Y, Sun B, Xie X et al (2015) DMH1 (4-[6-(4-isopropoxyphenyl)pyrazolo[1,5-a]pyrimidin-3-yl]quinoline) inhibits chemotherapeutic drug-induced autophagy. Acta Pharm Sin B 5:330–336
26. Hover LD, Owens P, Munden AL et al (2016) Bone morphogenetic protein signaling promotes tumorigenesis in a murine model of high-grade glioma. Neuro Oncol 18:928–938
27. Windhausen T, Squifflet S, Renn J et al (2015) BMP signaling regulates bone morphogenesis in zebrafish through promoting osteoblast function as assessed by their nitric oxide production. Molecules 20:7586–7601

Chapter 17

Bone Morphogenetic Proteins (BMPs) and Bone Regeneration

Deboleena Kanjilal and Jessica Ann Cottrell

Abstract

Many research methods exist to elucidate the role of BMP-2 during bone regeneration. This chapter briefly reviews important animal models used in these studies and provides details on the rat femur defect model. This animal model is frequently utilized to measure the efficacy of osteogenic factors like BMP-2. Detailed information about delivery methods, dose range, and dose duration used in BMP-2-related studies are provided.

Key words BMP-2, Bone, In vivo, Rat femur defect model, BMP-2 carriers, Chondrocytes, Osteoblasts

1 Introduction

Bone morphogenetic protein-2 (BMP-2) is an osteogenic growth factor [1–3], essential for development [4, 5], and necessary for bone regeneration [6–9]. Several BMP family members can induce bone formation, but only BMP-2 is FDA approved for clinical use [10, 11]. Purified recombinant human BMP-2 (rhBMP-2) is used clinically to treat open tibia fractures, promote spinal arthrodesis, and treat oral-maxillofacial bone defects. However, the osteogenic efficacy of BMP-2 in experimental animal models and in the clinic is dependent on preparation, dosage, carrier materials, graft use, and method of administration [2, 8, 12–15]. In the clinic, supraphysiological doses of rhBMP-2 are used to promote bone formation, although high doses of BMP-2 are not always as efficient as lower dosages [13, 16–18]. The use of BMP-2 has a risk of causing tissue toxicity as do many other growth factor treatments [19, 20]. BMP-2 in larger doses can cause ectopic bone formation outside of the desired target and can induce the formation of tumors [21, 22]. Research to reduce these risks through localized administration of lower BMP-2 doses paired with a better delivery method such as a controlled release matrix is an active area of investigation

Melissa B. Rogers (ed.), *Bone Morphogenetic Proteins: Methods and Protocols*, Methods in Molecular Biology, vol. 1891, https://doi.org/10.1007/978-1-4939-8904-1_17,

[17, 19, 21–23]. Other preclinical studies focus on finding optimal conditions for use of BMP-2 to treat skeletal injuries (e.g., tibia fractures) or promote arthrodesis (e.g., spinal fusions) [7, 12, 16, 24–28].

Several animal models have been used to study the use of BMP-2 and various application methods in preclinical studies. These models include rodents [29–31], rabbits [32], dogs [33], sheep [34], and nonhuman primates [35, 36]. In order to mimic specific orthopedic situations, different types of surgeries are performed on these animals such as spinal fusion [37], long bone fractures, calvarial defects [38], and maxillofacial injuries [39]. Most preliminary in vivo studies are done on rodents, as they are small in size and easy to handle [40]. Rodents can be manipulated in various ways to study the role of BMP-2 in bone formation [41]. For example, the rat femur segmental defect model can be used for dose-related studies [29], pharmacokinetic and retention studies [42], and efficacy of carrier-mediated delivery systems [7]. One of the most common surgeries performed on rodents, the rat femur defect surgery, to study BMP-2 delivery in collagen sponge is described below.

2 Materials

2.1 Rat Femur Defect

1. Rats: Rats of at least 12 weeks of age and weighing between 300 and 500 g are preferred, because application of the fixator onto the femur of smaller rats is more difficult. Some common strains used include Wistar, athymic, or Sprague Dawley.
2. List of drugs:
   (a) Acepromazine: Tranquilizer
   (b) Ketamine: Anesthesia
   (c) Xylazine: Anesthesia
   (d) Buprenorphine: Pain reliever
   (e) Enrofloxacin: Prophylactic antibiotic treatment for postoperative care.

   Rats can also be treated with a subcutaneous or intramuscular injection of acepromazine (1–2 mg/kg) approximately 10 min before ketamine/xylazine injection. This helps the rat react better to the ketamine/xylazine mixture. Ketamine and xylazine can be administered separately or as a cocktail. The xylazine is usually diluted ten times from its original concentration of 100 mg/mL to 10 mg/mL.
3. Fixator: Many different types of fixators exist such as internal and external fixators. The fixator we prefer to use is made up of polysulfone plate and two small stainless steel plates and consists of four holes for screw fixation to femur, two holes proximal and two holes distal of impending defect [43] (Fig. 1) (*see* **Note 1**).

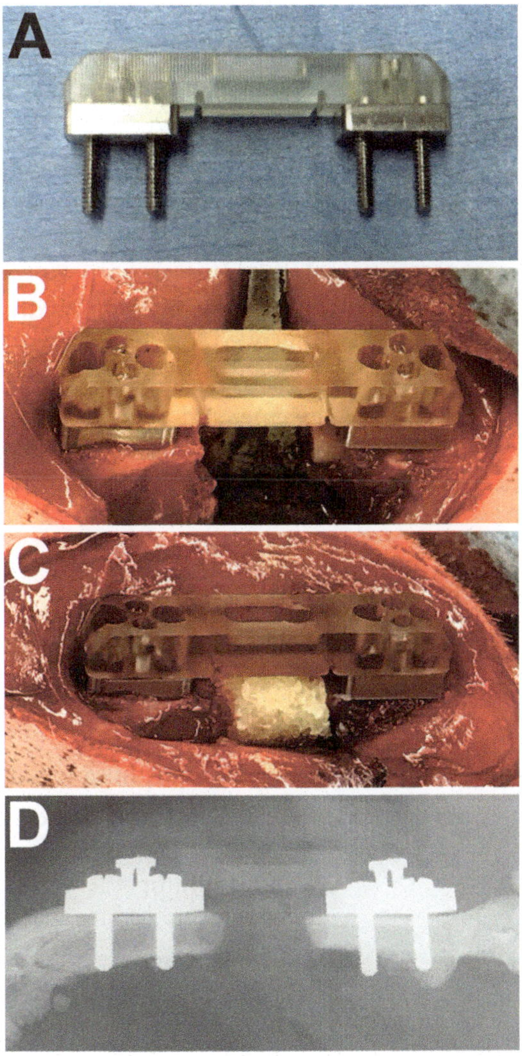

Fig. 1 Rat femur defect surgery using an internal fixator. (**a**) An assembled fixator prior to surgical placement in vivo. (**b**) The fixator attached to the femur and an 8 mm defect cut. (**c**) The test material press fit at the defect site. (**d**) A postsurgical radiograph image of the fixator on the defect site

4. Stainless steel screws: For the fixator mentioned above, four 1/8″-long 000-120 oval fillister head screws are needed to attach the polysulfone plate to the two stainless steel plates, and four ¼″ 00-90 oval fillister head screws are used to attach the fixator to the rat femur.
5. Saw: A small hand saw, Gigli saw, or an oscillating saw may be used to create the defect.
6. Sterile saline for irrigation.
7. 3-0 to 5-0 resorbable sutures.

8. Heating pad, maintained at 37 °C.
9. Drill bit: The drill bit diameter will be dependent on the fixator screws that are used. For the 00-90 screws, a 1 mm diameter drill bit is used.
10. Absorbable collagen hemostatic sponge cut to size to fill the intended defect space. Other carriers can be used depending upon defect size, animal model, and location of defect (*see* **Note 2**).
11. Infuse® rhBMP-2 (Medtronic Inc., Minneapolis, MN; FDA approved for clinical use).
12. BMP-2 buffer: 5 mM glutamate, 5 nM sodium chloride, 2.5% glycine, and 0.5% sucrose.

3 Methods

3.1 Rat Femur Segmental Defect Surgery (See Note 3)

1. Weigh the rat and determine the amount of acepromazine, ketamine, xylazine, buprenorphine, and enrofloxacin needed per rat based on weight (*see* Table 1) (*see* **Note 4**).
2. Inject buprenorphine subcutaneously at the dorsal scapula.
3. Inject enrofloxacin intramuscularly in the contralateral thigh.
4. Shave the right hind leg of the rat to be used.
5. Place the rat on a heated pad and prepare the right hind leg sterilely for surgery using standard aseptic techniques that includes multiple chlorhexidine washes, isopropanol rinses, and povidone-iodine application.
6. Make a 4 cm square hole in a surgical drape and place it over the rat, only exposing its right thigh.
7. Once the rat has obtained a deep level of anesthesia, make a 3–4 cm incision on the lateral side of the right leg centered between the hip and knee.

Table 1
Tranquilizer, anesthetic, and analgesic doses for rats

Drug	Dose	Purpose
Acepromazine	1–2 mg/kg	Tranquilizer
Ketamine	60–80 mg/kg	Anesthesia
Xylazine	5–10 mg/kg	Anesthesia
Buprenorphine	0.05 mg/kg	Pain relief
Enrofloxacin	22.7 mg/kg	Prophylactic antibiotic treatment for postoperative care

8. Bluntly dissect through the muscle bundles to expose the diaphysis of the femur.
9. Position the fixator on top of the femur, ensuring that no muscle is between the fixator and the bone and that the fixator is centered over the impending defect. Secure the fixator to the femur using locking bone forceps.
10. To secure the fixator, use the drill to make pin holes on the bone and screw in the pins into both cortices of the rat femora through the holes in the fixator and the newly made holes in the bone. Drill the most proximal hole first and then place a screw partially into the cortex to stabilize the fixator over the bone. Then, drill the most distal hole and partially insert the screw in one cortex. This will allow you to rotate the fixator on the bone to adjust the position. Once the proximal and distal holes have been screwed and fixed, drill the remaining two holes and fix it in place.
11. Irrigate the site of fixator with sterile saline.
12. Use the saw to remove a portion of the femur diaphysis, typically the length of the defect is between 4 and 8 mm (*see* **Note 5**).
13. After cutting the defect, irrigate the region again to remove any remaining bone dust or bone marrow contents.
14. Cut the collagen sponges into strips of $15 \times 15 \times 2.75$ mm (*see* **Note 6**).
15. Dilute the recombinant human BMP-2 with BMP-2 buffer to make 120 μg/mL of solution (*see* **Note 2**). Pipette 100 μL of solution onto cut collagen sponge to get final dose of 12 μg of rhBMP-2. Alternative BMP-2 carriers can be used (*see* **Note 2**).
16. Press fit the rhBMP-2-collage sponge into the defect.
17. Close the muscle over the defect using resorbable sutures.
18. Close the surgical incision using interrupted sutures.
19. Radiograph the rat femur to confirm successful defect production and application of the fixator and to provide a baseline to compare bone growth form. X-ray dosing used for rats are approximately 35 kVp for 90 s using MIN-R 2000 plus film (18×24 cm) with a Faxitron MX-20 (*see* **Notes** 7 and **8**).
20. Keep the rats on the heated pad till the rats regain consciousness.
21. Return rats to cages once the animals are sternal and alert.
22. As part of postoperative care, administer enrofloxacin via intramuscular injection once a day for 2 additional days post-surgery.

3.2 Microcomputed Tomography (Micro-CT)

1. Micro-CT can be used to measure bone volume within and outside the defect site.
2. Harvest the femur and fix it in 10% formalin.
3. Transfer the sample into saline or 70% ethanol before scanning to avoid formaldehyde contamination of the micro-CT scanner.
4. Prior to scanning, remove any metal components from the sample to prevent X-ray beam artifacts. Scanning conditions are dependent on the bone density (*see* **Note 9**).
5. Reconstruct the images into a 3D volume and analyze data using software available for the scanner used or other available software, such as ImageJ or Analyze 12.0 (Analyze direct) [44].

3.3 Histology and Histomorphometry for Animal Specimens

1. After micro-CT, embed samples in polymethyl methacrylate (PMMA) (methyl methacrylate solution treated with dehydrated benzoyl peroxide). The sample is dehydrated in 100% ethanol and xylene and then inserted into PMMA solution [45].
2. Cut 50–100 μm sagittal section through the target site for BMP-2 activity, polish, and glue to a 1/8″ (thick) × 1″ (wide) × 3″ (long) piece of plexiglass.
3. Stain the slide with Stevenel's blue (soft tissue) and van Gieson's picrofuchsin (mineralized tissue) [46].
4. Using digital images of the stained sections, measure cartilage and bone area or other morphometric parameters using image analysis software [7]. Alternatively, the undecalcified or decalcified samples can also be embedded in paraffin after fixing the sample in paraformaldehyde PFA and dehydrating in ethanol (*see* **Note 10**).

4 Notes

1. We prefer the custom fixator designed by Oest et al. [43] as it moves the plate away from the defect space and can be removed without affecting the bone for mechanical testing.
2. Identifying suitable carriers continues to pose challenges as conditions needed for carrier production may not be compatible with maintaining BMP-2 in a biological active state [12]. Other important factors to consider include potential host response to the material, porosity, and mechanical properties such as compressive and tensile strength, ease and cost of production, and the ability to control BMP-2 release. Because high doses of BMP-2 contribute to adverse effects such as ectopic bone formation [47] and osteoclast activation [48],

encapsulation [49] or entrapment of the BMPs may be more favorable as the required BMP-2 dose may lower [50]. Currently used carrier materials are briefly defined below.

(a) **Ceramics** (calcium phosphates and bioglass) have favorable mineral composition, microarchitecture, biodegradability, and bioresorbability properties [51–53], although brittleness limits their utility for treating defects in weight-bearing bones.

(b) **Synthetic polymers** (polycaprolactone (PCL), polyethylene glycol (PEG), poly(lactic-co-glycolic acid) (PLGA)) offer design flexibility and lower chances of disease transmission [36, 54–56]. However, cellular degradation may cause an inflammation, reduced mechanical integrity, and poor clearance of degraded products [57].

(c) **Natural polymers**: In addition to collagen, other natural polymer carriers include chitosan, silk fibroin, gelatin, and hyaluron (HA) [58–64]. Although promising either alone or combined with synthetic materials, disadvantages include sourcing, processing, and potential disease transmission [65].

(d) **Microparticles and nanoparticles**: All of these materials can be delivered in the form of microparticles (1–1000 μm) or nanoparticles (10–1000 nm) with the advantage of a high surface-to-volume ratio [66].

3. Thirty-eight percent of all fracture studies are done on rats [29, 67–69], including the rat femur segmental defect surgery [41]. This model involves cutting a critical sized defect in the diaphysis of the femur, with the proximal and distal ends held together by a fixator. The surgery is relatively simple to perform and can be manipulated in various ways to study different aspects of osteogenesis. For example, an external fixator can be used to manipulate mechanical loading conditions on the defect [70]. Age is a variable that can effect fracture healing; therefore it is important to select age-appropriate rats for each study. In most experiments, rats are used at 12 weeks of age [71], but studies have used rats that were anywhere from 6 to 20 weeks old [72]. The femur defect surgery has relatively lower risk than calvarial defect and spinal fusion surgeries. Calvarial defects are usually close to the brain, while spinal fusion surgeries are more complex and lengthy.

4. If using an oscillating saw, irrigate the site with saline while sawing to remove bone fragments and reduce thermal injury. Avoid any contact with the plate and saw.

5. Determine maximum volume of saline that can be absorbed by different-sized collagen sponge before proceeding to the next

step and adjust rhBMP-2 concentration and volumes accordingly.

6. In mesenchymal stem cells, low concentrations of rhBMP-2- stimulate adipocyte differentiation, while higher concentrations stimulate chondrocyte and osteoblast differentiation [73]. The delivery system influences the optimal rhBMP-2 dose [74]. For the rat femur segmental defect surgery described here, 12 μg of rhBMP-2 applied via absorbable collagen sponge appears to be the optimum dose for repair [29]. Doses using different carrier materials in different repair models typically range from 1 to 12 μg [7, 75–77]. BMP-2 can be released rapidly by adsorption of BMP-2 on the surface of the carrier material or more slowly by immobilization or entrapment of BMP-2 in the carrier material [78]. Rapid or burst release of rhBMP-2 is achieved by loading high concentrations of rhBMP-2 on absorbable collagen sponge [62]. Collagen, one of the most widely used natural polymers, has been FDA approved for delivering BMP-2 and BMP-7 in vivo. The major nonmineral component of bone, collagen has a high affinity for BMP-2 and high osteogenic properties [79] and can be used as a film, gel, or sponge.
7. These conditions can vary and are dependent on bone density and Faxitron capabilities.
8. X-rays of the affected limb can be used to assess healing during the course of the experiment and to gauge whether the defect has been bridged with bone (calcified tissue) at the endpoint.
9. We scan rat femur defect samples at 70 kVp, 142 mA, and 12 μm voxel size using a Bruker SkyScan 1172 scanner with 0.5 mm aluminum filter to reduce bone hardening.
10. Paraffin sections are more commonly used for bone enzyme histochemistry and immunohistochemistry [80]. Sections (5 μm) are applied to polylysine or similarly coated slides to promote section adhesion and incubated at 60 °C for 2 h [81]. Prior to staining, the sections are cooled to room temperature and deparaffinized in xylene and rehydrated through a descending ethanol series [82]. Stain the slides with hematoxylin and eosin (H&E) [11] and image under a light microscope.

References

1. Urist MR (1965) Bone: formation by autoinduction. Science 150(3698):893–899
2. Urist MR, Strates BS (1971) Bone morphogenetic protein. J Dent Res 50(6):1392–1406
3. Wozney JM et al (1988) Novel regulators of bone formation: molecular clones and activities. Science 242(4885):1528–1534
4. Kramer J et al (2000) Embryonic stem cell-derived chondrogenic differentiation in vitro: activation by BMP-2 and BMP-4. Mech Dev 92(2):193–205
5. Zhang H, Bradley A (1996) Mice deficient for BMP2 are nonviable and have defects in

amnion/chorion and cardiac development. Development 122(10):2977–2986

6. Sampath TK et al (1993) Drosophila transforming growth factor beta superfamily proteins induce endochondral bone formation in mammals. Proc Natl Acad Sci U S A 90 (13):6004–6008
7. Subramanian S et al (2015) Salicylic acid-based polymers for guided bone regeneration using bone morphogenetic protein-2. Tissue Eng Part A 21(13–14):2013–2024
8. Tsuji K et al (2006) BMP2 activity, although dispensable for bone formation, is required for the initiation of fracture healing. Nat Genet 38 (12):1424–1429
9. Wang EA et al (1988) Purification and characterization of other distinct bone-inducing factors. Proc Natl Acad Sci U S A 85 (24):9484–9488
10. Conway JD et al (2014) BMP-7 versus BMP-2 for the treatment of long bone nonunion. Orthopedics 37(12):e1049–e1057
11. Jo JY et al (2015) Sequential delivery of BMP-2 and BMP-7 for bone regeneration using a heparinized collagen membrane. Int J Oral Maxillofac Surg 44(7):921–928
12. Begam H et al (2017) Strategies for delivering bone morphogenetic protein for bone healing. Mater Sci Eng C Mater Biol Appl 70 (Pt 1):856–869
13. Carreira AC et al (2014) Bone morphogenetic proteins: structure, biological function and therapeutic applications. Arch Biochem Biophys 561:64–73
14. Carreira AC et al (2014) Bone morphogenetic proteins: facts, challenges, and future perspectives. J Dent Res 93(4):335–345
15. Carreira AC et al (2015) Bone morphogenetic proteins: promising molecules for bone healing, bioengineering, and regenerative medicine. Vitam Horm 99:293–322
16. Herberg S et al (2014) Low-dose bone morphogenetic protein-2/stromal cell-derived factor-1beta cotherapy induces bone regeneration in critical-size rat calvarial defects. Tissue Eng Part A 20(9–10):1444–1453
17. James AW et al (2016) A review of the clinical side effects of bone morphogenetic protein-2. Tissue Eng Part B Rev 22(4):284–297
18. Zhou J-JA (2008) Enhancing production of recombinant BMP-2 in mammalian cell culture systems by inhibition of pro-protein cleavage using 9DR peptides. Department of Dentistry, University of Toronto, Toronto. pp 1–121
19. Guillot R et al (2016) Assessment of a polyelectrolyte multilayer film coating loaded with BMP-2 on titanium and PEEK implants in the rabbit femoral condyle. Acta Biomater 36:310–322
20. Padgett RW, Wozney JM, Gelbart WM (1993) Human BMP sequences can confer normal dorsal-ventral patterning in the Drosophila embryo. Proc Natl Acad Sci U S A 90 (7):2905–2909
21. Jung MR et al (2012) Local BMP-7 release from a PLGA scaffolding-matrix for the repair of osteochondral defects in rabbits. J Control Release 162(3):485–491
22. Zhang L et al (2016) BMP signaling and its paradoxical effects in tumorigenesis and dissemination. Oncotarget 7(47):78206–78218
23. Shen J et al (2013) BMP2-induced inflammation can be suppressed by the osteoinductive growth factor NELL-1. Tissue Eng Part A 19 (21–22):2390–2401
24. Abbah SA et al (2013) Enhanced control of in vivo bone formation with surface functionalized alginate microbeads incorporating heparin and human bone morphogenetic protein-2. Tissue Eng Part A 19(3–4):350–359
25. Kim SE et al (2015) Fabrication of a BMP-2-immobilized porous microsphere modified by heparin for bone tissue engineering. Colloids Surf B Biointerfaces 134:453–460
26. Kuhn LT et al (2013) Fibroblast growth factor-2 and bone morphogenetic protein-2 have a synergistic stimulatory effect on bone formation in cell cultures from elderly mouse and human bone. J Gerontol A Biol Sci Med Sci 68(10):1170–1180
27. Suliman S et al (2015) Release and bioactivity of bone morphogenetic protein-2 are affected by scaffold binding techniques in vitro and in vivo. J Control Release 197:148–157
28. Zhang H et al (2010) Chemically-conjugated bone morphogenetic protein-2 on three-dimensional polycaprolactone scaffolds stimulates osteogenic activity in bone marrow stromal cells. Tissue Eng Part A 16 (11):3441–3448
29. Angle SR et al (2012) Healing of rat femoral segmental defect with bone morphogenetic protein-2: a dose response study. J Musculoskelet Neuronal Interact 12(1):28–37
30. Zara JN et al (2011) High doses of bone morphogenetic protein 2 induce structurally abnormal bone and inflammation in vivo. Tissue Eng Part A 17(9–10):1389–1399
31. Zwingenberger S et al (2013) Establishment of a femoral critical-size bone defect model in immunodeficient mice. J Surg Res 181(1): e7–e14
32. Zhang X et al (2012) Osteoconductive effectiveness of bone graft derived from antler

cancellous bone: an experimental study in the rabbit mandible defect model. Int J Oral Maxillofac Surg 41(11):1330–1337
33. Sato K, Urist MR (1985) Induced regeneration of calvaria by bone morphogenetic protein (BMP) in dogs. Clin Orthop Relat Res 197:301–311
34. Gerhart TN et al (1993) Healing segmental femoral defects in sheep using recombinant human bone morphogenetic protein. Clin Orthop Relat Res 293:317–326
35. Damien CJ et al (2002) Purified bovine BMP extract and collagen for spine arthrodesis: preclinical safety and efficacy. Spine (Phila Pa 1976) 27(16 Suppl 1):S50–S58
36. Seeherman HJ et al (2004) Recombinant human bone morphogenetic protein-2-delivered in an injectable calcium phosphate paste accelerates osteotomy-site healing in a nonhuman primate model. J Bone Joint Surg Am 86-A(9):1961–1972
37. Lee JW et al (2011) Improved spinal fusion efficacy by long-term delivery of bone morphogenetic protein-2 in a rabbit model. Acta Orthop 82(6):756–760
38. Cowan CM et al (2007) MicroCT evaluation of three-dimensional mineralization in response to BMP-2 doses in vitro and in critical sized rat calvarial defects. Tissue Eng 13(3):501–512
39. Boyne PJ (2001) Application of bone morphogenetic proteins in the treatment of clinical oral and maxillofacial osseous defects. J Bone Joint Surg Am 83-A Suppl 1(Pt 2):S146–S150
40. Li Y, Chen S, Li L, Qin L, Wang X (2015) Bone defect animal models for testing efficacy of bone substitute biomaterials. J Orthop Translat 3(3):95–104
41. O'Loughlin PF et al (2008) Selection and development of preclinical models in fracture-healing research. J Bone Joint Surg Am 90 (Suppl 1):79–84
42. Kempen DH et al (2009) Non-invasive monitoring of BMP-2 retention and bone formation in composites for bone tissue engineering using SPECT/CT and scintillation probes. J Control Release 134(3):169–176
43. Oest ME et al (2007) Quantitative assessment of scaffold and growth factor-mediated repair of critically sized bone defects. J Orthop Res 25 (7):941–950
44. Schneider CA, Rasband WS, Eliceiri KW (2012) NIH Image to ImageJ: 25 years of image analysis. Nat Methods 9(7):671–675
45. Bako P et al (2015) Methyl methacrylate embedding to study the morphology and immunohistochemistry of adult guinea pig and mouse cochleae. J Neurosci Methods 254:86–93
46. Maniatopoulos C et al (1986) An improved method for preparing histological sections of metallic implants. Int J Oral Maxillofac Implants 1(1):31–37
47. Wong DA et al (2008) Neurologic impairment from ectopic bone in the lumbar canal: a potential complication of off-label PLIF/TLIF use of bone morphogenetic protein-2 (BMP-2). Spine J 8(6):1011–1018
48. Kaneko H et al (2000) Direct stimulation of osteoclastic bone resorption by bone morphogenetic protein (BMP)-2 and expression of BMP receptors in mature osteoclasts. Bone 27 (4):479–486
49. Yilgor P, Hasirci N, Hasirci V (2010) Sequential BMP-2/BMP-7 delivery from polyester nanocapsules. J Biomed Mater Res A 93 (2):528–536
50. Boerckel JD et al (2011) Effects of protein dose and delivery system on BMP-mediated bone regeneration. Biomaterials 32 (22):5241–5251
51. Perez RA et al (2014) Therapeutic bioactive microcarriers: co-delivery of growth factors and stem cells for bone tissue engineering. Acta Biomater 10(1):520–530
52. van de Watering FC et al (2012) Differential loading methods for BMP-2 within injectable calcium phosphate cement. J Control Release 164(3):283–290
53. Akkineni AR et al (2015) 3D plotting of growth factor loaded calcium phosphate cement scaffolds. Acta Biomater 27:264–274
54. Kaito T et al (2005) Potentiation of the activity of bone morphogenetic protein-2 in bone regeneration by a PLA-PEG/hydroxyapatite composite. Biomaterials 26(1):73–79
55. Zhang HX et al (2016) In vitro and in vivo evaluation of calcium phosphate composite scaffolds containing BMP-VEGF loaded PLGA microspheres for the treatment of avascular necrosis of the femoral head. Mater Sci Eng C Mater Biol Appl 60:298–307
56. Cipitria A et al (2013) Polycaprolactone scaffold and reduced rhBMP-7 dose for the regeneration of critical-sized defects in sheep tibiae. Biomaterials 34(38):9960–9968
57. Wang Y, Kim YM, Langer R (2003) In vivo degradation characteristics of poly(glycerol sebacate). J Biomed Mater Res A 66 (1):192–197
58. Lee EJ, Kim HE (2016) Accelerated bony defect healing by chitosan/silica hybrid membrane with localized bone morphogenetic

protein-2 delivery. Mater Sci Eng C Mater Biol Appl 59:339–345

59. Zhou H et al (2009) Enhanced bioactivity of bone morphogenetic protein-2 with low dose of 2-N, 6-O-sulfated chitosan in vitro and in vivo. Biomaterials 30(9):1715–1724
60. Karageorgiou V et al (2006) Porous silk fibroin 3-D scaffolds for delivery of bone morphogenetic protein-2 in vitro and in vivo. J Biomed Mater Res A 78(2):324–334
61. Uebersax L, Merkle HP, Meinel L (2008) Insulin-like growth factor I releasing silk fibroin scaffolds induce chondrogenic differentiation of human mesenchymal stem cells. J Control Release 127(1):12–21
62. Bhakta G et al (2013) The influence of collagen and hyaluronan matrices on the delivery and bioactivity of bone morphogenetic protein-2 and ectopic bone formation. Acta Biomater 9(11):9098–9106
63. Takahashi Y, Yamamoto M, Tabata Y (2005) Enhanced osteoinduction by controlled release of bone morphogenetic protein-2 from biodegradable sponge composed of gelatin and beta-tricalcium phosphate. Biomaterials 26 (23):4856–4865
64. Suzuki K et al (2014) Effect of addition of hyaluronic acids on the osteoconductivity and biodegradability of synthetic octacalcium phosphate. Acta Biomater 10(1):531–543
65. Seeherman H, Wozney JM (2005) Delivery of bone morphogenetic proteins for orthopedic tissue regeneration. Cytokine Growth Factor Rev 16(3):329–345
66. Quinlan E et al (2015) Development of collagen-hydroxyapatite scaffolds incorporating PLGA and alginate microparticles for the controlled delivery of rhBMP-2 for bone tissue engineering. J Control Release 198:71–79
67. Liu F, Wells JW, Porter RM, Glatt V, Shen Z, Schinhan M, Ivkovic A, Vrahas MS, Evans CH, Ferreira E (2016) Interaction between living bone particles and rhBMP-2 in large segmental defect healing in the rat femur. J Orthop Res 34(12):2137–2145
68. Henslee AM et al (2011) Biodegradable composite scaffolds incorporating an intramedullary rod and delivering bone morphogenetic protein-2 for stabilization and bone regeneration in segmental long bone defects. Acta Biomater 7(10):3627–3637
69. Kolambkar YM et al (2011) Spatiotemporal delivery of bone morphogenetic protein enhances functional repair of segmental bone defects. Bone 49(3):485–492
70. Foo T et al (2013) External fixation of femoral defects in athymic rats: applications for human stem cell implantation and bone regeneration. J Tissue Eng 4:2041731413486368
71. Torricelli P et al (2002) Characterization of bone defect repair in young and aged rat femur induced by xenogenic demineralized bone matrix. J Periodontol 73(9):1003–1009
72. Jackson S, Andrews N, Ball D, Bellantuono I, Gray J (2016) Does age matter? The impact of rodent age on study outcomes. Lab Anim 1–9
73. Wang EA et al (1993) Bone morphogenetic protein-2 causes commitment and differentiation in C3H10T1/2 and 3T3 cells. Growth Factors 9(1):57–71
74. Luginbuehl V et al (2004) Localized delivery of growth factors for bone repair. Eur J Pharm Biopharm 58(2):197–208
75. Kolambkar YM et al (2011) An alginate-based hybrid system for growth factor delivery in the functional repair of large bone defects. Biomaterials 32(1):65–74
76. Kodera R et al (2014) Manipulation of anabolic and catabolic responses with bone morphogenetic protein and zoledronic acid in a rat spinal fusion model. Bone 58:26–32
77. Liu X et al (2013) Enhanced bone regeneration in rat calvarial defects implanted with surface-modified and BMP-loaded bioactive glass (13-93) scaffolds. Acta Biomater 9 (7):7506–7517
78. Lo KW et al (2012) Studies of bone morphogenetic protein-based surgical repair. Adv Drug Deliv Rev 64(12):1277–1291
79. Friess W et al (1999) Characterization of absorbable collagen sponges as recombinant human bone morphogenetic protein-2 carriers. Int J Pharm 185(1):51–60
80. Knabe C et al (2006) A method for immunohistochemical detection of osteogenic markers in undecalcified bone sections. Biotech Histochem 81(1):31–39
81. Goldschlager T et al (2010) Undecalcified bone preparation for histology, histomorphometry and fluorochrome analysis. J Vis Exp (35). https://doi.org/10.3791/1707
82. Wittenburg G et al (2009) Immunohistochemical comparison of differentiation markers on paraffin and plastic embedded human bone samples. J Physiol Pharmacol 60(Suppl 8):43–49

Chapter 18

Heterotopic Ossification in Mouse Models of Fibrodysplasia Ossificans Progressiva

Salin A. Chakkalakal and Eileen M. Shore

Abstract

Fibrodysplasia ossificans progressiva (FOP), a rare genetic disorder of progressive extra-skeletal ossification, is the most disabling form of heterotopic ossification (HO) in humans. Most people with FOP carry an activating mutation in a BMP type I receptor gene, $ACVR1^{R206H}$, that promotes ectopic chondrogenesis and osteogenesis and in turn HO. Advances in elucidating the cellular and molecular events and mechanisms that lead to the ectopic bone formation are being made through the use of genetically engineered mouse models that recapitulate the human disease. We describe methods for inducing heterotopic ossification in a mouse model that conditionally expresses the $Acvr1^{R206H}$ allele.

Key words Fibrodysplasia ossificans progressiva, FOP, Heterotopic ossification, HO, ACVR1, ALK2, Mouse model

1 Introduction

The BMP/TGFβ family of ligands and their receptors regulate many diverse and biologically critical cellular functions including cell differentiation, apoptosis, proliferation, migration, and stem cell reprogramming and are highly evolutionarily conserved in order to maintain precise control of these processes [1–6].

A human genetic disease, fibrodysplasia ossificans progressiva (FOP; MIM #135100, http://omim.org/entry/135100), is caused by rare mutations in the BMP type I receptor ACVR1 (also known as ALK2), with most cases caused by a recurrent heterozygous gain-of-function mutation c.617G>A; R206H [7–9]. People with FOP are born with no clinical manifestations of the disease except for a characteristic malformation of the great toe. However, during childhood and throughout adult life, extra-skeletal bone episodically forms in soft connective tissues such as skeletal muscle, often in response to injury. Nongenetic forms of such ectopic bone formation, known as heterotopic ossification (HO), are associated with a number of common conditions in

Melissa B. Rogers (ed.), *Bone Morphogenetic Proteins: Methods and Protocols*, Methods in Molecular Biology, vol. 1891, https://doi.org/10.1007/978-1-4939-8904-1_18,

adults that involve severe trauma such as spinal cord and head injuries, hip replacement surgery, and war-related blast injuries [10, 11].

In 2012, we published a detailed characterization of the first genetically engineered FOP mouse model with a knock-in allele of the Acvr1 R206H mutation [12]. This mouse model closely mimics the clinical features of FOP in humans and was the first demonstration that the ACVR1 R206H mutation is solely responsible for the FOP clinical phenotype and that FOP could be modeled in the mouse with high fidelity. However, genetic transmission of the $Acvr1^{R206H}$ allele in mice caused perinatal lethality; therefore, development of mouse models that conditionally activate the Acvr1 R206H mutation was necessary. Studies using a floxed conditional-on $Acvr1^{R206H}$ knock-in mouse were recently reported [13, 14].

Heterotopic ossification in FOP can occur in response to tissue trauma/injury or spontaneously in the absence of overt trauma. In our hands, postnatal expression of the mutant allele is not sufficient to allow robust and predictable HO formation in the absence of injury. However, we have found that global postnatal Cre recombinase activation and induction of $Acvr1^{R206H}$ expression in the conditional mouse model are sufficient to form HO in response to skeletal muscle injury [13]. Here, we describe our approach to activating and characterizing HO lesion formation in a conditional $Acvr1^{R206H}$ knock-in mouse by muscle injury.

2 Materials

2.1 Mouse Models

1. Conditional-on knock-in mouse model: $Acvr1^{[R206H]FlEx}$ was described in Hatsell et al., 2015. The floxed $Acvr1^{[R206H]FlEx}$ allele is expressed in response to Cre recombinase expression.
2. Mice double transgenic for R26-rtTA and tetO-Cre (heterozygous Gt(ROSA)26Sor$^{tm1(rtTA*M2)Jae}$ and hemizygous Tg (tetO-Cre)1Jaw; Jackson Laboratory) (*see* **Note 1**).

2.2 Doxycycline Chow

1. Sterilized doxycycline-embedded chow containing 625 mg/kg doxycycline.
2. Store doxycycline chow in sealed packets at 4 °C (*see* **Note 1**).

2.3 Cardiotoxin

1. 1× phosphate-buffered saline (PBS): 137 mM NaCl, 2.7 mM KCl, 110 mM Na_2HPO_4, 1.8 mM KH_2PO_4, pH 7.2. Weigh 8 g of NaCl, 0.2 g of KCl, 1.44 g of Na_2HPO_4, and 0.24 g KH_2PO_4; dissolve in 800 mL of deionized water. Adjust the pH to 7.2 with HCl, and add water to 1 L. Sterilize by autoclaving for 20 min at 15 psi.

2. 100 μM cardiotoxin stock solution: Cardiotoxin (from *Naja mossambica mossambica*) is obtained as lyophilized powder. Add 1 mg cardiotoxin to 1426 μL of sterile 1× PBS with gentle shaking. Store at −20 °C in 50 μL aliquots.
3. 10 μM cardiotoxin working solution: Just prior to use, dilute 100 μM cardiotoxin stock to 10 μM with sterile 1× PBS (10 μL of stock diluted with 90 μL of PBS). Keep on ice.
4. 28-gauge 1/2 U-100 insulin syringes.
5. Alcohol pads.
6. Gas vaporizer chamber for isoflurane-induced anesthesia.

2.4 Genotyping

1. Mouse genomic DNA.
2. PCR primers to detect the knock-in allele post-Cre recombination:

   Forward: 5′-TGTATTGCAGGACGCTGAAG-3′

   Reverse: 5′-CCCCTGAAGTGGAATAACCA-3′
3. 2× PCR master mix: 2.5 U of Taq DNA polymerase, 20 mM Tris–HCl (pH 9.0), 3 mM $MgCl_2$, 20 mM KCl, 16 mM $(NH_4)_2SO_4$, 0.1% NP-40, 1.6mM dNTPs.
4. 10× Tris acetate EDTA (TAE): 400 mM Tris acetate, 10 mM EDTA, pH 8.3. Dilute to 1× with deionized water for preparing agarose gels and gel electrophoresis running buffer.
5. Agarose.
6. Ethidium bromide at 10 mg/mL.

2.5 Histology (See Note 8)

1. 4% paraformaldehyde (PFA).
2. Decalcifying solution containing 5% formic acid.
3. Harris hematoxylin.
4. 0.5% (w/v) eosin Y in acidified 90% (w/v) ethanol (can be obtained commercially).
5. Acid alcohol: 5 mL of 12 N hydrochloric acid added to 500 mL 70% ethanol.
6. Ammonium water: Add 2.5 mL of 28.0–30.0% ammonium hydroxide stock solution to 500 mL deionized water.
7. Parafilm.
8. Scotch tape (or other similar tape).

3 Methods

Injury-induced heterotopic ossification in *Acvr1*$^{R206H/+}$;R26-rtTA;tetO-Cre (*Acvr1*$^{R206H/+}$) mice

3.1 Postnatal Global Induction of the Acvr1 R206H Mutation

1. Cross *Acvr1*$^{[R206H]FlEx/+}$ mice with mice double transgenic for R26-rtTA and tetO-Cre (heterozygous Gt(ROSA)26Sor$^{tm1(rtTA*M2)Jae}$ and hemizygous Tg(tetO-Cre)1Jaw; Jackson Laboratory) to generate *Acvr1*$^{[R206H]FlEx/+}$;Gt(ROSA)26Sor$^{tm1(rtTA*M2)Jae}$; Tg(tetO-Cre)1Jaw mice (referred to as *Acvr1*$^{cR206H/+}$*;rt-tetO-Cre* below) (*see* **Note 2**).
2. Induce Cre recombinase expression in *Acvr1*$^{cR206H/+}$*;rt-tetO-Cre* mice at 4–6 weeks old (*see* **Note 3**) by replacing normal food chow with doxycycline chow for 3 days (*see* **Note 4**).
3. PCR amplification to detect Cre recombination: Add 40 ng mouse genomic DNA (in 2–3 μL) to 10 μL of 2× PCR master mix and 0.5 μL each of forward and reverse primers (50 μM); add water to a total volume of 20 μL. PCR amplification: initial denaturation 94 °C (3 min); followed by 35 cycles of denaturation 94 °C (30 s), annealing 55 °C (45 s), and extension 72 °C (30 s); then final elongation at 72 °C for 10 min.
4. Agarose gel electrophoresis: Electrophorese 10 μL of each PCR-amplified sample through 4% (w/v) agarose gels (4 g of agarose added to 100 mL of 1× TAE with 0.5 μL of ethidium bromide stock solution) (Fig. 1).

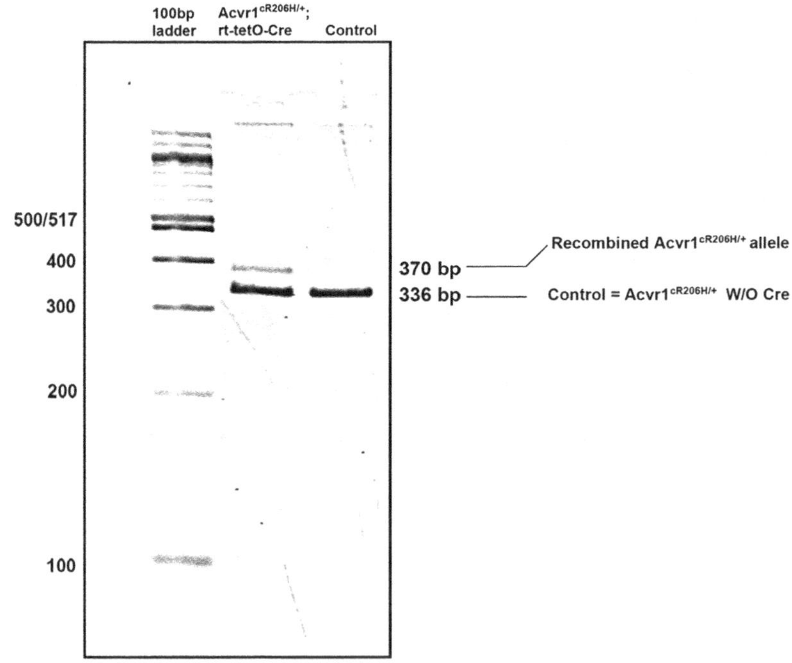

Fig. 1 Agarose gel electrophoresis of PCR products to verify Cre recombination of the *Acvr1*^{R206H} allele. Genomic DNA samples were isolated from a *Acvr1*$^{cR206H/+}$*; rt-tetO-Cre* mouse and a control *Acvr1*$^{+/+}$ mouse after 3 days on doxycycline chow. Following PCR amplification to detect *Acvr1*, samples were electrophoresed through a 4% agarose/TAE/ethidium gel. PCR products from wild-type *Acvr1* (336 bp) and recombined *Acvr1*^{cR206H} (370 bp) alleles are indicated

3.2 Hind Limb Injury by Cardiotoxin to Induce Heterotopic Ossification

1. Restrain mice by isoflurane-induced anesthesia (3–5% isoflurane) using a gas vaporizer chamber (*see* **Note 5**).
2. Load sterile syringes aseptically with 50 μL of 10 μM cardiotoxin working solution; place on ice. Clean hind limbs of mice with alcohol pad. Inject cardiotoxin aseptically into the quadriceps muscle (*see* **Notes 6** and 7). Inject contralateral limbs with 1× PBS as controls.
3. Place mice into cages and confirm recovery from anesthesia by noting that mice are awake with normal movement.

3.3 Tissue Harvest and Fixation

1. Euthanize mice at the desired assay time point for analysis (*see* **Note** 7). Hind limbs are evaluated by standard radiography for the presence of HO formation (Fig. 2a). HO volume can be quantified by microCT analyses (Fig. 2b). Progression and stages of HO can be examined by histology (Fig. 2c).

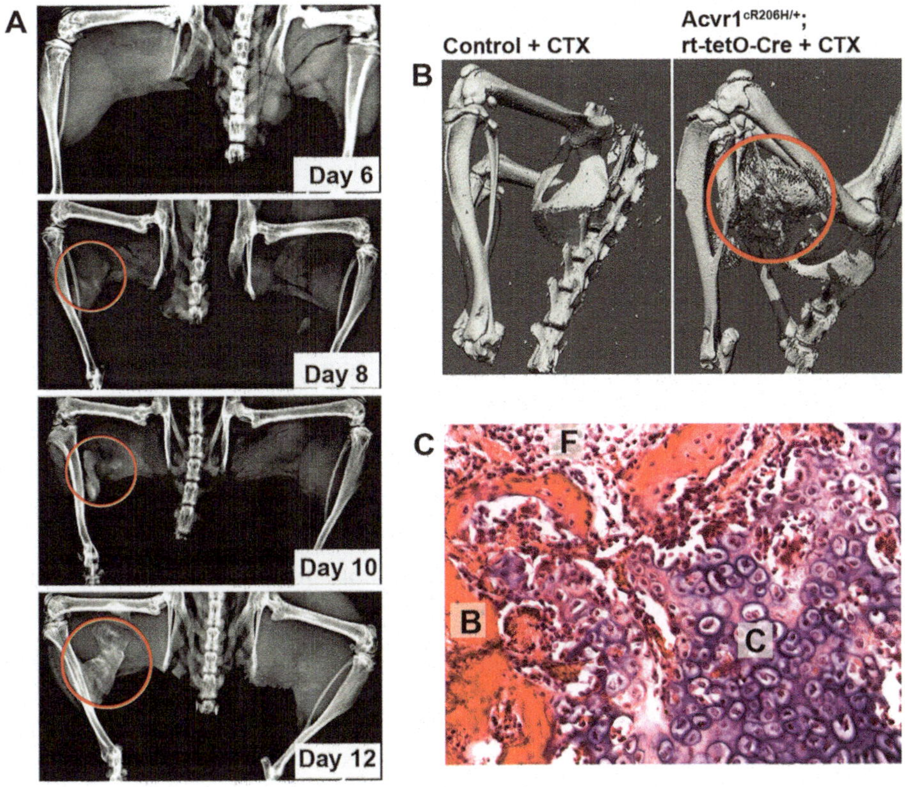

Fig. 2 Cardiotoxin-injured skeletal muscle in *Acvr1*$^{cR206H/+}$ mice induces heterotopic ossification. *Acvr1*$^{R206H/+}$ mice (at 4 weeks old) were provided doxycycline chow for 3 days to induce global expression of the *Acvr1*^{cR206H} allele, followed by cardiotoxin injury to quadriceps muscles. (**a**) X-ray images of the hind limbs at 6, 8, 10, and 12 days post-cardiotoxin injection are shown; circled areas indicate HO formation. (**b**) MicroCT images of a control and a mutant leg showing extensive HO formation at the site of cardiotoxin injection in the mutant (circle). (**c**) Decalcified tissue section from a hind limb of a cardiotoxin-injected mouse was stained with hematoxylin and eosin. Evidence of heterotopic lesion formation is shown by the presence of cartilage (C) and bone (B); fibroproliferative cells are also present (F)

2. Dissect hind limbs from euthanized mice. Place in 50 mL tubes and fix by filling tubes with 4% paraformaldehyde (4% PFA) to fully immerse sample. Incubate at room temperature with gentle shaking/rocking overnight (12–16 h).
3. Pour off PFA. Wash the PFA-fixed samples by filling tube with 1× PBS and rock gently for 5 min. After three washes, remove hind limb from the tube and process for imaging and/or histologic analyses.

3.4 Imaging Analyses

1. Following PFA and washes, carefully blot excess PBS from fixed hind limb with paper towels. Place tissue on a square of Parafilm that has been taped to scanning surface to prevent movement. Image using a standard small animal X-ray instrument to detect HO.
2. For more detailed analysis of HO, acquire microcomputed tomography (μCT) volumetric data at the following parameters: 80 kVp and 80 μA X-ray tube voltage and current, 250 μm aluminum filter, 1.7 s integration time, 400 views at 0.5° increments, 2 × 2 detector bin mode, and 4 averages. Reconstruct image data at a resolution of 40.5 μm isotropic voxels using a Feldkamp cone beam algorithm. Analyze and determine volume from the reconstructed 3D data using software such as OsiriX (www.osirix-viewer.com).

3.5 Histological Evaluation of Heterotopic Ossification

1. Following PFA and washes (*see* **Note 8**), decalcify samples by filling 50 mL tube containing the sample with decalcifying solution containing 5% formic acid; shake gently at room temperature for 3 days, replacing with fresh decalcifying solution each day. Pour off decalcifying solution. Wash the decalcified samples by filling tube with 1× PBS and rock gently for 5 min. After three washes, remove hind limb from the tube and process for paraffin embedding and section at 7 μm. Place sections on slides for histological staining.
2. Follow standard protocols to remove paraffin using three changes of xylene, 5 min each.
3. Rehydrate sections by immersing slides in three changes each of 100%, 95%, and 70% ethanol for 3 min each followed by immersion in deionized water twice.
4. Immerse hydrated sections on slides in Harris hematoxylin for 5 min. Rinse gently in running deionized water until excess stain no longer leaches from tissue. Dip slides three times in acid alcohol and then rinse by immersing in three changes of deionized water. Place slides in ammonium water for 30 s for developing the blue stain (bluing).
5. Dip slides in three changes of deionized water, and then place in 95% ethanol for 1 min. Immerse in 1% eosin Y solution for

5 min. Dehydrate in three changes of 95% ethanol and three changes of 100% ethanol, and then clear by immersing in xylene and three changes of 5 min each.

6. Cover slip and mount tissue sections and dry overnight. Observe microscopically for HO formation.

4 Notes

1. The rtTA-tetO "on" system [15] is responsive to doxycycline to activate Cre expression. Use fresh doxycycline chow for induction of Cre recombinase. Moisture and increased temperature can cause hydrolysis and inactivation of doxycycline.

2. We (and others [14]) have also induced global expression of the floxed *Acvr1*^{R206H} allele using tamoxifen and ERT2-Cre (B6.129-*Gt(ROSA)26Sor*$^{tm1(cre/ERT2)Tyj}$/J). Since tamoxifen, an estrogen analog, could plausibly affect bone formation and growth, we preferred to use the rt-tetO system in our initial studies [13].

   To induce global expression, tamoxifen is administered by intraperitoneal injection. Dissolve tamoxifen in corn oil to 20 mg/mL by gentle rocking overnight at 37 °C. For adult mice, 100 μL (75 mg tamoxifen/kg body weight) is administered via intraperitoneal injection. Repeat for 3 consecutive days (https://www.jax.org/research-and-faculty/tools/cre-repository/tamoxifen).

   Alternatively, to allow the expression of the *Acvr1*^{R206H} allele during embryonic development but avoid the lethality that is associated with global expression of *Acvr1*^{R206H} during embryogenesis, we have induced *Acvr1*^{R206H} expression prenatally in a limited population of skeletal progenitor cells (Prrx1^{+}). During embryonic development, Prrx1 is expressed in a population of skeletal progenitor cells that gives rise to lateral plate mesoderm-derived limb mesenchymal cells. *Acvr1*^{R206H}*;Prrx1-Cre* mice are born at the expected Mendelian frequency. *Acvr1*^{R206H}*;Prrx1-Cre* mice form extensive spontaneous extra-skeletal bone by 1 month of age [13].

3. For cardiotoxin-induced heterotopic ossification, mice are used within 6 weeks of age. We have found that the amount of HO formation by injury decreases as the mouse ages.

4. This diet delivers a daily dose of 2–3 mg of doxycycline based on an average 4–5 g/day of food consumption by a mouse.

5. When injecting mice, immobilization using isoflurane anesthesia is recommended. Avoid injecting into blood vessels, as this can be fatal to mice. Care should be taken during injections to avoid needle contact with skeletal bone; injury to periosteum can stimulate orthotopic ossification even in wild-type mice.

6. Vigorous shaking/vortexing of cardiotoxin can cause denaturation. Prepare diluted working solutions of cardiotoxin in pre-loaded syringes and keep on ice.
7. Progression to heterotopic bone occurs through endochondral ossification and forms mineralized bone by ~14 days post-cardiotoxin muscle injury. Cartilage formation is detected after 5–7 days. Impaired hind limb movement from day 7 post-injury is typically observed. Note that mice must be handled gently; pinching or squeezing appears to result in sufficient tissue trauma to stimulate HO.
8. Histological analysis can use PFA-fixed samples that have been processed specifically for histology or can use the same fixed sample used for imaging analysis without further processing.

All animal procedures were reviewed and approved by the Institutional Animal Care and Use Committee at the University of Pennsylvania.

Acknowledgments

We thank the International Fibrodysplasia Ossificans Progressiva Association (IFOPA), the Center for Research in FOP and Related Disorders, the Ian Cali Endowment for FOP Research, the Whitney Weldon Endowment for FOP Research, the Ashley Martucci FOP Research Fund, the Penn Center of Musculoskeletal Disorders (NIH P30-AR06919), the Cali-Weldon Professorship of FOP Research (EMS), and the National Institutes of Health (NIH R01-AR41916) for supporting our work. We also thank Regeneron Pharmaceuticals for developing the conditional Acvr1 R206H mouse model.

References

1. Elliott RL, Blobe GC (2005) Role of transforming growth factor beta in human cancer. J Clin Oncol 23(9):2078–2093. https://doi.org/10.1200/jco.2005.02.047
2. Hayashi Y, Hsiao EC, Sami S, Lancero M, Schlieve CR, Nguyen T, Yano K, Nagahashi A, Ikeya M, Matsumoto Y, Nishimura K, Fukuda A, Hisatake K, Tomoda K, Asaka I, Toguchida J, Conklin BR, Yamanaka S (2016) BMP-SMAD-ID promotes reprogramming to pluripotency by inhibiting p16/INK4A-dependent senescence. Proc Natl Acad Sci U S A 113(46):13057–13062
3. Miyazawa K, Shinozaki M, Hara T, Furuya T, Miyazono K (2002) Two major Smad pathways in TGF-beta superfamily signalling. Genes Cells 7(12):1191–1204
4. Watabe T, Miyazono K (2009) Roles of TGF-beta family signaling in stem cell renewal and differentiation. Cell Res 19(1):103–115
5. Wu MY, Hill CS (2009) Tgf-beta superfamily signaling in embryonic development and homeostasis. Dev Cell 16(3):329–343
6. Wu M, Chen G, Li YP (2016) TGF-β and BMP signaling in osteoblast, skeletal development, and bone formation, homeostasis and disease. Bone Res 4:16009. https://doi.org/10.1038/boneres.2016.9
7. Kaplan FS, Xu M, Seemann P, Connor JM, Glaser DL, Carroll L, Delai P, Fastnacht-Urban E, Forman SJ, Gillessen-Kaesbach G, Hoover-Fong J, Koster B, Pauli RM, Reardon W, Zaidi S-A, Zasloff M, Morhart R, Mundlos S, Groppe J, Shore EM (2009)

Classic and atypical fibrodysplasia ossificans progressiva (FOP) phenotypes are caused by mutations in the bone morphogenetic protein (BMP) type I receptor ACVR1. Hum Mutat 30(3):379–390

8. Shore EM, Xu MQ, Feldman GJ, Fenstermacher DA, Cho T-J, Choi IH, Connor JM, Delai P, Glaser DL, Le Merrer M, Morhart R, Rogers JG, Smith R, Triffitt JT, Urtizberea JA, Zasloff M, Brown MA, Kaplan FS (2006) A recurrent mutation in the BMP type I receptor ACVR1 causes inherited and sporadic fibrodysplasia ossificans progressiva. Nat Genet 38 (5):525–527
9. Shore EM, Kaplan FS (2010) Inherited human diseases of heterotopic bone formation. Nat Rev Rheumatol 6(9):518–527. https://doi.org/10.1038/nrrheum.2010.122
10. Pignolo RJ, Foley KL (2005) Nonhereditary heterotopic ossification. Clin Rev Bone Miner Metab 3(3–4):261–266
11. Edwards DS, Kuhn KM, Potter BK, Forsberg JA (2016) Heterotopic ossification: a review of current understanding, treatment, and future. J Orthop Trauma 30:S27–S30
12. Chakkalakal SA, Zhang DY, Culbert AL, Convente MR, Caron RJ, Wright AC, Maidment ADA, Kaplan FS, Shore EM (2012) An Acvr1 R206H knock-in mouse has fibrodysplasia ossificans progressiva. J Bone Miner Res 27 (8):1746–1756. https://doi.org/10.1002/jbmr.1637
13. Chakkalakal SA, Uchibe K, Convente MR, Zhang DY, Economides AN, Kaplan FS, Pacifici M, Iwamoto M, Shore EM (2016) Palovarotene inhibits heterotopic ossification and maintains limb mobility and growth in mice with the human ACVR1(R206H) fibrodysplasia ossificans progressiva (FOP) mutation. J Bone Miner Res 31(9):1666–1675. https://doi.org/10.1002/jbmr.2820
14. Hatsell SJ, Idone V, Wolken DMA, Huang L, Kim HJ, Wang LL, Wen XL, Nannuru KC, Jimenez J, Xie LQ, Das N, Makhoul G, Chernomorsky R, D'Ambrosio D, Corpina RA, Schoenherr CJ, Feeley K, Yu PB, Yancopoulos GD, Murphy AJ, Economides AN (2015) ACVR1(R206H) receptor mutation causes fibrodysplasia ossificans progressiva by imparting responsiveness to activin A. Sci Transl Med 7(303). https://doi.org/10.1126/scitranslmed.aac4358
15. Lewandoski M (2001) Conditional control of gene expression in the mouse. Nat Rev Genet 2 (10):743–755. https://doi.org/10.1038/35093537

Chapter 19

Double-Humanized Mouse Model to Study Bone Morphogenetic Protein (BMP) Signaling in Tumor Xenografts

Jenna Newman, Rachel NeMoyer, Andrew Zloza, and John Langenfeld

Abstract

The activation of the bone morphogenic protein (BMP) signaling pathway in cancer cells has been shown to enhance migration and tumor angiogenesis and promote survival. The BMP signaling pathway regulates benign cells in the tumor microenvironment and is a known regulator of immune cells. The development of BMP receptor inhibitors has allowed the study of tumor xenografts in mice. We describe a double-humanized mouse model with adoptively transferred human immune and human tumor cells that can be used to assess the effects of BMP inhibitors on these human cells in vivo.

Key words Bone morphogenic protein, BMP inhibitor, Xenografts, Double-humanized mouse model, Peripheral blood mononuclear cells, Cell culture, Flow cytometry, ELISA, Western blotting, Immunohistochemistry, NOD *scid* gamma mice

1 Introduction

There is now significant evidence that the BMP signaling cascade promotes tumorigenesis in many types of cancers. Bone morphogenetic proteins are embryonic proteins that are part of the transforming growth factor (TGF-β) superfamily, which are aberrantly expressed in many carcinomas. Inhibition of BMP receptors with small molecule inhibitors decreases growth and induces death of cancer cells, specifically lung cancer cells [1].

Importantly, BMP signaling promotes survival of cancer cells through the regulation of potent anti-apoptotic proteins, which include X-linked inhibitor of apoptosis (XIAP), TGF-β-activated kinase (TAK1), and inhibitor of differentiation proteins (Id1–Id3) [1–6]. The regulation of XIAP, TAK1, and Id1 by the BMP signaling cascade also occurs during embryogenesis in *Xenopus*, demonstrating that these are evolutionary conserved signaling pathways. Both the BMP and TGF-β signaling pathways are known to

Melissa B. Rogers (ed.), *Bone Morphogenetic Proteins: Methods and Protocols*, Methods in Molecular Biology, vol. 1891,
https://doi.org/10.1007/978-1-4939-8904-1_19,

regulate the differentiation and activation of immune cells [1–6]. Tumor xenograft studies are essential for the development of therapeutic strategies to treat patients with cancer. With the advent of improved outcomes in patients treated with immunotherapy for lung cancer and melanoma, it is important to examine responses to both the immune cells within the tumor microenvironment and the tumors cells themselves. Several BMP inhibitors have been developed (LDN, DMH1, DMH2), some of which have demonstrated antitumor responses in tumor xenografts [7, 8]. With the development of more potent BMP inhibitors for in vivo use, we anticipate finding an even greater response either as a single agent or in combination with other therapeutics. Below we describe the process of adoptively transferring human immune cells, establishing tumor xenografts from cultured human cancer cells, harvesting and injecting tumor cells into mice, injecting mice with BMP inhibitor, and harvesting tumors for analyses.

2 Materials

1. NOD *scid* gamma mice.
2. Sterile phosphate-buffered saline (PBS): 155 mM NaCl, 3 mM Na_2HPO_4–$7H_2O$, 1 mM KH_2PO_4, pH 7.4.
3. Sterile 0.25% trypsin-EDTA.
4. Incomplete DMEM (e.g., Gibco™, catalog number 11965092).
5. Complete DMEM (cDMEM): 10% heat-inactivated fetal bovine serum, 100 U/mL penicillin, 100 μg/mL streptomycin, 2 mM glutamine.
6. Trypan blue, 0.4%.
7. Cryovial with an internal anti-adhesive coating.
8. 1 mL insulin syringe (26 gauge).
9. Electronic calipers.
10. BMP inhibitor (please refer to Augeri et al. [1] for a brief discussion of relative activities of current BMP inhibitors).
11. Isoflurane inhalation anesthetic.
12. Anesthesia machine (with O_2 flush valve) and vaporizer.
13. CO_2 inhalation apparatus.
14. BMP-sensitive lung cancer tumor cells (including but not limited to H1299 and A549 cell lines).
15. 1.077 ± 0.001 g/mL Ficoll-Paque solution.
16. Vacutainer sodium heparin tubes.
17. Hanks' Balanced Salt Solution (HBSS).
18. Flow cytometer.

3 Methods

Unless otherwise noted, all procedures use sterile solutions and supplies and follow procedures to minimize worker exposure to blood borne pathogens.

3.1 Adoptively Transferring Immune Cells

3.1.1 Preparing Ficoll-Paque Solution with Whole Blood

1. Add 15 mL sterile Ficoll-Paque (density 1.077 ± 0.001 g/mL) to a 50 mL conical tube.
2. In a separate 50 mL conical tube, combine two Vacutainer sodium heparin tubes of blood (up to 20 mL total) and dilute with sterile PBS to a total volume of 35 mL.
3. Pipet the blood-PBS mixture slowly onto the Ficoll-Paque layer in the other 50 mL conical tube, without disturbing the Ficoll-Paque layer (*see* **Note 1**). Repeat process to accommodate as many tubes of blood as required.

3.1.2 Isolation of Human Peripheral Blood Mononuclear Cells (PBMCs)

1. Place conical tubes layered with Ficoll and blood into a centrifuge, set at 30 min, 1450 rpm/410 × *g*, room temperature (no brake). Subsequent to centrifugation, extract the layer of PBMCs (Fig. 1) from the turbid interface (buffy coat fraction) of the serum layer (very top) and the Ficoll layer (atop the erythrocyte pellet at the bottom of the tube).
2. Wash PBMCs with PBS.
3. Collect the PBMCs by centrifugation for 5 min at 1450 rpm/410 × *g*. The brake can be applied for this centrifugation.
4. Count PBMCs using a hemocytometer.

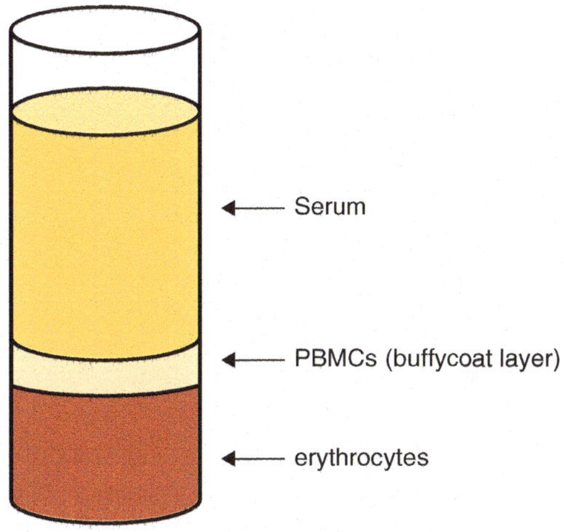

Fig. 1 Diagram of human blood separated by Ficoll-Paque solution. Human blood is separated into three layers following mixture with Ficoll-Paque and subsequent centrifugation. Serum, peripheral blood mononuclear cells (PBMCs), and erythrocytes form three distinct layers as shown. PBMCs are collected to be used for humanization of mice

3.1.3 Humanizing the Immune System of Mice with PMBCs

1. After centrifugation, pour off the supernatant by inverting the tube.
2. Resuspend one million PBMCs in a volume of ~100 μL sterile PBS for each mouse.
3. Inject NOD *scid* gamma (NSG) mice intraperitoneally.
4. Tumor cells may be injected 2–3 weeks following administration of human PBMCs (*see* **Note 2**).

3.2 Establishing Tumor Xenografts in Mice

Recipient mice should be shaved at the site of injection prior to preparing cells for injection.

3.2.1 Harvesting Cancer Cell Lines

1. Grow a T175 flask of BMP-sensitive human lung cancer cells (H1299) to 80–90% confluence (in the exponential growth phase) in a 37 °C, 5% CO_2 incubator in cDMEM.
2. Dilute the residual media with 10 mL of PBS prior to aspirating the media.
3. Remove the PBS and media by aspiration.
4. Add 3 mL of sterile 0.25% trypsin-EDTA (1×) to the flask.
5. Place the flask back into the 37 °C incubator for 1–5 min to activate the trypsin and detach cells from the plate (*see* **Note 3**).
6. Dilute trypsin with the addition of 7 mL of cDMEM (*see* **Note 4**).
7. Transfer the suspension to a 50 mL conical tube.

3.2.2 Preparing Cells to Be Injected

1. Mix 50 μL of this suspension with 50 μL of trypan blue to count live, trypan blue-excluding cells with a hemocytometer [9].
2. Centrifuge the remaining trypsin/DMEM suspension at 1450 rpm/410 × g at 5 °C for 5 min.
3. Decant the supernatant and resuspend the pellet in DMEM in a cryovial harboring an internal anti-adhesive coating on ice (*see* **Note 4**). Two million H1299 cells should be resuspended in a volume of 100–200 μL DMEM per mouse.
4. Carry the cells on ice to the animal facility for injection into mice (*see* **Note 5**).

3.2.3 Injecting Cells into Mice

1. Mix the cryovial containing cells well to resuspend the cells that settle at the bottom of the cryovial. Load a 1 mL insulin syringe with two million H1299 cells per mouse in a volume of 100–200 μL.
2. Inject cells, bevel side up, subcutaneously into the right flank of the mouse (*see* **Note 6**).

3.3 Injecting Mice with BMP Inhibitor

One week following the injection of H1299 cells, mice are treated twice daily with BMP inhibitor or a vehicle control.

1. Anesthetize mice with the inhalation anesthetic isoflurane, administered using an anesthesia machine with O_2 flush valve and vaporizer.
2. Administer 6 mg/mL BMP inhibitor or control vehicle in a total volume of 100 μL intraperitoneally per mouse twice daily for approximately 2 weeks.
3. Monitor tumor growth via measurement with electronic calipers every 2–3 days throughout the duration of the experiment.

3.4 Analysis of Tumors

1. At the termination of the experiment, euthanize mice by CO_2 inhalation followed by cervical dislocation.
2. Dissect tumors from mice, place in HBSS, and store on ice during transport of the tissues from the mouse facility to laboratory for analyses by flow cytometry, Western blotting, ELISAs, or immunohistochemistry.

4 Notes

1. Overlay slowly (~1 min per blood-PBS tube) to ensure that the Ficoll and blood form distinct layers within the conical tube.
2. Recipient mice should be an immunodeficient inbred laboratory strain [10]. Compromised antigen presentation and the lack of functional adaptive immunity [10] observed in NSG mice enable reconstitution of the human immune system subsequent to transfer of PBMCs, providing a model in which human disease can be studied in the context of a recapitulated human immune system. Humanization status can be confirmed by collecting mouse blood via facial vein bleed, and flow cytometry can be conducted following the processing of blood.
3. Cells must be carefully monitored and immediately removed from the incubator when the cell monolayer has detached from the flask.
4. It is essential that cDMEM be added to trypsin-treated cells. Fetal bovine serum inactivates trypsin, minimizing toxicity that prolonged exposure to trypsin causes.

   Conversely, DMEM without serum must be used to resuspend the pellet for injection into mice. Fetal bovine serum is immunogenic and may inhibit growth of the cell line in vivo.
5. Cells must be stored on ice during transport to the mouse room to prevent cell death. Injection of a cell mixture with a

high ratio of dead/live cells is immunogenic and negatively impacts tumor growth.

6. The needle should be oriented parallel to the skin upon entry into the subcutaneous space to avoid penetration into the muscle beneath the subcutaneous space.

References

1. Augeri DJ, Langenfeld E, Castle M, Gilleran JA, Langenfeld J (2016) Inhibition of BMP and of TGFβ receptors downregulates expression of XIAP and TAK1 leading to lung cancer cell death. Mol Cancer 15(1):27
2. Langenfeld E, Deen M, Zachariah E, Langenfeld J (2013) Small molecule antagonist of the bone morphogenetic protein type I receptors suppresses growth and expression of Id1 and Id3 in lung cancer cells expressing Oct4 or nestin. Mol Cancer 12(1):129
3. Langenfeld E, Hong CC, Lanke G, Langenfeld J (2013) Bone morphogenetic protein type I receptor antagonists decrease growth and induce cell death of lung cancer cell lines. PLoS One 8(4):e61256
4. Langenfeld EM, Kong Y, Langenfeld J (2006) Bone morphogenetic protein 2 stimulation of tumor growth involves the activation of Smad-1/5. Oncogene 25(5):685–692
5. Langenfeld EM, Langenfeld J (2004) Bone morphogenetic protein-2 stimulates angiogenesis in developing tumors11NIH K22 grant CA91919-01A1 and UMDNJ foundation to J. Langenfeld. Mol Cancer Res 2 (3):141–149
6. Langenfeld EM, Calvano SE, Abou-Nukta F, Lowry SF, Amenta P, Langenfeld J (2003) The mature bone morphogenetic protein-2 is aberrantly expressed in non-small cell lung carcinomas and stimulates tumor growth of A549 cells. Carcinogenesis 24(9):1445–1454
7. Hao J, Ho JN, Lewis JA, Karim KA, Daniels RN, Gentry PR et al (2010) In vivo structure–activity relationship study of dorsomorphin analogues identifies selective VEGF and BMP inhibitors. ACS Chem Biol 5(2):245–253
8. Langenfeld EM, Bojnowski J, Perone J, Langenfeld J (2005) Expression of bone morphogenetic proteins in human lung carcinomas. Ann Thorac Surg 80(3):1028–1032
9. Counting cells using a hemocytometer. Abcam. http://www.abcam.com/protocols/counting-cells-using-a-haemocytometer. Accessed 19 Jan 2017
10. Patton J, Vuyyuru R, Siglin A, Root M, Manser T (2015) Evaluation of the efficiency of human immune system reconstitution in NSG mice and NSG mice containing a human HLA.A2 transgene using hematopoietic stem cells purified from different sources. J Immunol Methods 422:13–21

INDEX

Melissa B. Rogers (ed.), *Bone Morphogenetic Proteins: Methods and Protocols*, Methods in Molecular Biology, vol. 1891, https://doi.org/10.1007/978-1-4939-8904-1, © Springer Science+Business Media, LLC, part of Springer Nature 2019

R

S

T

V

W

X

Y

Z

Printed by Printforce, the Netherlands